Mathematical Handbook
for Electrical Engineers

For a listing of recent titles in the *Artech House Technology Management and Professional Development Library,* turn to the back of this book.

Mathematical Handbook
for Electrical Engineers

Sergey A. Leonov
Alexander I. Leonov

ARTECH
HOUSE

BOSTON | LONDON
artechhouse.com

Library of Congress Cataloging-in-Publication Data
A catalog record of this book is available from the Library of Congress.

British Library Cataloguing in Publication Data
A catalogue record of this book is available from the British Library.

Cover design by Igor Valdman

© 2005 ARTECH HOUSE, INC.
685 Canton Street
Norwood, MA 02062

International Standard Book Number: 1-58053-779-0

10 9 8 7 6 5 4 3 2 1

CONTENTS

Part II
MATHEMATICAL ALGORITHMS TO SOLVE
COMMON PROBLEMS IN ELECTRICAL ENGINEERING

PART III REFERENCE DATA

PREFACE

Mathematics is immense. Mathematics is complicated. Mathematics is frightening.

The purpose of this handbook is to refute all three statements above. The authors have tried to make the handbook concise, simple, and clear. The focus is to provide a single-volume reference source on what are the most common mathematical techniques used to solve problems in electrical engineering, and how to use them for obtaining analytical or computer-aided solutions. Thus, the focus is on providing clear and simple answers "how to..." for the practical engineers and university students.

The handbook consists of three parts.

Part I, *Fundamentals of Engineering Mathematics*, has a dual purpose. On the one hand, it contains definitions, laws and rules, which can be used by electrical engineers to solve a majority of the problems that require math involvement. It is unlikely that in practical computation tasks an electrical engineer will require knowledge of mathematical rules and formulas beyond the scope of those presented in the handbook.

On the other hand, it can be used by the engineers and university students for systematic reading. It does not pretend to be a textbook, since it does not contain any derivations (those who are interested in derivations can find them in the referenced bibliography) but once Part I is read thoroughly from the beginning to the end (and it is only about 250 pages), it will equip the reader with knowledge of mathematical techniques sufficient to solve the majority of electrical engineering problems.

The question university students and electrical engineers often ask themselves looking at the brick-size volumes of mathematical textbooks and handbooks is: "Do I really have to know all this?" The answer is: a modern electrical engineer should know basic math within these 250 pages. It should be quite sufficient to solve practical tasks of analysis, calculations, and computer simulation he or she faces in his or her daily work.

Each paragraph is short, written in clear language, and in most cases contains simple illustrative examples to enhance understanding of the subject.

Part II, *Mathematical Algorithms to Solve Common Problems in Electrical Engineering*, contains examples of how to solve typical problems in electrical engineering that require involvement of mathematics. Major applied disciplines such as electrical circuits, devices and systems, antennas and propagation, waveforms and signal processing, stochastic radio engineering are covered.

Students can use them to enhance their understanding how to apply mathematical methods to solve electrical engineering problems, and engineers can use them as the guidelines to solve similar problems or use cited algorithms as the building blocks to create their own analysis programs.

Part III, *Reference Data,* provides the single-volume reference source for useful mathematical formulas and equations frequently used in electrical engineering: constants, units and conversion factors, sets of coordinates, common equalities, identities, series, differentiation and integration formulas, random functions parameters and distribution laws, stochastic simulation algorithms, and so forth that are typically scattered over numerous books and papers.

Nowadays for a practical engineer or a student, there is no necessity to browse through the thick volumes with the tables for derivatives, integrals, and so forth, or even to know how to write algorithms for numerical computation methods (that is why they are not included in the handbook), since there is a variety of mathematical software packages that can provide an answer at the touch of the button on the computer keyboard. One of them is MATHCAD, which is referenced as the basic software package in the examples through the handbook. It is a convenient tool because in order to find a derivative, an integral or a solution of the equation, a user does not have to do any programming. He only writes it in the worksheet in the manner he would write it in the notebook and lets the software provide the answer. We wish we were that lucky when we were university students 30–50 years ago in the era of brick-size folios with the tables of derivatives/integrals/special functions and slide rulers!

Dr. Alexander Leonov,
Dr. Sergey Leonov,
Waterloo, Canada, April 2005

PART I

FUNDAMENTALS OF ENGINEERING MATHEMATICS

Chapter 1

ELEMENTARY MATHEMATICS

1.1 Fundamentals of Algebra

1.1.1 Numbers

- The *natural numbers* are the set of numbers N = {1, 2, 3, 4, 5,...}
- The *integers* are the set of numbers N = {0, 1, -1, 2, -2, 3, -3,...}
- The *real numbers* are the set of all points on the number line (integers and fractions)
- The *rational numbers* are the real numbers that can be written as the ratio of

  two integers (e.g., $\dfrac{1}{2}$; $-\dfrac{8}{115}$)
- The *irrational numbers* are the real numbers that cannot be written as the ratio

  of two integers (e.g., $\sqrt{2} = 1.41...$; $\pi = 3.14...$)

1.1.2 Fundamental Rules of Arithmetic

Basic Binary Operations:

- Addition: $a+b=c$ • Subtraction: $c-b=a$
- Multiplication: $a\cdot b=c$ • Division: $c/b=a$

Basic Rules of Binary Operations:

- $a+b=b+a$ • $a+b+c=(a+b)+c=a+(b+c)$
- $a+c \succ b+c$ for any c if $a \succ b$
- $a\cdot b=b\cdot a$ • $a\cdot b\cdot c=(a\cdot b)\cdot c=a\cdot(b\cdot c)$
- $(a+b)\cdot c=a\cdot c+b\cdot c$
- $a\cdot c \succ b\cdot c$ for any c if $a \succ b$

3

Important Rules of Binary Operations:

- $a+(b+c) = a+b+c$ • $a-(b+c) = (a-b)-c$
- $(a+b)+c = (a+c)+b = a+(b+c)$
- $(a+b)-c = (a-c)+b = a+(b-c)$
- $(a-b)-c = (a-c)-b = a-(b+c)$
- $a\cdot(b\cdot c) = (a\cdot b)\cdot c$ • $(a\cdot b)\cdot c = (a\cdot c)\cdot b = a\cdot(b\cdot c)$
- $(a\cdot b)/c = (a/c)\cdot b = a\cdot(b/c)$ • $(a+b)/c = (a/c)+(b/c)$
- $(a-b)/c = (a/c)-(b/c)$

Note. A common mistake:

$$a/(b+c) \neq a/b + a/c \qquad a/(b-c) \neq a/b - a/c$$

Null (0) is less than any natural number. The series $(0, 1,2,3,4, \ldots)$ is called an extended series of natural numbers. For any natural number $a \neq 0$:

- Summation: $a+0 = 0+a = a$; $0+0 = 0$
- Subtraction: $a-0 = a$; $a-a = 0$; $0-0 = 0$
- Multiplication: $a\cdot 0 = 0\cdot a = 0$; $0\cdot 0 = 0$
- Division: $0/a = 0$; $a/0$ is not defined for natural numbers

The product of n numbers a is written as a^n and is called a number a in nth power.

Basic Rules: • $a^n \times a^k = a^{n+k}$ • $a^n \div a^k = a^{n-k}$

• $(a^n)^k = a^{n \times k}$ • $\sqrt[n]{a} = a^{1/n}$ • $a^0 = 0^0 = 1$

1.1.3 Fractions and Percents

A *positive fraction* is a pair of natural numbers a (a *numerator*) and b (a *denominator*) written as $\dfrac{a}{b}$. A denominator shows in how many parts a unity quantity is divided, and a numerator shows how many of these parts are taken.

♦ <u>Example 1.1</u> The fraction $\dfrac{5}{7}$ means that a unity quantity is divided into 7 equal parts and 5 parts are taken.

Basic Rules: • $\dfrac{a}{b} = \dfrac{c}{d}$ if $a \cdot d = b \cdot c$ • $\dfrac{a}{b} \succ \dfrac{c}{d}$ if $a \cdot d \succ b \cdot c$ • $\dfrac{a \cdot c}{b \cdot c} = \dfrac{a}{b}$

• $\dfrac{a/c}{b/c} = \dfrac{a}{b}$ • $\dfrac{a}{b} + \dfrac{c}{d} = \dfrac{a \cdot d + c \cdot b}{b \cdot d}$ • $\dfrac{a}{b} - \dfrac{c}{d} = \dfrac{a \cdot d - c \cdot b}{b \cdot d}$ • $\dfrac{a}{b} \cdot \dfrac{c}{d} = \dfrac{a \cdot c}{b \cdot d}$

• $\dfrac{a}{b} \Big/ \dfrac{c}{d} = \dfrac{a \cdot d}{b \cdot c}$

A *decimal fraction* is a fraction $\dfrac{a}{b}$ when a denominator b is equal to a natural power of 10.

♦ <u>Example 1.2</u> $\dfrac{4}{10} = 0.4$; $\dfrac{12}{100} = 0.12$; $\dfrac{3451}{1000} = 3.451$

A *percent* is ones hundredth of a number (quantity). Any real number can be represented in terms of percentage by multiplying it by 100.

♦ <u>Example 1.3</u> $0.20 = 20\%$; $3 = 300\%$; $\dfrac{3}{4} = 75\%$

Operations with percents are typically based on the ratio $\dfrac{a}{b} = \dfrac{p}{100}$.

♦ <u>Example 1.4</u> How many percent is a number $a = 14$ from $b = 28$?

$\dfrac{14}{28} = \dfrac{p}{100}$. Thus $p = \dfrac{14}{28} \cdot 100 = 50\%$

• What number a is 25% from $b = 300$?

$\dfrac{a}{300} = \dfrac{25}{100}$. Thus $a = \dfrac{300}{100} \cdot 25 = 75$

• What is number b if 4% of this number is $a = 50$?

$\dfrac{50}{b} = \dfrac{4}{100}$. Thus $b = \dfrac{50}{4} \cdot 100 = 1250$

1.1.4 Modulus and Intervals

Modulus of the positive real number is equal to the number. Modulus of the negative real number is equal to the opposite number.

$$|a| = \begin{cases} a, & \text{if } a \ge 0 \\ -a, & \text{if } a < 0 \end{cases}$$

♦ <u>Example 1.5</u> $|3| = 3; |-3| = 3$.

Basic rules:

- $|a| \geq 0$; $|-a| = |a|$; $a \leq |a|$; $|x| \leq a$ is equal to $-a \leq x \leq a$
- $|x| > a$ is equal to $x < a$ and $x > a$; $|a+b| \leq |a| + |b|$
- $|a-b| \geq |a| - |b|$; $|a \cdot b| = |a| \cdot |b|$; $\left|\dfrac{a}{b}\right| = \dfrac{|a|}{|b|}$

An *open interval* is the set of numbers between two numbers a and b (does not include end points). It satisfies the inequality

$a < x < b$

and is denoted by (a, b).

A *closed interval* is the set of numbers between two numbers a and b that includes the end of points

$a \leq x \leq b$

and is denoted by $[a, b]$.

1.1.5 A Monomial

A *monomial* is a product of a number (monomial factor) and powers of numbers, denoted by letters.

♦ <u>Example 1.6</u> $5a^2b$; $\dfrac{1}{6}xyz^3$; 3.75; 0 are monomials.

Basic rules:

- Addition/Subtraction: $3ab^2 - 4ab^2 + 6ab^2 = (3 - 4 + 6)ab^2 = 5ab^2$
- Multiplication: $2ab \cdot (-4a^2b^3) = -8a^3b^4$
- Division: $8a^3b^2 : 4ab = 2a^2b$
- Expansion: $a + [b - (c - d)] = a + [b - c + d] = a + b - c + d$
- Factorization: $a - b + c - d = a + (-b + c - d) = a - (b - c + d)$

1.1.6 A Polynomial

A *polynomial* is an algebraic sum of a finite number of monomials.

♦ <u>Example 1.7</u> $2x^2 + 3y^2 + 6xy$; $a_n x^n + a_{n-1} x^{n-1} + \ldots a_1 x + a_0$ are polynomials.

Basic rules:

- Addition/Subtraction: $(3ab^2 + 5ab) + (4ab^2 - 3ab) = 7ab^2 + 2ab$
- Multiplication:

$(3ab^2 + 5ab) \cdot (4ab^2 - 3ab) = 12a^2 b^4 - 9a^2 b^3 + 20a^2 b^3 - 15a^2 b^2$

$= 12a^2 b^4 + 11a^2 b^3 - 15a^2 b^2 = a^2 b^2 (12b^2 + 11b - 15)$

- Division: $(16a^4 b^4 + 6a^2 b^2) : 2ab = 8a^3 b^3 + 3ab = ab(8a^2 b^2 + 3)$
- Factorization:

$ax - ay + bx - by = a(x - y) + b(x - y) = (x - y) \cdot (a + b)$;

$a^2 + 2ab + b^2 = a^2 + ab + b^2 + ab = a(a + b) + b(b + a) = (a + b) \cdot (b + a) = (a + b)^2$

- Expansion:

$(a + b)^2 = (a + b) \cdot (a + b) = a^2 + 2ab + b^2$

$(a + b)^3 = (a + b) \cdot (a + b) \cdot (a + b) = a^3 + 3a^2 b + 3ab^2 + b^3$

$(a + b)^4 = (a + b)^2 \cdot (a + b)^2 = a^4 + 4a^3 b + 6a^2 b^2 + 4ab^3 + b^4$

1.1.7 The Greatest Common Divisor

The *greatest common divisor* (g.c.d.) of two or several polynomials is their common divisor of the highest power.

♦ <u>Example 1.8</u> $x^3 + 3x^2 + 4x + 12 = (x + 3) \cdot (x^2 + 4)$;

$x^3 + 4x^2 + 4x + 3 = (x + 3) \cdot (x^2 + x + 1)$;   g.c.d $= x + 3$

1.1.8 The Least Common Multiple

The *least common multiple* (l.c.m.) is the polynomial of the lowest power that can be divided by each of given polynomials.

♦ <u>Example 1.9</u> For monomials $2x^2 y, 12xy^2 z, 6x^3 yz$

l.c.m $= x^2 y \times yz \times x = x^3 y^2 z$

If P and Q are polynomials with g.c.d $= u$ and l.c.m $= v$, then

$$v = \frac{P \cdot Q}{u}; u = \frac{P \cdot Q}{v}$$

1.1.9 The Algebraic Fraction

An *algebraic fraction* is the ratio of two polynomials P and Q, where P is called a *numerator*, and Q is called a *denominator*.

♦ Example 1.10 $\dfrac{x}{y+z}; \dfrac{a^2 + 5a - 17}{a + 3}; \dfrac{1}{b}$

Basic operations:

- Addition/Subtraction: $\dfrac{P}{Q} \pm \dfrac{R}{S} = \dfrac{PS \pm QR}{QS}$

- Multiplication: $\dfrac{P}{Q} \cdot \dfrac{R}{S} = \dfrac{PR}{QS}$

- Division: $\dfrac{P}{Q} : \dfrac{R}{S} = \dfrac{P}{Q} \cdot \dfrac{S}{R} = \dfrac{PS}{QR}$

Basic rules:

- $\left(\dfrac{P}{Q} + \dfrac{R}{S} \right) + \dfrac{U}{V} = \dfrac{P}{Q} + \left(\dfrac{R}{S} + \dfrac{U}{V} \right)$

- $\dfrac{P}{Q} \cdot \dfrac{R}{S} = \dfrac{R}{S} \cdot \dfrac{P}{Q}$

- $\dfrac{P}{Q} \cdot \left(\dfrac{R}{S} \cdot \dfrac{U}{V} \right) = \left(\dfrac{P}{Q} \cdot \dfrac{R}{S} \right) \cdot \dfrac{U}{V}$

- $\left(\dfrac{P}{Q} + \dfrac{R}{S} \right) \cdot \dfrac{U}{V} = \dfrac{P}{Q} \cdot \dfrac{U}{V} + \dfrac{R}{S} \cdot \dfrac{U}{V}$

1.1.10 A Proportion

A *proportion* is the identity of two algebraic polynomials:

$$\frac{A}{B} \equiv \frac{C}{D}, \quad B \neq 0; D \neq 0$$

◆ <u>Example 1.11</u> $\dfrac{8}{x} = \dfrac{a}{6}$; $\dfrac{z+3}{z^2+y+4} = \dfrac{z}{y+3}$

Basic rules:

If $\dfrac{A}{B} = \dfrac{C}{D}$, then

- $A \cdot D = B \cdot C$
- $\dfrac{A+B}{B} = \dfrac{C+D}{D}$
- $\dfrac{A-B}{B} = \dfrac{C-D}{D}$
- $\dfrac{A+B}{A-B} = \dfrac{C+D}{C-D}$

1.1.11 A Root

A *root* of nth power from a number a is the number b for which the nth power is equal to a.

$\sqrt[n]{a} = b$ if $b^n = a$.

◆ <u>Example 1.12</u> $\sqrt[4]{16} = 2$ since $2^4 = 16$; $\sqrt[4]{16} = -2$ since $(-2)^4 = 16$
$\sqrt[3]{9} = 3$ since $3^3 = 9$; $\sqrt[3]{-9} = -3$ since $(-3)^3 = -9$; $\sqrt[n]{0} = 0$; $\sqrt[n]{1} = 1$

Basic rules:
- Multiplication: $\sqrt[n]{a} \cdot \sqrt[n]{b} = \sqrt[n]{ab}$, $a \geq 0, b \geq 0$
- Division: $\dfrac{\sqrt[n]{a}}{\sqrt[n]{b}} = \sqrt[n]{\dfrac{a}{b}}$, $a \geq 0, b \geq 0$
- Powering: $\left(\sqrt[n]{a}\right)^m = \sqrt[n]{a^m}$, $a \geq 0$
- A root from a root: $\sqrt[m]{\sqrt[n]{a}} = \sqrt[mn]{a}$, $a \geq 0$
- A root of odd power from a negative number:

$$\sqrt[2n+1]{-a} = -\sqrt[2n+1]{a}, \quad a \geq 0$$

Important Rules:

- $\sqrt[2k]{a^{2k} \cdot b} = |a|\sqrt[2k]{b}$
- $\sqrt[2k+1]{a^{2k+1} \cdot b} = a\sqrt[2k+1]{b}$
- $a\sqrt[2k]{b} = \begin{cases} \sqrt[2k]{a^{2k} \cdot b}, & \text{if } a \geq 0 \\ -\sqrt[2k]{a^{2k} \cdot b}, & \text{if } a < 0 \end{cases}$
- $a\sqrt[2k+1]{b} = \sqrt[2k+1]{a^{2k+1} \cdot b}$

$$\bullet \quad \sqrt{a \pm \sqrt{b}} = \sqrt{\frac{a + \sqrt{a^2 - b}}{2}} \pm \sqrt{\frac{a - \sqrt{a^2 - b}}{2}}$$

◆ Example 1.13 $\sqrt{8} = \sqrt{2^2 \cdot 2} = 2\sqrt{2}$; $\sqrt{(x-4)^3} = (x-4)\sqrt{(x-4)}$;

$\sqrt{11 + \sqrt{40}} = 1 + \sqrt{10}$

1.2 Functions

1.2.1 Definitions

The *function* is the relationship between several variable quantities that is typically denoted as:

$$u = f(x, y, z, t, \ldots)$$

In case there are only two quantities whose values are related, the typical notation is:

$$y = f(x)$$

The variable x is called *independent* or *free* variable (*argument* of the function), and variable y is called a *dependent* variable (*value* of the function), (i.e., the value the function f depends on the value of its argument x).

◆ Example 1.14 The relationship between the temperature measured in degrees Celsius $T°C$ and degrees Fahrenheit $T°F$ is as follows:

$$T°F = \frac{9}{5} T°C + 32$$

Here $T°C$ is the independent variable x and $T°F$ is the dependent variable y. This function also can be written as:

$$y = \frac{9}{5} x + 32$$

For a particular value of x, (e.g., $T°C = 30°$), there is only one value of y $T°F = 86°$.

The *inverse* function with respect to the function f is the function that reverses the operations carried out by f; that is if $y = f(x)$, the inverse function is $x = f^{-1}(y)$.

♦ <u>Example 1.15</u>   Find the inverse function for the function in the previous example.

Solution.   Since $y = \dfrac{9}{5}x + 32$, then $x = \dfrac{5}{9}(y - 32)$. Since y becomes an

argument,   this   function   can   also   be   written   as: $y = \dfrac{5}{9}(x - 32)$ or

$T°C = \dfrac{5}{9}(T°F - 32)$. If $T°F = 86°$, the corresponding $T°C = 30°$.

The *composite* function is a function of a function, for example if $y = f(x)$ and $z = \varphi(y)$, it can be written as:

$$z = f[\varphi(x)]$$

The *even* and *odd* functions are ones that satisfy the equations:

$$f(-x) = f(x); \quad f(-x) = -f(x)$$

The *periodic* function is such that its image values are repeated at regular intervals in its domain with period *T*:

$$f(x + nT) = f(x), \quad n = 0, \pm 1, \pm 2, \ldots$$

Parameter T is called the *period* of the function. Parameter $f = 1/T$ is called the *frequency* of a periodic function, and parameter $\omega = 2\pi \cdot f$ is called the *circular* frequency of a periodic function.

If the function $y = f(x)$ can be represented as an analytic expression, it is called an *elementary* one. The common elementary functions are *polynomial, rational, trigonometric (circular), inverse trigonometric, exponential, logarithmic, hyperbolic,* and *inverse hyperbolic* functions, and the functions that can be represented as the sums, differences, products, or ratios of these functions.

1.2.2 Algebraic Functions

The most common algebraic functions are *polynomial* functions and *rational* functions.

A *polynomial* function has the general form:

$$y = f(x) = \sum_{k=0}^{n} a_k \cdot x^k = a_0 + a_1 x + a_2 x^2 + \ldots a_n x^n$$

For $k = 1$, the polynomial function reduces to a *linear* function:

$y = a_0 + a_1 x$

For $k = 2$, the polynomial function becomes a *quadratic* function:

$y = a_0 + a_1 x + a_2 x^2$

Rational functions have the general form:

$$y = f(x) = \frac{P(x)}{Q(x)} = \frac{\displaystyle\sum_{k=0}^{n} a_k x^k}{\displaystyle\sum_{l=0}^{m} b_l x^l}$$

If the degree of $P(x)$ is less than degree of $Q(x)$ $(n < m)$, $f(x)$ is called a *strictly proper rational function*. If $n = m$, it is called a *proper rational function*, and if $n > m$, it is called an *improper rational function*.

The real number x_u is called a *null* of the rational function, if $P(x_u) = 0$ and $Q(x_u) \neq 0$.

The real number x_g is called a *pole* of the rational function, if $P(x_g) \neq 0$ and $Q(x_g) = 0$.

Thus, in order to find the nulls of the rational functions, we have to solve the equation

$P(x) = 0$

and in order to find the poles of the rational function, we have to solve the equation

$Q(x) = 0$

◆ Example 1.16 Find the nulls and the poles of the function

$$y = \frac{x^2 - 4}{x^2 + 10x + 9}$$

Solution. The nulls are found from the equation $x^2 = 4$, thus nulls are $x_1 = 2, x_2 = -2$.

The poles are found from the equation:

$$x^2 + 10x + 9 = 0$$

Thus, poles are $x_3 = -1, x_4 = -9$.

The class of *irrational* algebraic functions are the functions of the type $(n = 1, 2, \ldots)$

$$y = \sqrt[n]{x}$$

1.2.3 Transcendental Functions

The functions that do not fall into the class of algebraic ones are called *transcendental* functions. Basic transcendental functions are:

- Trigonometric (circular) functions (see Section 1.7.2);
- Exponential functions;
- Logarithmic functions;
- Hyperbolic functions.

1.2.4 Exponential and Logarithmic Functions

An *exponential* function is the function of the type

$$y = a^x$$

where a is a positive constant and x is the independent variable.
When $a = e$, where

$$e = \lim_{n \to \infty} \left(1 + \frac{1}{n} \right)^n \approx 2.718\ldots$$

we obtain the *standard exponential function*:

$$y = e^x$$

A *logarithmic* function is one inverse to the exponential function:

$$y = \log_a x$$

Thus, if $y = a^x$, then $x = \log_a y$. For $a = e$ and $a = 10$, the functions are typically denoted:

$$y = \log_e x = \ln x \quad y = \log_{10} x = \lg x$$

This means that if $y = e^x$, then $x = \ln y$, and $y = 10^x$, then $x = \lg y$.
The basic properties of logarithms are:

- $\log_a(x_1 x_2) = \log_a x_1 + \log_a x_2$

- $\log_a\left(\dfrac{x_1}{x_2}\right) = \log_a x_1 - \log_a x_2$

- $\log_a x^n = n \cdot \log_a x$. Thus, $\log_a \sqrt[c]{x} = \dfrac{1}{c} \cdot \log_a x$

- $\log_{a^k} x = \dfrac{1}{k} \cdot \log_{|a|} x$, if $a^k \succ 0$

- $\log_{ab} x = \dfrac{\log_{|a|} x}{1 + \log_{|a|}|b|}$, if $ab \succ 0$

- $x = a^{\log_a x}$

- $\log_a x = \dfrac{\log_b x}{\log_b a}$. Thus, if $x = b$, $\log_a b = \dfrac{1}{\log_b a}$

1.2.5 Decibels

The fundamental division of a logarithmic scale for expressing the ratio of two amounts P_1 and P_2 plays a very important role in electrical engineering. The number of bels B describing the ratio of powers P_1/P_2 is:

$$B = \log_{10}\left(\frac{P_1}{P_2}\right)$$

A *decibel* (dB) is one-tenth of a bel. Thus the ratio of two powers or voltages $P_1 = V_1^2$, $P_2 = V_2^2$ in decibel is expressed as:

$$dB = 10\log_{10}\left(\frac{P_1}{P_2}\right) = 20\log_{10}\left(\frac{V_1}{V_2}\right)$$

The convenience of expressing powers in decibel is that powers coming from different sources can be added (subtracted) rather then multiplied (divided).

♦ <u>Example 1.17</u> The gain of a directional radar antenna is 30 dB with respect to the isotropic antenna on transmit and 40 dB on receive. Find the total gain of the signal transmitted to the object and received back.

Solution. The ratio of the power P_t transmitted in the direction of the object with respect to the power that would be transmitted by isotropic antenna P_0 is:

$$G_t = P_t/P_0$$

On receive, it is

$$G_r = P_r/P_0$$

where P_r is the received power. The total gain of directional antenna G is:

$$G = G_t \cdot G_r$$

In decibel notation,

$$10\log_{10} G[dB] = 10\log_{10}(G_t \cdot G_r) =$$

$$10\log_{10} G_t[dB] + 10\log_{10} G_r[dB] = 30[dB] + 40[dB] = 70[dB]$$

Thus, total gain

$10\log_{10} G = 70[dB]$ or $G = 10^{\frac{70}{10}} = 10^7$ [*times*]   with respect to isotropic antenna.

1.2.6 Hyperbolic and Inverse Hyperbolic Functions

Hyperbolic functions are defined as follows:

- The hyperbolic sine $\sinh x = shx = \dfrac{e^x - e^{-x}}{2}$

- The hyperbolic cosine $\cosh x = chx = \dfrac{e^x + e^{-x}}{2}$

- The hyperbolic tangent $\tanh x = thx = \dfrac{e^x - e^{-x}}{e^x + e^{-x}}$

- The hyperbolic cotangent $\coth x = cthx = \dfrac{e^x + e^{-x}}{e^x - e^{-x}}$

- The hyperbolic secant $\operatorname{sec} h\, x = \dfrac{2}{e^x + e^{-x}}$

- The hyperbolic cosecant $\operatorname{cosech} x = \dfrac{2}{e^x - e^{-x}}$

The inverse hyperbolic functions are defined as:

- $y = \operatorname{Arsinh} x = Arshx = \ln\left(x + \sqrt{x^2 + 1}\right)$

- $y = \operatorname{Arcosh} x = Archx = \ln\left(x + \sqrt{x^2 - 1}\right)$ $|x \geq 1|$

- $y = \operatorname{Artanh} x = Arthx = \dfrac{1}{2}\ln\dfrac{1+x}{1-x}$, $|x| < 1$

- $y = \operatorname{Arcoth} x = Arcthx = \dfrac{1}{2}\ln\dfrac{1-x}{1+x}$, $-\infty \prec x \prec -1$ and $1 \prec x \prec \infty$

Other notations correspondingly are:

- $y = \sinh x^{-1}$ • $y = \cosh x^{-1}$ • $y = \tanh x^{-1}$ • $y = \coth x^{-1}$

Basic relationships and identities for hyperbolic and inverse hyperbolic function are given in Appendix 2.

1.2.7 Piecewise Defined Functions

There are a number of piecewise defined functions that are used in electrical engineering. The most common are:

- The *signum* function [Figure 1.1(a)]

$$y = \operatorname{sgn}(x) = \begin{cases} 1, & x > 0 \\ -1, & x < 0 \\ 0, & x = 0 \end{cases}$$

- The *unit step* function [Heaviside function, Figure 1.1(b)]

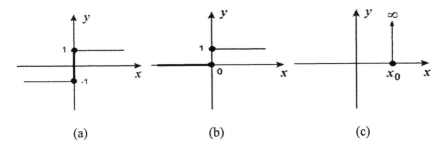

Figure 1.1 (a) The signum function, (b) unit step function, and (c) delta function.

$$y = H(x) = \begin{cases} 0, & x < 0 \\ 1, & x > 0 \end{cases}$$

- The *delta* function [Dirac function, Figure 1.1(c)]

$$y = \delta(x) = \begin{cases} \infty, & x = x_0 \\ 0, & x \neq x_0 \end{cases}$$

The major properties of the delta function are given in <u>Appendix 3</u>.

- The *floor* function

$y = \lfloor x \rfloor$ = the greatest integer not greater than x

- The *ceiling* function

$y = \lceil x \rceil$ = the least integer not less then x

- The *fractional-part* function

$$y = FRACPT(x) = x - \lfloor x \rfloor$$

1.3 Equations

1.3.1 Definitions

An *equation* is an equality that contains unknown variables (typically denoted $x, y, z, \ldots$), which have to be found.

If both sides of the equation contain algebraic expressions with respect to unknown variables, it is called an *algebraic equation*.

If at least one side of the equation contains a transcendental function (e.g., exponential, trigonometric, logarithmic), it is called a *transcendental equation*.

The generic form of the *rational algebraic* equation is:

$$a_n x^n + a_{n-1} x^{n-1} + \ldots + a_1 x + a_0 = 0$$

where x is the unknown variable, $a_n, a_{n-1}, \ldots$ are the coefficients, and n is a natural number, indicating the order of the equation.

For $n = 1$, the equation is called a *linear* one; for $n = 2$, it is called a *quadratic* one; for $n = 3$, it is called a *cubic* one; for $n = 4$, it is called the equation of the *fourth order,* and so forth.

If at least one of the factors of the algebraic equation is irrational with respect to the unknown variable, the algebraic equation is called the *irrational* one. For example,

$$\sqrt{x+3} + x = 4.$$

1.3.2 How to Solve a Rational Linear Equation

The rational linear equation with a single unknown variable x has the form $(n = 1)$:

$$a_1 x + a_0 = 0, \quad a_1 \neq 0$$

It has a single solution:

$$x = -\frac{a_0}{a_1}$$

♦ Example 1.18 Solve the equation $3x - 9 = 0$.
Solution. $x = 9/3 = 3$.

1.3.3 How to Solve a Rational Quadratic Equation

The rational quadratic equation with a single unknown variable x has the form $(n = 2)$:

$$a_2 x^2 + a_1 x + a_0 = 0$$

It has two solutions (roots):

$$x_1 = \frac{-a_1 - \sqrt{a_1^2 - 4a_2a_0}}{2a_2}; \quad x_2 = \frac{-a_1 + \sqrt{a_1^2 - 4a_2a_0}}{2a_2}$$

When $a_2 = 1$, the equation is typically written as:

$$x^2 + px + q = 0$$

and the solutions are

$$x_{1,2} = -\frac{p}{2} \pm \sqrt{\left(\frac{p}{2}\right)^2 - q}$$

♦ Example 1.19 Solve the equation $x^2 + 10x + 9 = 0$.

Solution. The roots are $x_{1,2} = -5 \pm \sqrt{25 - 9}$. Thus, $x_1 = -1; x_2 = -9$.

1.3.4 How to Solve a Rational Cubic Equation

The rational cubic equation with a single unknown variable x has the form $(n = 3)$:

$$a_3x^3 + a_2x^2 + a_1x + a_0 = 0$$

In general, the cubic equation has three roots x_1, x_2, x_3 and typically can be solved by numeric methods with the aid of a computer (see Chapter 9). However, there are two situations when an analytical solution can be obtained using the following rules.

Rule #1. If the linear factor can be found by expansion

$$a_3x^3 + a_2x^2 + a_1x + a_0 = a(x - b)(x^2 + px + q)$$

Then the first root $x_1 = b$, and $x_{2,3}$ are found from the quadratic equation

$$x^2 + px + q = 0$$

Rule #2. If the cubic equation can be represented in the form

$$x^3 + px + q = 0$$

Then the roots are:

$$x_1 = u+v; \qquad x_2 = \frac{u+v}{2} + \frac{u-v}{2}j\sqrt{3}; \qquad x_3 = \frac{u+v}{2} - \frac{u-v}{2}j\sqrt{3};$$

$$u = \sqrt[3]{-\frac{q}{2} + \sqrt{\left(\frac{p}{3}\right)^3 + \left(\frac{q}{2}\right)^2}}; \qquad v = \sqrt[3]{-\frac{q}{2} - \sqrt{\left(\frac{p}{3}\right)^3 + \left(\frac{q}{2}\right)^2}}$$

♦ Example 1.20 Solve the cubic equation $x^3 - 6x^2 + 21x - 52 = 0$.
Solution. Assuming $x = t+2$, the equation can be written as

$$t^3 + 9t - 26 = 0$$

Thus, $p = 9$, $q = -26$. Using the formulas above, we obtain the solutions as follows ($u = 3$, $v = -1$):

$$t_1 = 2, \ t_2 = -1 + 2\sqrt{3}j, \ t_3 = -1 - 2\sqrt{3}j$$

Thus, the roots of the equation are as follows:

$$x_1 = 4, \ x_2 = 1 + 2\sqrt{3}j, \ x_3 = 1 - 2\sqrt{3}j$$

1.3.5 How to Solve the Rational Equations of Higher Orders

Equations of the orders higher than third typically do not have analytical solutions, and have to be solved with the aid of a computer (see Chapter 9). However, here are some cases when equations of higher order can be reduced to the quadratic equation.

- Biquadratic equation

$$a_4 x^4 + a_2 x^2 + a_0 = 0;$$

$$x_{1,2} = \pm\sqrt{\frac{-a_2 - \sqrt{a_2^2 - 4a_4 a_0}}{2a_4}} \ ; \qquad x_{3,4} = \pm\sqrt{\frac{-a_2 + \sqrt{a_2^2 - 4a_4 a_0}}{2a_4}}$$

- Equations with mutually reciprocal variables:

$$a\frac{f_1(x)}{f_2(x)} + b\frac{f_2(x)}{f_1(x)} = c$$

Denoting

$$u = \frac{f_1(x)}{f_2(x)}, \quad au + b\frac{1}{u} = c, \text{ or } au^2 - cu + b = 0$$

that takes a form of quadratic equation

$$a_2 u^2 + a_1 u + a_0 = 0, \quad a_2 = a; \quad a_1 = -c; \quad a_0 = b$$

- Equations of the fourth order convertible to the formula of a perfect square.

♦ Example 1.21 Solve the equation $x^4 + 6x^3 + 5x^2 - 12x + 3 = 0$ [1].

Solution . This equation can be represented as

$$x^4 + 6x^3 + 9x^2 - 4x^2 - 12x + 3 = 0 \text{ or}$$

$$\left(x^2 + 3x\right)^2 - 4\left(x^2 + 3x\right) + 3 = 0 \text{ or}$$

$$u^2 - 4u + 3 = 0, \text{ where } u = x^2 + 3x$$

Thus $u_1 = 1$, $u_2 = 3$ and $x_{1,2} = \dfrac{-3 \pm \sqrt{13}}{2}$; $x_{3,4} = \dfrac{-3 \pm \sqrt{21}}{2}$

- The equations that take the form:

$$\frac{ax}{px^2 + nx + q} + \frac{bx}{px^2 + mx + q} = c$$

are reduced to the quadratic equations with substitution:

$$u = px + \frac{q}{x}$$

This results in equation:

$$\frac{a}{u+n} + \frac{b}{u+m} = c \text{ or } \left(\text{if } u \neq -n, \quad u \neq -m\right)$$

$$cu^2 + \left(mc + nc - a - b\right)u + mnc - am - bn = 0$$

♦ <u>Example 1.22</u> Solve the equation $\dfrac{2x}{2x^2 - 5x + 3} + \dfrac{13x}{2x^2 + x + 3} = 6$ [1].

Solution. In this equation,

$a = 2, b = 13, p = 2, n = -5, m = 1, q = 3, c = 6$.

Thus, the final equation takes the form

$6u^2 - 39u + 33 = 0$, or $2u^2 - 13u + 11 = 0$

with the roots $u_1 = 1$, $u_2 = \dfrac{11}{2}$.

For $u_1 = 1$, the original equation does not have real roots. For $u_2 = \dfrac{11}{2}$,

$2x + \dfrac{3}{x} = \dfrac{11}{2}$ or $4x^2 - 11x + 6 = 0$

with the roots $x_1 = \dfrac{3}{4}$, $x_2 = 2$.

1.3.6 How to Solve Irrational Equations

The typical approach to finding analytical solutions for irrational equations is to raise them to the required power until the expressions with radixes are eliminated.

♦ <u>Example 1.23</u> Solve the equation $\sqrt{5 + x} + \sqrt{5 - x} = \dfrac{x}{2}$ [1].

Solution. After we quadrate both sides of the equation, it takes the form

$5 + x + 5 - x + 2\sqrt{25 - x^2} = \dfrac{x^2}{4}$ or $2\sqrt{25 - x^2} = \dfrac{x^2}{4} - 10$

After we quadrate it the second time, we obtain

$100 - 4x^2 = \dfrac{x^4}{16} - 5x^2 + 100$ or $x^4 - 16x^2 = x^2\left(x^2 - 16\right) = 0$

Thus, the roots are $x_1 = x_2 = 0$; $x_3 = 4$; $x_4 = -4$.

1.3.7 How to Solve Transcendental Equations

In general, transcendental equations can be solved only with the aid of a computer. However, there are some cases that can lead to analytical solutions.

    A. The Exponential Equations

• The unknown variable x occurs only in the powers of expressions that are not added or subtracted. Then the logarithm of the equation should be taken.

♦ Example 1.24 Solve the equation [2]:

$$3^x = 4^{x-2} \cdot 2^x$$

Solution. Taking the logarithm of the equation, we obtain:

$$x \cdot \log 3 = (x-2) \cdot \log 4 + x \cdot \log 2$$

Thus, $x = \dfrac{2\log 4}{\log 4 - \log 3 + \log 2} = \dfrac{\log 16}{\log(8/3)}$

• The unknown variable x occurs only in the power indices of the expressions with the bases, which are integer powers of the same number a. Then often the substitution $y = a^x$ leads to the algebraic equation with respect to y.

♦ Example 1.25 Solve the equation [2]:

$$2^{x-1} = 8^{x-2} - 4^{x-2}$$

Solution. The equation can be written as

$$\frac{2^x}{2} = \frac{8^x}{64} - \frac{4^x}{16} \quad \text{or} \quad 8^x - 4 \cdot 4^x - 32 \cdot 2^x = 0$$

Substituting $y = 2^x$, we obtain

$$y^3 - 4y^2 - 32y = 0 \quad \text{or} \quad y\left(y^2 - 4y - 32\right) = 0$$

This equation has the roots

$$y_1 = 0, \quad y_2 = 8, \quad y_3 = -4.$$

The only real root of the equation corresponds to $y = 8$. Thus, in this case $x = 3$.

B. Logarithmic Equations

• Unknown variable x occurs only in the logarithm argument or the equation contains logarithms of the same polynomial $P(x)$. Then, substituting $y = \log_a P(x)$, we can obtain the algebraic equation with respect to y.

♦ Example 1.26 Solve the equation [2]:

$$4 - \lg\left(\frac{5}{2}x\right) = 3\sqrt{\lg\left(\frac{5}{2}x\right)}$$

Solution. Substituting $y = \sqrt{\lg\left(\frac{5}{2}x\right)}$, we obtain the equation

$y^2 + 3y - 4 = 0$ with roots $y_1 = 1$, $y_2 = -4$.

The only real solution corresponds to $y = 1$.

From $\sqrt{\lg\left(\frac{5}{2}x\right)} = 1$, we obtain $\frac{5}{2}x = 10$. Thus, $x = 4$.

• The unknown variable x occurs only in the arguments of the logarithm and the argument of the logarithm in the equation is the same, although the bases are different. Often, such an equation can be solved after the logarithms are transformed to the same base.

♦ Example 1.27 Solve the equation [2]:

$\log_2 x + \log_3 x + \log_4 x = 3 + \log_3 4$

Solution. Let us transform the equation into the form

$$\log_4 x\left(\frac{\log_2 x}{\log_4 x} + \frac{\log_3 x}{\log_4 x} + 1\right) = 3 + \log_3 4$$

Using the properties of the logarithms we can further transform the equation as:

$$\log_4 x\left(\log_2 4 + \log_3 4 + 1\right) = 3 + \log_3 4$$

that results in $\log_4 x = 1$. Thus, $x = 4$.

C. *Trigonometric Equations*

● The unknown variable x occurs only in the arguments of the trigonometric functions. Then, using trigonometric identities (see Appendix 4), it is often possible to transform the equation into the form that contains a single trigonometric function of x. This function is substituted with y that leads to the algebraic equation with respect to y.

♦ Example 1.28 Solve the equation [2]:

$$4 \sin x = 4 \cos^2 x - 1$$

Solution. Using formula $\cos^2 x = 1 - \sin^2 x$, we transform the equation to the form

$$4 \sin x = 4\left(1 - \sin^2 x\right) - 1 \quad \text{or} \quad 4 \sin^2 x + 4 \sin x - 3 = 0$$

Substituting $y = \sin x$, we obtain the equation:

$4 y^2 + 4 y - 3 = 0$ with the roots $y_1 = \dfrac{1}{2}$, $y_2 = -\dfrac{3}{2}$.

The only real solution corresponds to $y = \dfrac{1}{2}$ or $\sin x = \dfrac{1}{2}$.

Thus, there are two solutions of the latter equation:

$$x_1 = \frac{\pi}{6} + 2\pi \cdot n, \quad x_2 = \frac{5\pi}{6} + 2\pi \cdot n, \quad n = 0, \pm 1, \pm 2, \ldots$$

1.3.8 How to Solve a Set of Linear Equations

In a generic case, the set of m linear equations with n unassigned variables $x_1, x_2, \ldots, x_n$ has the form:

$$a_{11} x_1 + a_{12} x_2 + \ldots a_{1n} x_n = b_1$$
$$a_{21} x_1 + a_{22} x_2 + \ldots a_{2n} x_n = b_2$$
$$\ldots\ldots\ldots\ldots\ldots\ldots\ldots\ldots\ldots$$
$$a_{m1} x_1 + a_{m2} x_2 + \ldots a_{mn} x_n = b_n$$

The solution when $x_1, x_2, \ldots, x_n = 0$ is called the *trivial solution*. If at least one of the roots is not equal to zero, it is called the *nontrivial solution*. When $n = 3$, the variables x_1, x_2, x_3 are typically denoted as x, y, z.

There are two major ways to solve the system of linear equations.

• *Elimination method.* For a simple set of linear equations (typically, when $n = 2$), one of the unassigned variables can be expressed via another one using the first equation, and then this expression is put into the second equation, which becomes the equation with a single unassigned variable only.

♦ Example 1.29 Solve a set of equations:

$$x + 2y = 4 \qquad (1.1)$$
$$2x + y = 5 \qquad (1.2)$$

Solution. From the first equation we find:

$$x = 4 - 2y \qquad (1.3)$$

Putting this expression to (1.2) we obtain:

$2(4 - 2y) + y = 5$ or $8 - 4y + y = 5$ or $3y = 3$. Thus, $y = 1$.

Putting $y = 1$ in (1.3), we find $x = 2$. The solutions are: $x = 2$, $y = 1$.

• *Matrix method.* When $n > 2$ the most efficient method to solve a set of equations is the matrix method (see Chapter 6).

1.4 Theory of Combinations

1.4.1 Factorial

The n-factorial is the function that satisfies the conditions:

$$f(0) = 1, \; f(n+1) = (n+1) \cdot f(n) \text{ for all integers } n \geq 0.$$

For any natural number n:

$$n! = 1 \cdot 2 \cdot 3 \cdot \ldots n$$

♦ Example 1.30 $1! = 1; \; 2! = 1 \cdot 2 = 2; \; 3! = 1 \cdot 2 \cdot 3 = 6; \; 6! = 1 \cdot 2 \cdot 3 \cdot 4 \cdot 5 \cdot 6 = 720.$

1.4.2 A Binomial Coefficient

A *binomial coefficient* C_n^k or in another notation $\binom{n}{k}$ is the function

$$C_n^k = \begin{cases} \dfrac{n!}{k!(n-k)!} & \text{if } 0 \le k \le n \\ 0 & \text{if } 0 \le n \prec k \end{cases} \quad \text{for any integer } n \ge 0.$$

♦ <u>Example 1.31</u> $\quad C_5^3 = \dfrac{5!}{3!(5-3)!} = 10$

For any real a and integer $k \ge 0$, the binomial coefficient C_a^k is the function

$$C_a^k = \begin{cases} \dfrac{a(a-1)(a-2)...(a-k+1)}{k!} & \text{if } k > 0 \\ 1 & \text{if } k = 0 \end{cases}$$

♦ <u>Example 1.32</u> $\quad C_{-2}^3 = \dfrac{-2 \cdot (-2-1) \cdot (-2-3)}{3!} = -4$

The properties of the binomial coefficients

• For integer $n \ge 0$, $k \ge 0$ $\quad C_n^k = C_n^{n-k}$

• For real a, b

$$\binom{a}{k} + \binom{a}{k+1} = \binom{a+1}{k+1}; \quad \binom{a}{0} + \binom{a+1}{1} + \binom{a+2}{2} + ... + \binom{a+k}{k} = \binom{a+k+1}{k}$$

$$\binom{a}{0}\binom{b}{k} + \binom{a}{1}\binom{b}{k-1} + ... + \binom{a}{k}\binom{b}{0} = \binom{a+b}{k}$$

• If integer $a = b = n \ge 0$

$$\left(C_n^0\right)^2 + \left(C_n^1\right)^2 + ... + \left(C_n^n\right)^2 = C_{2n}^n$$

1.4.3 A Polynomial Coefficient

A *polynomial coefficient* is the function $C_n(k_1, k_2, ... k_m)$ or in another notation

$$\binom{n}{k_1, k_2, ... k_m}$$

$$C_n(k_1, k_2, \ldots k_m) = \frac{n!}{k_1! k_2! \ldots k_m!}$$

defined for all integers $n \geq 0$ and all successions of integers

$$k_1 \geq 0, k_2 \geq 0, \ldots, k_m \geq 0$$

for which

$$\sum_{i=1}^{m} k_j = n$$

♦ Example 1.33 $C_6(2,1,3) = \dfrac{6!}{2! \cdot 1! \cdot 3!} = \dfrac{1 \cdot 2 \cdot 3 \cdot 4 \cdot 5 \cdot 6}{1 \cdot 2 \cdot 1 \cdot 1 \cdot 2 \cdot 3} = 60$

1.4.4 Newton's Binomial

For all real $a \neq 0, b \neq 0$ and all natural numbers n, the following formula of *Newton's binomial* is valid:

$$(a+b)^n = \sum_{k=0}^{n} C_n^k a^{n-k} b^k = C_n^0 a^n b^0 + C_n^1 a^{n-1} b^1 + \ldots + C_n^n a^0 b^n$$

$$(a-b)^n = \sum_{k=0}^{n} (-1)^k C_n^k a^{n-k} b^k = C_n^0 a^n b^0 - C_n^1 a^{n-1} b^1 + \ldots + C_n^n a^0 b^n$$

For $a = b = 1$ correspondingly

$$\sum_{k=0}^{n} C_n^k = 2^n$$

For $a = 1, b = -1$

$$\sum_{k=0}^{n} (-1)^k C_n^k = 0$$

♦ Underline Example 1.34

$$(a+b)^3 = C_3^0 a^3 + C_3^1 a^2 b + C_3^2 a^1 b^2 + C_3^3 b^3 = a^3 + 3a^2 b + 3ab^2 + b^3$$

1.4.5 Polynomial Formula

For any real $a_1 \neq 0, a_2 \neq 0, \ldots, a_m \neq 0$ and any natural number n,

$$(a_1 + a_2 + \ldots a_m)^n = \sum_{k_1 + k_2 + \ldots k_m = n} C_n(k_1, k_2, \ldots k_m) \times a_1^{k_1} a_2^{k_2} \ldots a_m^{k_m}$$

for a succession $(k_1, k_2, \ldots, k_m)$ for which

$$\sum_{i=1}^{m} k_m = n$$

For $a_1 = a_2 = \ldots a_m = 1$

$$\sum_{k_1 + k_2 + \ldots k_m = n} C_n(k_1, k_2, \ldots, k_m) = m^n$$

♦ Underline Example 1.35

$$(x+y+z)^3 = C_3(3,0,0)x^3 + C_3(2,1,0)x^2 y + C_3(2,0,1)x^2 z + C_3(1,2,0)xy^2 + C_3(1,1,1) \cdot$$
$$xyz + C_3(1,0,2)xz^2 + C_3(0,3,0)z^3 + C_3(0,2,1)y^2 z + C_3(0,1,2)yz^2 + C_3(0,0,3)z^3 =$$
$$x^3 + 3x^2 y + 3x^2 z + 3xy^2 + 6xyz + 3xz^2 + y^3 + 3y^2 z + 3yz^2 + z^3$$

1.4.6 Commutation

Commutation of n elements is a set of groups, which contains n elements and the succession of elements is taken into account. The number of commutations of n elements P_n is equal:

$$P_n = n!$$

♦ Underline Example 1.36   How many different combinations exist to put 10 books on a bookshelf?
Solution. Ten books on a bookshelf, taking into account the succession they follow each other, can be put in $P_{10} = 10! = 3,628,800$ different combinations.

1.4.7 Occupancy

The *occupancy* of n elements is k groups $(k \leq n)$ is a combination of groups of k out of n elements which differ in elements (at least one) or a succession of elements in the group.

The number of different groups in an occupancy A_n^k is:

$$A_n^k = \frac{n!}{(n-k)!}$$

♦ Example 1.37 There are six different numbers 3, 5, 7, 11, 13, and 17 given. How many different fractions can be composed from these numbers if each fraction consists of two numbers?

Solution. Since the fractions $\dfrac{a}{b}$ and $\dfrac{b}{a}$ composed from two numbers a, b are different, the number of possible fractions is [1]:

$$A_6^2 = \frac{6!}{4!} = 5 \cdot 6 = 30$$

1.4.8 Combination

The *combination* of n elements in k groups $(k \leq n)$ is a combination of groups of k out of n elements which differ in elements (at least one) but a succession of elements in the group does not matter. The number of different groups in a combination C_n^k is:

$$C_n^k = \frac{n!}{k!(n-k)!}$$

♦ Example 1.38 What is a probability of winning a 7-out-of-49 lotto when you fill in the card with one combination?

Solution. The number of different combinations is C_{49}^7. Thus, your chance of winning is 1 in $C_{49}^7 = 85{,}900{,}584$ (probability of winning is 1 over 85,900,584 or $P = 0.00000001$). Good luck!

1.5 Plane Geometry

1.5.1 Triangle

Basic Rules (Figure 1.2):
• The sum of two sides in a triangle is always greater than the third side:

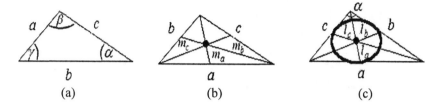

Figure 1.2 (a) A general triangle, (b) medians, and (c) bisectors.

$b + c \geq a.$

- The sum of the angles is equal to $180°$:

$\alpha + \beta + \gamma = 180° = \pi.$

- The *median* is a line, connecting the apex of a triangle and a middle of the opposite side. All medians intersect in a single point, which is the center of gravity of a triangle. The length of the median belonging to a side a is:

$$m_a = \frac{\sqrt{2(b^2 + c^2) - a^2}}{2}$$

- A *bisector* is a line that bisects the interior angle. All bisectors intersect in a single point, which is in the center of the triangle. The length of a bisector of the angle α is equal to:

$$l_a = \frac{\sqrt{bc[(b + c)^2 - a^2]}}{b + c}$$

- Area of the triangle:

$$A = \frac{(ab \sin \gamma)}{2} = \sqrt{p(p - a)(p - b)(p - c)} \quad \text{where} \quad p = \frac{(a + b + c)}{2}$$

- For a *rectangular* triangle $a^2 + b^2 = c^2$ (*Pythagoras' theorem*, Figure 1.3).

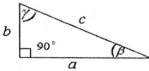

Figure 1.3 A rectangular triangle.

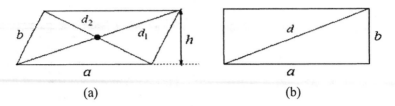

(a)   (b)

Figure 1.4 (a) A parallelogram and (b) a rectangle.

The area of a rectangular triangle is: $A = \dfrac{ab}{2} = \dfrac{\left(a^2 \tan \beta\right)}{2} = \dfrac{\left(c^2 \sin 2\beta\right)}{4}$

1.5.2 Parallelogram

Basic Rules [Figure 1.4(a)]:
- The opposite sides are equal.
- The opposite sides are parallel.
- The diagonals in the intersection point are divided into equal parts.
- The opposite angles are equal.
- The sides and diagonals are related as: $d_1^2 + d_2^2 = 2\left(a^2 + b^2\right)$.
- The area is: $A = ah$.

1.5.3 Rectangle

A *rectangle* [Figure 1.4(b)] is a parallelogram if all angles are equal to $90°$ and diagonals are equal: $d_1 = d_2 = d$. The area is $A = ab$.

1.5.4 Square

A *square* is a rectangle if $a = b$. Basic formulas [Figure 1.5(a)]:

- $d = a\sqrt{2}$ • $a = \dfrac{d\sqrt{2}}{2}$ • The area is $A = a^2 = \dfrac{d^2}{2}$.

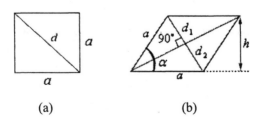

(a)   (b)

Figure 1.5 (a) A square and (b) a rhombus.

1.5.5 Rhombus

A *rhombus* [Figure 1.5(b)] is a parallelogram if the following conditions are met: all sides are equal; diagonals are mutually perpendicular; and diagonals bisect the angles.

The area is $A = ah = a^2 \sin \alpha = \dfrac{d_1 d_2}{2}$.

1.5.6 Trapezoid

A *trapezoid* [Figure 1.6 (a)] is a quadrangle with two parallel sides.

Area is $A = \dfrac{(a+b)h}{2}$. If $d = c$, $A = (a - c \cdot \cos \gamma) \cdot c \sin \gamma = (b + c \cdot \cos \gamma) \cdot c \sin \gamma$

1.5.7 Quadrangle

Sum of all the angles of a quadrangle is equal to $360°$ [Figure 1.6(b)].

Area is $A = \dfrac{(d_1 d_2 \sin \alpha)}{2}$

1.5.8 Polygon

If the number of sides is equal to n, the sum of the interior angles is equal to $180°(n-2)$. The sum of exterior angles is equal to $360°$. For a *regular* polygon, the following formulas are valid: the central angle $\alpha = \dfrac{360°}{n}$ and the exterior angle $\beta = \dfrac{360°}{n}$.

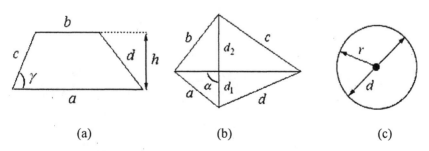

(a)   (b)   (c)

Figure 1.6 (a) A trapezoid, (b) a quadrangle, and (c) a circle

If R is the circumscribed radius, and r is the radius of incircle, then the side a is:

$$a = 2\sqrt{R^2 - r^2} = 2R\sin\left(\frac{\alpha}{2}\right) = 2r\tan\left(\frac{\alpha}{2}\right)$$

$$\text{Area } A = \frac{\pi a r}{2} = \pi r^2 \tan\left(\frac{\alpha}{2}\right) = \frac{\left(\pi R^2 \sin\alpha\right)}{2} = \frac{\left(\pi a^2 \cot\left(\frac{\alpha}{2}\right)\right)}{4}$$

1.5.9 Circle

A *circle* is the planar curve whose points are all equidistant from a fixed point called the center of the circle [Figure 1.6(c)]. The ratio of the length of the circumference c to the diameter d is constant and equal to $\pi = 3.14159265\ldots$

$$\pi = \frac{c}{d}$$

Basic formulas:

$$c = 2\pi r = \pi d = 2\sqrt{\pi A}\,; \quad \text{Area } A = \pi r^2 = \frac{\pi d^2}{4} = \frac{cd}{4}$$

1.6 Solid Geometry

1.6.1 Angles

• The *dihedral angle* is formed by two semiplanes starting at one straight line. The dihedral angle is measured by its linear angle ABC [Figure 1.7(a)].

• A *polyhedral angle OABCDE* is formed by several planes, which have the common point (apex) and intersect through the sequence of straight lines OA, OB, $OC\ldots$[edges, Figure 1.7(b)].

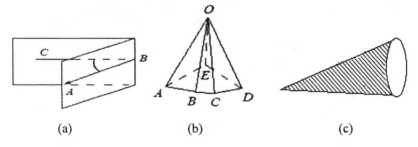

(a)   (b)   (c)

Figure 1.7 (a) A dihedral angle, (b) a polyhedral angle, and (c) a solid angle.

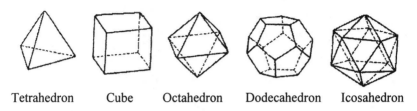

| Tetrahedron | Cube | Octahedron | Dodecahedron | Icosahedron |

Figure 1.8 Regular polyhedrons.

• A *solid angle* is a portion of the space enclosed with straight lines starting in a single point and leading to all points of a closed curve [Figure 1.7(c)].

1.6.2 Polyhedron

A *polyhedron* is a body limited by the planes (faces). The following notations are used to describe parameters of polyhedrons hereinafter:

V – volume
A – full area of the surface
M – side surface area
h – height
F – area of the base

A *regular polyhedron* is a polyhedron in which all faces are regular polygons and all polyhedral angles are equal. There are only five regular polyhedrons (Figure 1.8). Their parameters are shown in Table 1.1 (a is the length of the edge).

Table 1.1
Parameters of the Regular Polyhedrons

Regular Polyhedrons	Number of Faces and Its Shape	Number of Faces	Apexes	Full Surface	Volume
Tetrahedron	4 triangles	6	4	$1.73\,a^2$	$0.12\,a^3$
Cube	6 squares	12	8	$6\,a^2$	a^3
Octahedron	8 triangles	12	6	$3.46\,a^2$	$0.47\,a^3$
Dodecahedron	12 fiveangles	30	20	$20.65\,a^2$	$7.66\,a^3$
Icosahedron	20 triangles	30	12	$8.66\,a^2$	$2.18\,a^3$

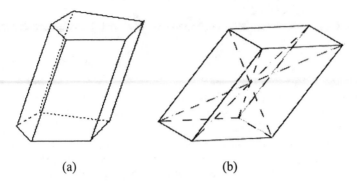

(a)   (b)

Figure 1.9 (a) A prism and (b) a parallelepiped.

1.6.3 Prism

A *prism* [Figure 1.9(a)] is a polyhedron with regular polygons in the base and parallelograms as the faces. If the edges are perpendicular to the base, the prism is called a *right* one. If the prism is a right one and its bases are the regular polygons, the prism is called the *regular* one.

Basic formulas:
- $M = pl$ • $A = M + 2F$ • $V = Fh$

where l is the length of an edge and p is the perimeter of a section by a plane perpendicular to an edge.

1.6.4 Parallelepiped

A *parallelepiped* is a prism whose bases are parallelograms [Figure 1.9(b)]. If parallelepiped is a right one and its bases are rectangles, it is called a *rectangular* one [Figure 1.10(a)]. For a rectangular parallelepiped with edges a, b, c, and diagonal d, the following formulas are valid:

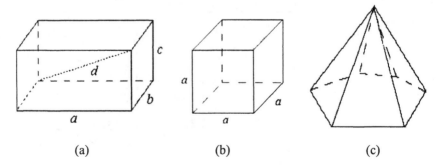

(a)   (b)   (c)

Figure 1.10 (a) A rectangular parallelepiped, (b) a cube, and (c) a pyramid.

- $d^2 = a^2 + b^2 + c^2$ • $V = abc$ • $A = 2(ab + bc + ca)$

1.6.5 Cube

A *cube* [Figure 1.10 (b)] is a rectangular parallelepiped with equal edges: $a = b = c$.

- $d^2 = 3a^2$ • $V = a^3$ • $A = 6a^2$

1.6.6 Pyramid

A *pyramid* has a polygon as a base and triangles as faces that come to a single apex [Figure 1.10 (c)]. The volume is $V = Fh/3$.

1.6.7 Cylinder

A *cylinder* is a body limited by a straight line (generatrix) moving along some curve parallel to itself (a cylindrical surface) and two parallel planes, which are the bases of the cylinder [Figure 1.11(a)].

- A *circular right cylinder* has a circle as a base and its generatrix is perpendicular to the plane of the base [Figure 1.11(b)]. If r is the radius of the base:

$$M = 2\pi r h; \quad A = 2\pi r(r + h); \quad V = \pi r^2 h.$$

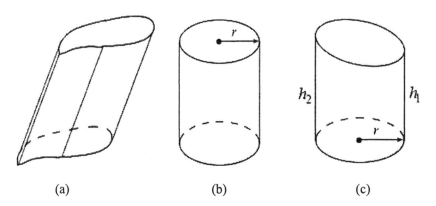

(a) (b) (c)

Figure 1.11 (a) A cylinder, (b) a circular right cylinder, and (c) a truncated circular cylinder.

- For a *truncated circular cylinder* [Figure 1.10(c)]:

$$M = \pi r (h_1 + h_2);$$

$$A = \pi r \left[h_1 + h_2 + r + \sqrt{r^2 + \frac{(h_2 - h_1)^2}{2}} \right];$$

$$V = \pi r^2 \frac{h_1 + h_2}{2}$$

1.6.8 Cone

A *cone* is a body limited by a straight line (generatrix) moving along a curve and having a fixed point (apex), and a base plane [Figure 1.12(a)]. For any cone

$$V = \frac{hF}{3}.$$

- A *circular right cone* [Figure 1.12(b)] has a circle as a base and its height goes through the center of the circle (*l* is a length of a generatrix, *r* is the radius of the circle).

$$M = \pi r l = \pi r \sqrt{r^2 + h^2}; \quad A = \pi r (r + l); \quad V = \frac{\pi r^2 h}{3}.$$

- A *truncated right cone* [Figure 1.12(c)].

$$l = \sqrt{h^2 + (R - r)^2}; \quad M = \pi l (R + r); \quad V = \frac{\pi h (R^2 + r^2 + Rr)}{3}$$

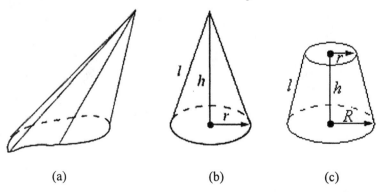

(a) (b) (c)

Figure 1.12 (a) A cone, (b) a circular right cone, and (c) a truncated right cone.

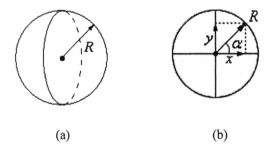

(a)   (b)

Figure 1.13 (a) A sphere and (b) definition of trigonometric functions.

1.6.9 Sphere

A *sphere* [Figure 1.13(a)] is a body each section of which is a circle. R is a radius, $D = 2R$ is a diameter.

Area is $A = 4\pi R^2 = \pi D^2 = \sqrt[3]{36\pi V^2}$. Volume is $V = \dfrac{4\pi R^3}{3} = \dfrac{\pi D^3}{6} = \dfrac{\sqrt{A^3/\pi}}{6}$.

1.7 Planar Trigonometry

1.7.1 Angles

The planar angles are typically measured in degrees or radians:

$$\alpha[\text{rad}] = \frac{\pi}{180} \cdot \alpha[\text{deg}]; \quad \alpha[\text{deg}] = \frac{180}{\pi} \cdot \alpha[\text{rad}]$$

Thus, 1 rad is $\dfrac{180}{\pi} = 57°17'44''$; $1° = \dfrac{\pi}{180} = 0.01745\,\text{rad}$.

A complete angle in degrees is equal to $360° = 2\pi$ radians.

1.7.2 Trigonometric Functions

Basic trigonometric functions are defined as follows [Figure 1.13(b)]:

- $\sin \alpha = \dfrac{y}{R}$; $\cos \alpha = \dfrac{x}{R}$ • $\tan \alpha = \dfrac{y}{x}$ if $x \neq 0$; $\cot \alpha = \dfrac{x}{y}$ if $y \neq 0$

- $\sec \alpha = \dfrac{R}{x}$ if $x \neq 0$; $\csc \alpha = \dfrac{R}{y}$ if $y \neq 0$

Basic formulas for trigonometric functions of multiple angles $(n = 0, \pm 1, \pm 2, \ldots)$

- $\sin(\alpha + 360° \cdot n) = \sin \alpha$
- $\cos(\alpha + 360° \cdot n) = \cos \alpha$
- $\tan(\alpha + 360° \cdot n) = \tan \alpha$
- $\cot(\alpha + 360° \cdot n) = \cot \alpha$
- $\sec(\alpha + 360° \cdot n) = \sec \alpha$
- $\csc(\alpha + 360° \cdot n) = \csc \alpha$

Basic formulas for trigonometric function conversion:

- $\sin\left(\dfrac{\pi}{2} \pm \alpha\right) = \cos \alpha ; \quad \sin(\pi \pm \alpha) = \mp \sin \alpha$

- $\sin\left(\dfrac{3\pi}{2} \pm \alpha\right) = -\cos \alpha ; \quad \sin(2\pi \pm \alpha) = \pm \sin \alpha$

- $\cos\left(\dfrac{\pi}{2} \pm \alpha\right) = \mp \sin \alpha ; \quad \cos(\pi \pm \alpha) = -\cos \alpha$

- $\cos\left(\dfrac{3\pi}{2} \pm \alpha\right) = \pm \sin \alpha ; \quad \cos(2\pi \pm \alpha) = \cos \alpha$

- $\tan\left(\dfrac{\pi}{2} \pm \alpha\right) = \mp \cot \alpha ; \quad \tan(\pi \pm \alpha) = \pm \tan \alpha$

- $\cot\left(\dfrac{\pi}{2} \pm \alpha\right) = \mp \tan \alpha ; \quad \cot(\pi \pm \alpha) = \pm \cot \alpha$

Even and odd trigonometric functions

- $\sin(-\alpha) = -\sin \alpha$ • $\cos(-\alpha) = \cos \alpha$
- $\tan(-\alpha) = -\tan \alpha$ • $\cot(-\alpha) = -\cot \alpha$
- $\sec(-\alpha) = \sec \alpha$ • $\csc(-\alpha) = -\csc \alpha$

The major trigonometric identities are given in Appendix 4.

1.7.3 How to Solve Triangles

- *Rectangular* triangles. Basic formulas [Figure 1.14(a)]: $\alpha + \beta = 90°$;

$$\sin \alpha = \cos \beta = \frac{a}{c} ; \quad \cos \alpha = \sin \beta = \frac{b}{c} ; \quad \tan \alpha = \cot \beta = \frac{a}{b} ; \quad \cot \alpha = \tan \beta = \frac{b}{a}$$

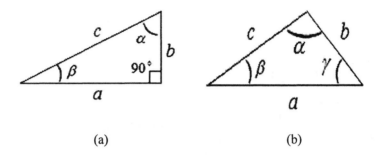

(a)   (b)

Figure 1.14 (a) A right triangle and (b) a general triangle.

- *General* triangles [Figure 1.14(b)]. Hereinafter, $p = \dfrac{a+b+c}{2}$ is half-perimeter, R is a circumradius, r is a radius of incircle, A is an area.

  Basic formulas:

$$\alpha + \beta + \gamma = 180°$$

$$\frac{a}{\sin \alpha} = \frac{b}{\sin \beta} = \frac{c}{\sin \gamma} = 2R$$

$$c^2 = a^2 + b^2 - 2ab \cos \gamma$$

$$\frac{a-b}{a+b} = \frac{\tan[(\alpha - \beta)/2]}{\tan[(\alpha + \beta)/2]} = \frac{\tan[(\alpha - \beta)/2]}{\cot[\gamma/2]}$$

$$\tan \frac{\gamma}{2} = \sqrt{\frac{(p-a)(p-b)}{p(p-c)}}$$

$$\sin \frac{\gamma}{2} = \sqrt{\frac{(p-a)(p-b)}{ab}} \; ; \quad \cos \frac{\gamma}{2} = \sqrt{\frac{p(p-c)}{ab}}$$

$$\frac{a+b}{c} = \frac{\cos[(\alpha - \beta)/2]}{\sin(\gamma/2)} = \frac{\cos[(\alpha - \beta)/2]}{\cos[(\alpha + \beta)/2]}$$

$$\frac{a-b}{c} = \frac{\sin[(\alpha - \beta)/2]}{\cos(\gamma/2)} = \frac{\sin[(\alpha - \beta)/2]}{\sin[(\alpha + \beta)/2]}$$

$$c = a\cos\beta + b\cos\alpha \; ; \; \tan\gamma = \frac{c\sin\alpha}{b - c\cdot\cos\alpha} = \frac{c\cdot\sin\beta}{a - c\cdot\cos\beta}$$

$$R = \frac{p}{4\cos(\alpha/2)\cos(\beta/2)\cos(\gamma/2)}$$

$$r = \sqrt{\frac{(p-a)(p-b)(p-c)}{p}} = p\tan\frac{\alpha}{2}\tan\frac{\beta}{2} = \tan\frac{\gamma}{2}$$

$$= 4R\sin\frac{\alpha}{2}\sin\frac{\beta}{2}\sin\frac{\gamma}{2} = (p-c)\tan\frac{\gamma}{2}$$

A height h_c on the side c: $h_c = a\sin\beta = b\sin\alpha$

A median m_c on the side c: $m_c = \frac{1}{2}\sqrt{a^2 + b^2 + 2ab\cos\gamma}$

A bisector l_γ of the angle γ:

$$l_\gamma = \frac{2ac\cos(\beta/2)}{a+c} = \frac{2bc\cos(\alpha/2)}{b+c}$$

$$A = \frac{1}{2}ab\sin\gamma = 2R^2\sin\alpha\sin\beta\sin\gamma = c^2\frac{\sin\alpha\sin\beta}{2\sin\gamma} = c^2\frac{\sin\alpha\sin\beta}{2\sin(\alpha+\beta)}$$

$$= \sqrt{p(p-a)(p-b)(p-c)} = rp$$

• Example 1.39 For a general triangle, the following parameters are given:

1. One side and two angles (e.g., c, α, β). Find the third angle and the sides.

Solution. The angle $\gamma = 180° - \alpha - \beta$. Sides are: $a = c\dfrac{\sin\alpha}{\sin\gamma}$; $b = c\dfrac{\sin\beta}{\sin\gamma}$

2. Two sides and angle between them (e.g., a, b, γ). Find the third side and angles.

Solution. $c = \sqrt{a^2 + b^2 - 2ab\cos\gamma}$, $\cos\alpha = \dfrac{b^2 + c^2 - a^2}{2bc}$, $\beta = 180° - \alpha - \gamma$

3. Two sides and the angle opposite to one of them (e.g., a, b, α).
Solution. The third side is found as in the previous example, an angle β from

$\sin\beta = \dfrac{b}{a}\sin\alpha$ and $\gamma = 180° - \alpha - \beta$

4. Three sides (e.g., a, b, c). Find the angles.

Solution. $\tan\dfrac{\alpha}{2} = \sqrt{\dfrac{(p-b)(p-c)}{p(p-a)}}$; $\tan\dfrac{\beta}{2} = \sqrt{\dfrac{(p-a)(p-c)}{p(p-b)}}$

$\tan\dfrac{\gamma}{2} = \sqrt{\dfrac{(p-a)(p-b)}{p(p-c)}}$

1.7.4 Inverse Trigonometric Functions

- $y = a\sin x$, if $\sin y = x$ and $-\pi/2 \le y \le \pi/2$
- $y = a\cos x$, if $\cos y = x$ and $0 \le y \le \pi$
- $y = a\tan x$, if $\tan y = x$ and $-\pi/2 < y < \pi/2$
- $y = a\cot x$, if $\cot y = x$ and $0 < y < \pi$

Basic formulas for multiple angles $(n = 0, \pm 1, \pm 2, \ldots)$:

- $A\sin x = (-1)^n a\sin x + \pi n$ for all angles α satisfying $\sin\alpha = x$, $-1 \le x \le 1$
- $A\cos x = \pm a\cos x + 2\pi n$ for all angles α satisfying $\cos\alpha = x$, $-1 \le x \le 1$
- $A\tan x = a\tan x + \pi n$ for all angles α satisfying $\tan\alpha = x$, $-\infty \le x \le \infty$
- $A\cot x = a\cot x + \pi n$ for all angles α satisfying $\cot\alpha = x$, $-\infty \le x \le \infty$

Basic identities and conversion formulas for inverse trigonometric functions are given in Appendix 4.

1.8 Solid Trigonometry

1.8.1 How to Solve Generic Spherical Triangles

For *Eulerian* triangles (Figure 1.15) basic formulas are as follows:

$a + b > c$, $\left|a - b\right| < c$; $\alpha + \beta < \gamma + \pi$; $\alpha < \beta$ if $a < b$;

$\alpha = \beta$ if $a < b$; $\pi < \alpha + \beta + \gamma < 3\pi$; $0 < a + b + c < 2\pi$;

For a spherical triangle, the sum of angles is always more than π. A difference $\alpha + \beta + \gamma - \pi = \varepsilon$ is called a *spherical excess* and is used to calculate the area of the spherical triangle:

$A = R^2 \cdot \varepsilon$

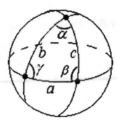

Figure 1.15 A Eulerian triangle.

where R is a radius of a sphere. Basic identities and conversion formulas are given in Appendix 5.

1.8.2 How to Solve the Right Spherical Triangles

For the *right spherical triangle, a* and *b* are legs, *c* is a hypotenuse, and α and β are angles alternate to sides *a* and *b* correspondingly.

Basic formulas:

- $\sin a = \cos(90° - a) = \sin \alpha \sin c$
- $\sin b = \cos(90° - b) = \sin \beta \sin c$
- $\cos c = \sin(90° - \alpha)\sin(90° - b) = \cos a \cos b$
- $\cos \alpha = \sin(90° - \alpha)\sin \beta = \cos a \sin \beta$
- $\cos \beta = \sin(90° - \beta)\sin \alpha = \cos b \sin \alpha$
- $\sin \alpha = \cos(90° - \alpha) = \cot(90° - b)\cot \beta = \tan b \cot \beta$
- $\sin b = \cos(90° - b) = \cot(90° - \alpha)\cot \alpha = \tan a \cot \alpha$
- $\cos c = \cot \alpha \cdot \cot \beta$
- $\cos \alpha = \cot(90° - b)\cot c = \tan b \cot c$
- $\cos \beta = \cot(90° - a)\cot c = \tan a \cot c$

1.9 Coordinates

1.9.1 Two-Dimensional (2D) Coordinates

Basic *planar* (two-dimensional) coordinates include the *generic* planar system of coordinates, *Cartesian* coordinates, and *polar* coordinates.

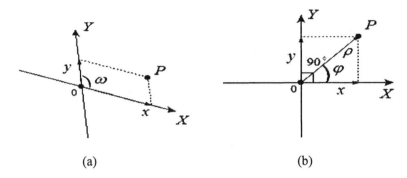

Figure 1.16 (a) Generic 2D coordinates and (b) Cartesian 2D coordinates.

The generic system with arbitrary angle between axes w is shown in Figure 1.16(a). Cartesian coordinates $(w = 90°)$ and polar coordinates are shown in Figure 1.16(b). In the generic and Cartesian coordinates, the position of a point P is described by its projections $P = (x, y)$ on axes X and Y. In the polar coordinates, a point P is described by polar radius ρ and angle φ, $P = (\rho, \varphi)$. Formulas of 2D coordinates conversion are given in Appendix 6.

1.9.2 Three-Dimensional (3D) Coordinates

Basic three-dimensional coordinates include a *generic* spatial system of coordinates [Figure 1.17(a)], *Cartesian* coordinates [Figure 1.17(b)], *spherical* coordinates [Figure 1.17(b)], and *cylindrical* coordinates [Figure 1.17(c)].

A generic (oblique) and Cartesian (rectangular) system of coordinates describe the position of a point P by its projections on the axes X, Y, Z, $P = (x, y, z)$. In spherical coordinates, a point P is described as $P = (\rho_s, \varphi, \theta)$ and in cylindrical coordinates $P = (\rho_c, \varphi, z)$. Formulas of 3D coordinates conversion are given in Appendix 6.

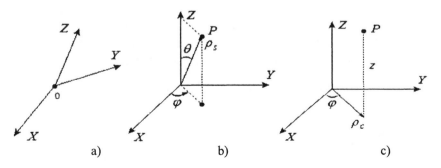

Figure 1.17 (a) Oblique, (b) Cartesian, and (c) cylindrical 3D coordinates.

1.10 Planar Analytic Geometry

1.10.1 The Distance Between Two Points

The *distance d* between two points $P_1 = (x_1, y_1)$ and $P_2 = (x_2, y_2)$ is:

$$d = \sqrt{(x_2 - x_1)^2 + (y_2 - y_1)^2 + 2(x_2 - x_1)(y_2 - y_1)\cos w}$$

Thus, for Cartesian coordinates ($w = 90^\circ$):

$$d = \sqrt{(x_2 - x_1)^2 + (y_2 - y_1)^2}$$

and for polar coordinates $P_1 = (\rho_1, \varphi_1)$, $P_2 = (\rho_2, \varphi_2)$

$$d = \sqrt{\rho_1^2 + \rho_2^2 - 2\rho_1\rho_2 \cos(\varphi_2 - \varphi_1)}$$

1.10.2 The Equation of a Straight Line

A *straight line* passing through two points $P_1 = (x_1, y_1)$ and $P_2 = (x_2, y_2)$ is described by the equation:

$$Ax + By + C = 0$$

If $B \neq 0$ the equation can be written as:

$$y = kx + b$$

where $k = \dfrac{-A}{B}$, $b = \dfrac{-C}{B}$.

If $A = 0$, the line is parallel to axis X; if $B = 0$, the line is parallel to axis Y; if $C = 0$, the line passes through the center of coordinates.
Other common forms of the equations for a straight line are:

- Equation in polar coordinates (ρ, φ):

$$\rho = \frac{p}{\cos(\varphi - \alpha)}$$

where p is a distance from a pole to a straight line; α is the angle between a polar axis and a normal to a straight line.

- Normal equation:

$$x\cos\alpha + y\sin\alpha - p = 0$$

where p is a distance from a line to the center of coordinates; α is the angle between a normal to a straight line and axis OX.

- Equation in sections:

$$\frac{x}{a} + \frac{y}{b} = 1$$

The line intersects axis X in the point $A = (a,0)$ and axis Y in the point $B = (0,b)$.

1.10.3 The Equation of an Ellipse

An *ellipse* (Figure 1.18) is a locus $M = (x,y)$ for which a sum of the distances to two fixed points $F_1 = (+c,0)$ and $F_2 = (-c,0)$ (focuses) is constant (equal to $2a$).

Basic equations are:

- Canonical equation $\dfrac{x^2}{a^2} + \dfrac{y^2}{b^2} = 1$.

- Parametric equation $x = a\cos t, \ y = b\sin t$.

- Equation in polar coordinates $\rho = \dfrac{p}{1 + e\cos\varphi}, \quad e < 1$.

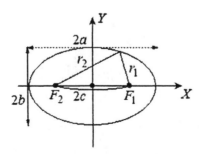

Figure 1.18 An ellipse.

Basic parameters: $\varepsilon = \dfrac{c}{a}; \ p = \dfrac{b^2}{a}; \ d = \dfrac{a}{e}.$

For any point of the ellipse $M = (x, y)$: $\dfrac{r_1}{d_1} = \dfrac{r_2}{d_2} = e.$

Area of the ellipse: $A = \pi a b$.

Length of the ellipse: $L = 4aE(e)$, where $E(e) = E(e, \pi/2)$ is a complete elliptical integral of second order. Approximate formula:

$$L \approx \pi[1.5(a + b) - \sqrt{ab}]$$

1.10.4 The Equation of a Circle

A *circle* is an ellipse for which $a = b$. Both focuses coincide with a center of a circle $(c = 0)$. Basic equations are:
- Canonical equation (R is a radius)

$$x^2 + y^2 = R^2$$

for a circle with a center in the center of coordinates. If a center is at point $C = (x_0, y_0)$, the equation is:

$$(x - x_0)^2 + (y - y_0)^2 = R^2$$

- Parametric equation

$$x = x_0 + R\cos t; \quad y = y_0 + R\cos t$$

where t is an angle of the instantaneous radius and positive direction of the axis OX (Figure 1.19).

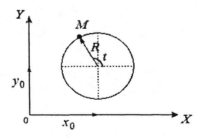

Figure 1.19 A circle.

- Equation in polar coordinates

$$\rho^2 - 2\rho\rho_0 \cos(\varphi - \varphi_0) + \rho_0^2 = R^2$$

where $C = (\rho_0, \varphi_0)$ are the polar coordinates of the circle center. If the center is located on the polar axis and coincides with a pole, the equation is:

$$\rho = 2R\cos\varphi$$

1.10.5 The Equation of a Hyperbola

A *hyperbola* [Figure 1.20(a)] is a locus $M = (x, y)$ for which modulus of the difference of the distances from two fixed points $F_1 = (+c, 0)$ and $F_2 = (-c, 0)$ (focuses) is constant (equals $2a$).
    Basic equations are:

- Canonical equation $\dfrac{x^2}{a^2} - \dfrac{y^2}{b^2} = 1$

- Parametric equation $x = a \cdot cht$, $y = b \cdot sht$

- Equation in polar coordinates

$$\rho = \frac{p}{1 + e\cos\varphi}, e > 1$$

Basic parameters: $e = \dfrac{c}{a} > 1$; $p = \dfrac{b^2}{a}$; $d = \dfrac{a}{e}$.

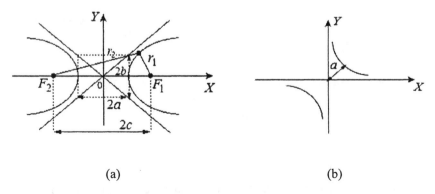

(a)   (b)

Figure 1.20 (a) A hyperbola and (b) an equilateral hyperbola.

For any point of hyperbola, $M = (x, y)$: $\dfrac{r_1}{d_1} = \dfrac{r_2}{d_2} = e.$

For equilateral hyperbola $a = b$ equation is: $x^2 - y^2 = a.$

When axes X and Y are used as the reference frames [Figure 1.20(b)], the equation of an equilateral hyperbola is:

$$xy = \frac{a^2}{2}$$

1.10.6 The Equation of a Parabola

A *parabola* [Figure 1.21(a)] is a locus, each point of which $M = (x, y)$ is equidistant to a fixed point (focus) $F = (p/2, 0)$ and a given straight line (directrix). Basic equations are:

- Canonical equation $y^2 = 2px$, p is a focal parameter.

- Equation in polar coordinates $\rho = \dfrac{p}{1 + \cos \varphi}.$

Basic parameters: $e = 1$; $R = \dfrac{(p + 2x_1)^{3/2}}{\sqrt{p}} = \dfrac{p^2}{\sin^3 u} = \dfrac{n^2}{p^2},$

where R is a parabola curvature radius at the point $M = (x_1, y_1)$, and n is the length of the normal MN [Figure 1.21(b)].

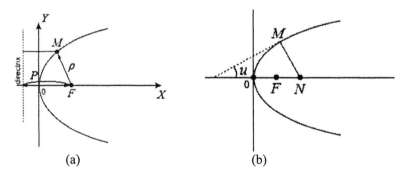

(a) (b)

Figure 1.21 A parabola representation for (a) canonical equation and (b) equation in polar coordinates.

1.11 Solid Analytic Geometry

1.11.1 The Distance Between Two Points
The *distance* d between two points $P_1 = (x_1, y_1, z_1)$ and $P_2 = (x_2, y_2, z_2)$ in Cartesian coordinates is:

$$d = \sqrt{(x_2 - x_1)^2 + (y_2 - y_1)^2 + (z_2 - z_1)^2}$$

1.11.2 The Equation of a Plane

The *plane* going through the point M with coordinates x_0, y_0, z_0 perpendicular to the vector $\bar{N}$ with coordinates A, B, C [Figure 1.22(a)] is described by the linear equation:

$$A(x - x_0) + B(y - y_0) + C(z - z_0) = 0 \tag{1.4}$$

If we rearrange (1.4) and denote

$$D = -(Ax_0 + By_0 + Cz_0)$$

then it takes the form

$$Ax + By + Cz + D = 0$$

♦ Example 1.40 The plane going through point $M(2; 1; -1)$ and perpendicular to the vector $N(-2; 4; 3)$ is given by the equation [3]:

$$-2(x - 2) + 4(y - 1) + 3(z + 1) = 0 \text{ or } -2x + 4y + 3z + 3 = 0$$

Two planes are with the equations:

$$A_1 x + B_1 y + C_1 z + D_1 = 0 \text{ and } A_2 x + B_2 y + C_2 z + D_2 = 0$$

are parallel to each other, if $\dfrac{A_2}{A_1} = \dfrac{B_2}{B_1} = \dfrac{C_2}{C_1}$.

Two planes with the same equations are perpendicular to each other if

$$A_1 A_2 + B_1 B_2 + C_1 C_2 = 0$$

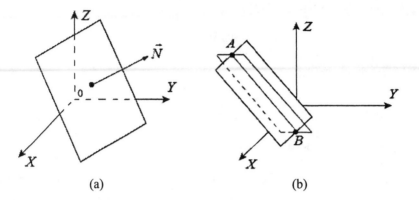

Figure 1.22 (a) A plane and (b) a straight line.

Two planes with equations above form four dihedral angles $\varphi_1 = \varphi_2 = \varphi$, $\theta_1 = \theta_2 = \theta$. The angles φ and θ can be found from equations:

$$\cos \varphi = \pm \frac{A_1 A_2 + B_1 B_2 + C_1 C_2}{\sqrt{A_1^2 + B_1^2 + C_1^2} \cdot \sqrt{A_2^2 + B_2^2 + C_2^2}} \; ; \quad \theta = \pi - \varphi$$

♦ <u>Example 1.41</u> Find the angles between two planes with equations

$x - y + \sqrt{2}z + 2 = 0$ and $x + y + \sqrt{2}z - 3 = 0$ [3].

Solution.

$$\cos \varphi = \frac{1 \cdot 1 + (-1) \cdot 1 + \sqrt{2} \cdot \sqrt{2}}{\sqrt{1 + 1 + 2} \cdot \sqrt{1 + 1 + 2}} = \pm \frac{1}{2}$$

Thus, $\varphi_1 = \varphi_2 = 60°$ or $\varphi_1 = \varphi_2 = 120°$ depending on what sign is taken.

Other common forms of the equations for the plane are:
• Normal equation

$$x \cos \alpha + y \cos \beta + z \cos \gamma - p = 0$$

where

$$\cos \alpha = \mp \frac{A}{\sqrt{A^2 + B^2 + C^2}} \; ; \quad \cos \beta = \mp \frac{B}{\sqrt{A^2 + B^2 + C^2}} \; ;$$

$$\cos\gamma = \mp \frac{C}{\sqrt{A^2+B^2+C^2}} \quad (-\text{ if } D>0, +\text{ if } D \prec 0); \quad p = \frac{|D|}{\sqrt{A^2+B^2+C^2}}$$

- Equation in sections

$$\frac{x}{a}+\frac{y}{b}+\frac{z}{c}=1$$

This is the equation of the plane going through the points $(a,0,0)$, $(0,b,0)$, $(0,0,c)$.

1.11.3 The Equation of a Straight Line

Any *straight line AB* can be represented as the intersection of two planes [Figure 1.22(b)] given by equations:

$$A_1 x + B_1 y + C_1 z + D_1 = 0$$

$$A_2 x + B_2 y + C_2 z + D_2 = 0$$

This set of equations represents the equation of a straight line in the case where A_1, B_1, C_1 are not proportional to A_2, B_2, C_2 (planes are not parallel to each other).

Any nonzero vector $\vec{L}(l,m,n)$ lying on the straight line or parallel to it with coordinates l,m,n is called a *direction vector,* and l,m,n are called the *direction coefficients* [Figure 1.23(a)].

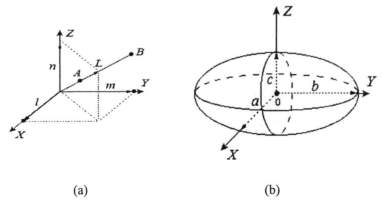

(a)   (b)

Figure 1.23 (a) A direction vector and (b) an ellipsoid.

The angles between the straight line AB and the axes of the coordinates are defined by the *direction cosines*:

$$\cos\alpha = \frac{l}{\sqrt{l^2 + m^2 + n^2}}$$

$$\cos\beta = \frac{m}{\sqrt{l^2 + m^2 + n^2}}$$

$$\cos\gamma = \frac{n}{\sqrt{l^2 + m^2 + n^2}}$$

The angle φ between two straight lines $L\left(l,m,n\right)$ and $K\left(l',m',n'\right)$ is found from the formula:

$$\cos\varphi = \frac{l\cdot l' + m\cdot m' + n\cdot n'}{\sqrt{l^2 + m^2 + n^2}\cdot\sqrt{l'^2 + m'^2 + n'^2}}$$

1.11.4 The Equation of an Ellipsoid and Sphere

The equation of the *ellipsoid* is:

$$\frac{x^2}{a^2} + \frac{y^2}{b^2} + \frac{z^2}{c^2} = 1$$

where a,b,c are semi-axes [Figure 1.23(b)].

If $a = b > c$, this leads to the *oblate ellipsoid of revolution* that results from rotation of the ellipse

$$\frac{x^2}{a^2} + \frac{z^2}{c^2} = 1$$

around its minor axis.

If $a = b \prec c$, this leads to the *elongated ellipsoid of revolution* that results from rotation of the same ellipse around its major axis.

Any section of an ellipsoid is an ellipse. The volume of ellipsoid $V = \frac{3}{4}\pi abc$.

When $a = b = c = R$, the ellipsoid reduces to the *sphere* with radius R:

$$x^2 + y^2 + z^2 = R^2$$

The volume of the sphere $V = \dfrac{4}{3}\pi R^3$.

1.11.5 The Equation of a Hyperboloid

The equation of the *hyperboloid* of *one sheet* [Figure 1.24 (a)] is:

$$\frac{x^2}{a^2} + \frac{y^2}{b^2} - \frac{z^2}{c^2} = 1$$

The equation of the *hyperboloid* of *two sheets* [Figure 1.24 (b)] is:

$$\frac{x^2}{a^2} + \frac{y^2}{b^2} - \frac{z^2}{c^2} = -1$$

In both cases the sections parallel to the axis Z are hyperbolas.

1.11.6 The Equation of a Cone

The equation of a *cone* with the apex placed in the center of the coordinates [Figure 1.24(c)]:

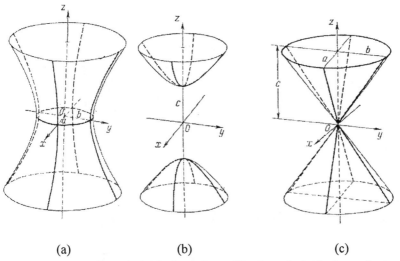

(a)   (b)   (c)

Figure 1.24 (a) A hyperboloid of one sheet, (b) a hyperboloid of two sheets, and (c) a cone.

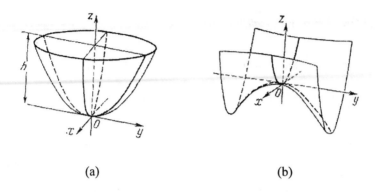

Figure 1.25 (a) An elliptic and (b) hyperbolic paraboloid.

$$\frac{x^2}{a^2} + \frac{y^2}{b^2} - \frac{z^2}{c^2} = 0$$

If $a = b$, it results in the *circular cone*.

1.11.7 The Equation of a Paraboloid

The equation of the *elliptic paraboloid* [Figure 1.25(a)] is:

$$z = \frac{x^2}{a^2} + \frac{y^2}{b^2}$$

The sections parallel to the axis Z are parabolas, and the sections parallel to the plane XOY are ellipses.

If $a = b$, this leads to a *paraboloid of revolution* that results from rotation of parabola

$$z = \frac{x^2}{a^2}$$

in the plane XOZ around its axis.

The equation of a *hyperbolic paraboloid* [Figure 1.25(b)] is:

$$z = \frac{x^2}{a^2} - \frac{y^2}{b^2}$$

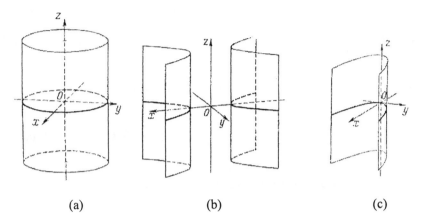

Figure 1.26 (a) An elliptic, (b) hyperbolic, and (c) parabolic cylinder.

1.11.8 The Equation of a Cylinder

The shape of a cylinder is defined by its direction vector. If we assume that it is lying in the plane *XOY* and generatrices are parallel to axis *OZ* (Figure 1.26), then the equations are as follows:

- *Elliptic cylinder* [Figure 1.26(a)]: $\dfrac{x^2}{a^2} + \dfrac{y^2}{b^2} = 1$

- *Circular cylinder* $(a = b = R)$: $x^2 + y^2 = R^2$

- *Hyperbolic cylinder* [Figure 1.26(b)]: $\dfrac{x^2}{a^2} - \dfrac{y^2}{b^2} = 1$

- *Parabolic cylinder* [Figure 1.26(c)]: $y^2 = 2px$

References

[1] Yaremchuk, F. P., and P. A. Rudenko, *Algebra and Elementary Functions*, (in Russian), Kiev: Naukova Dumka, 1976.

[2] Bronshtein I. N., and K. A. Semendyaev, *Mathematical Handbook for Engineers and Students*, (in Russian), Moscow: Nauka, 1980.

[3] Vigodskiy, M. Ya., *Mathematical Handbook*, (in Russian), Moscow: Nauka, 1972.

Selected Bibliography

Devine, D. F., and J. E. Kaufmann, *Elementary Mathematics*, New York: John Wiley and Sons, 1977.

Ellis, A. J., *Basic Algebra and Geometry for Scientists and Engineers*, New York: John Wiley and Sons, 1982.

James, G. (ed.), *Modern Engineering Mathematics*, Harlow, England: Pearson Education Ltd., 2001.

Kurtz, M., *Handbook of Applied Mathematics for Engineers and Scientists*, New York: McGraw-Hill, 1992.

Mizrahi, A., and M. Sullivan, *Calculus and Analytic Geometry*, Belmont, CA: Wadsworth Publishing Co., 1986.

Chapter 2

SEQUENCES, SERIES, AND LIMITS

2.1 Definitions

Let us consider a function that has a domain of the set of whole numbers $(0,1,2,\ldots,n)$. The *sequence* is the set of values of this function

$$f(0), f(1), f(2)\ldots$$

If the sequence does not end (that is, denoted by dots ...), it is called an *infinite* sequence; otherwise, it is called a *finite* or *terminating* sequence.

A *series* is an extended sum of terms, for example,

$$S_N = 1 + 2 + 3 + \ldots N = \sum_{n=1}^{N} n$$

2.2 Finite Sequences and Series

• An *arithmetic sequence* is one in which the difference between successive terms is a constant number (e.g., 1, 6, 11, 16, 21...) that can be written as

$$y_n = a + nd, \quad n = 0,1,2,\ldots$$

In the sequence above, $a = 1$, $d = 5$. The old name for such sequences was *arithmetic progression.*

An *arithmetic series* is the sum of the terms of an arithmetic sequence

$$S_N = a + (a+d) + (a+2d) + \ldots [a + (N-1)d] = \sum_{n=0}^{N-1} (a + nd)$$

• A *geometric sequence* is one in which the ratio of successive terms is a constant number (e.g., 2, 6, 18, 54...) that can be written as

$$y_n = a \cdot r^n, \quad n = 0, 1, 2,\ldots$$

59

where a is the first term, and r is the ratio. In the sequence above, $a = 2, r = 3$. The old name for such a sequence is *geometric progression*.

A *geometric series* is the sum of the terms of a geometric sequence. The general form is:

$$S_N = a + ar + ar^2 + \ldots ar^{N-1} = \sum_{n=0}^{N-1} ar^n$$

If $r = 1$, $S_N = aN$

If $r \neq 1$, $S_N = \dfrac{a\left(1 - r^N\right)}{1 - r}$

2.3 A Limit of a Sequence

If the term of the sequence y_n tends to a value b when n increases (tends to infinity), then the number b is called a *limit of a sequence* $y_1, y_2, \ldots, y_n$:

$$\lim_{n \to \infty} y_n = b$$

Another form of the definition of a limit of a sequence is as follows: a number b is called a *limit of a sequence* $y_1, y_2, \ldots, y_n$, if the modulus of a difference $y_n - b$ is always smaller than the chosen small positive number ε when n is larger than some number N:

$$|y_n - b| < \varepsilon \text{ for } n \geq N$$

(N depends on the choice of ε).

♦ Example 2.1 For a sequence $y_1 = 0.3$, $y_2 = 0.33$, $y_3 = 0.333\ldots$ the term y_n tends to $\dfrac{1}{3}$; thus, $\lim y_n = \dfrac{1}{3}$. The difference $y_n - \dfrac{1}{3} = -\dfrac{1}{3 \cdot 10^n}$ [1].
We can always choose the number ε so that starting from some number N, this difference will be smaller than ε. For example, if $\varepsilon = 0.01$, $N = 2$; if $\varepsilon = 0.001$, $N = 3$; and so forth.

2.4 Infinite Series

Infinite series do not end (that is, do not have a last term in the series). For example,

$$1 + 3 + 5 + 7 + \ldots + (2n+1) + \ldots$$

The infinite series is called a *convergent* one when the sum S_n of a series of n terms tends to a limit as $n \to \infty$. Otherwise, it is called a *divergent* series. If a series has both negative and positive terms, for example,

$$S = \sum_{n=0}^{\infty} y_n \quad \text{and the associated series} \quad S_1 = \sum_{n=0}^{\infty} |y_n|$$

is convergent, then S is convergent and said to be *absolutely convergent*.

The series that frequently occurs in the solution of practical electrical engineering problems are *power series*

$$a_0 + a_1 x + a_2 x^2 + \ldots + a_n x^n + \ldots$$

with four special cases:

- The geometric series

$$\frac{1}{1+x} = 1 - x + x^2 - x^3 + \ldots (-1)^n x^n + \ldots \quad (1 \prec x \prec 1)$$

- The binominal series

$$(1+x)^k = 1 + \binom{k}{1}x + \binom{k}{2}x^2 + \binom{k}{3}x^3 + \ldots + \binom{k}{n}x^n + \ldots \quad (1 \prec x \prec 1)$$

where $\binom{k}{n} = \dfrac{k(k-1)\ldots(k-n+1)}{n!}$ is the binominal coefficient (see Section 1.4.2).

- The exponential series

$$e^x = 1 + \frac{x}{1!} + \frac{x^2}{2!} + \frac{x^3}{3!} + \ldots + \frac{x^n}{n!} + \ldots \quad (\text{all } x)$$

- The logarithmic series

$$\ln(1+x) = x - \frac{x^2}{2} + \frac{x^3}{3} - \frac{x^4}{4} + \ldots (-1)^n \frac{x^{n+1}}{n+1} + \ldots \quad (1 \prec x \leq 1)$$

Some useful finite and infinite series are given in <u>Appendix 8.</u>

2.5 Limit of a Function

The definition of the limit can be extended to include a function of real variables.

A function $f(x)$ is said to approach a limit b as x approaches the value a if, given any small positive quantity ε, it is possible to find a positive number δ such that

$$|f(x) - b| < \varepsilon$$

for all x satisfying $0 < |x - a| < \delta$.

In other words, we can make the value of $f(x)$ as close as we want to b by taking x sufficiently close to a.

The limit of a *constant value c* is equal to the value c.

2.6 Infinitesimal and Nonterminating Values

The value is called an *infinitesimal* one if its limit is equal to zero.

♦ <u>Example 2.2</u>   The function $x^2 - 4$ is an infinitesimal value for $x \to 2$ and $x \to -2$. For $x \to 0$, this function is not the infinitesimal value.

Basic properties of the infinitesimal values are as follows:
- The sum of any given number of the infinitesimal values is the infinitesimal value.
- The product of the limited value (a value whose modulus does not exceed a positive number M) and an infinitesimal value is the infinitesimal value.
- The fraction with numerator equal to an infinitesimal value and denominator of a value with a limit not equal to zero is the infinitesimal value.
- Two infinitesimal values are called the equivalent ones if the limit of their ratio is equal to unity, and the limit of the quotient of two infinitesimal values will not change if one of the values is substituted by the equivalent one.

♦ <u>Example 2.3</u>   Find the limit $\lim\limits_{x\to 0}\dfrac{\sin 2x}{x}$ [1].

Solution. Since

$$\lim_{x\to 0}\frac{\sin 2x}{2x}=1$$

the value $\sin 2x$ is equivalent to $2x$.
Thus,

$$\lim_{x\to 0}\frac{\sin 2x}{x}=\lim_{x\to 0}\frac{2x}{x}=2$$

A *nonterminating* value is the value for which its modulus is greater than any given positive number M starting from some number of the term of the sequence. Sometimes a nonterminating value is defined as a value with the limit that tends to plus or minus infinity $S\to\infty$ or $S\to-\infty$.

♦ <u>Example 2.4</u>   The function $\dfrac{1}{x}$ is the nonterminating value for $x\to 0$ since

$$\lim_{x\to\infty}\frac{1}{x}=\infty \ [1].$$

If x is the infinitesimal value, then $\dfrac{1}{x}$ is the nonterminating value and vice versa.

♦ <u>Example 2.5</u>   For $x\to 0$, the function $\tan x$ is the infinitesimal value, and the function $\cot x=\dfrac{1}{\tan x}$ is the nonterminating value [1].

2.7 Basic Properties of a Limit

• The limit of the sum of any given numbers of the factors is equal to the sum of the limits of each factor

$$\lim(s_1+s_2+\ldots+s_n)=\lim s_1+\lim s_2+\ldots+\lim s_n$$

• The limit of the product of any given numbers of the factors is equal to the product of the limits of each factor

$$\lim(s_1\cdot s_2\cdot\ldots\cdot s_n)=\lim s_1\cdot\lim s_2\cdot\ldots\cdot\lim s_n$$

• The constant multiplier c can be moved out of the limit sign

$$\lim c \cdot s = c \lim s$$

- The limit of the quotient is equal to the quotient of the limits if the limit of the devisor is not equal to zero

$$\lim \frac{s_1}{s_2} = \frac{\lim s_1}{\lim s_2}$$

◆ Example 2.6 Find the limit $\lim\limits_{x \to 5} \frac{x+4}{x-2}$ [1].

Solution.

$$\lim_{x \to 5} \frac{x+4}{x-2} = \lim_{x \to 5}(x+4) : \lim_{x \to 5}(x-2) = 9 : 3 = 3$$

2.8 Two Important Limits

- $\lim\limits_{x \to 0} \dfrac{\sin x}{x} = 1$ if x is measured in radians

- $\lim\limits_{n \to \infty} \left(1 + \dfrac{1}{n}\right)^n = e \approx 2.71828\dots$ (e is the first letter of the name Euler).

2.9 Continuous and Discontinuous Functions

The function $f(x)$ is called a *continuous* one at the point $x = x_0$ if $f(x) \to f(x_0)$ as $x \to x_0$. Otherwise, the function is called a *discontinuous* at this point. The function shown in Figure 2.1(a) is a continuous one at the point $x = 0$, and the function shown in Figure 2.1(b) is a discontinuous at this point.

2.10 Properties of the Continuous Functions

If function $f(x)$ is continuous on the interval $[a,b]$, then it has the following

Euler, Leonard (1707–1783). The distinguished Swiss mathematician and the most prolific mathematical writer of all time who published over 800 papers and who made major contributions to mathematics, optics, mechanics, electricity, and magnetism. He worked at the St. Petersburg Academy, and the Berlin Academy of Science and won the Paris Academy Prize 12 times.

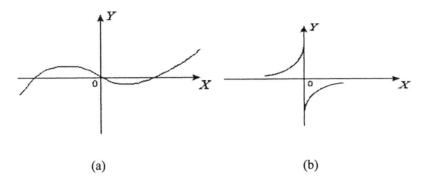

Figure 2.1 (a) Continuous and (b) discontinuous functions at $x = 0$.

basic properties:

- $f(x)$ is a bounded function; that is the numbers n and N exist such that for all x within $[a,b]$

$$M < f(x) < N$$

- $f(x)$ has the least value (*minimum*) and the largest value (*maximum*) on $[a,b]$, and the difference between the two is called the *oscillation* of $f(x)$ on $[a,b]$.

- $f(x)$ takes every value between its minimum and maximum somewhere between $x = a$ and $x = b$.

- If $a \leq x_1 \leq x_2 \leq ... x_n \leq b$ (the interval $[a,b]$ is divided into n parts), there is an $X \in [a,b]$ such as

$$f(X) = \frac{f(x_1) + f(x_2) + ... + f(x_n)}{n}$$

- Given $\varepsilon > 0$, $f(x)$ can be approximated on $[a,b]$ by a polynomial of suitable degree such that

$$|f(x) - P_n(x)| < \varepsilon \text{ for } x \in [a,b] \quad (Weierstrass's\ theorem)$$

- If two functions $f(x)$ and $g(x)$ are continuous functions, so are the functions:

$c \cdot f(x)$ where c is a constant; $f(x)+g(x)$; $f(x) \cdot g(x)$; and

$f(x)/g(x)$, except where $g(x) = 0$.

- All basic elementary functions

$y = c$; $y = a^x$; $y = x$; $y = \log_a x$; $y = x^n$; $y = e^x$;

$y = \sin x$; $y = \text{asin } x$; $y = \cos x$; $y = \text{acos } x$;

$y = \tan x$; $y = \text{atan } x$; $y = \cot x$; $y = \text{acot } x$

are continuous functions within its entire domain.

Reference

[1] Vigodskiy, M. Ya., *Mathematical Handbook*, (in Russian), Moscow: Nauka, 1972.

Selected Bibliography

Binmore, K. G., *Mathematical Analysis: A Straightforward Approach*, London: Cambridge Press, 1976.

Bronshtein, I.N., and K.A. Semendyaev, *Mathematical Handbook for Engineers and Students*, (in Russian), Moscow: Nauka, 1980.

James, G. (ed.), *Modern Engineering Mathematics*, Harlow, England: Pearson Education Ltd., 2001.

Moss, R. M. F., and G. T. Roberts, *A Preliminary Course in Analysis*, London: Chapman and Hall Ltd., 1968.

Chapter 3

DIFFERENTIAL AND INTEGRAL CALCULUS

3.1 Fundamentals of Differentiation

3.1.1 Historical Background

The differential calculus evolved from tasks dealing with variation and motion, and primarily from two basic tasks:

- To determine a tangent at a point on a curve;
- To determine a velocity of an object.

In general, the task is to find a function $f(t)$ based on the underlying function $f(t)$. The function $\dot{f}(t)$, which typically is a velocity of the function $f(t)$ variation with respect to its argument, later was termed "a derivative." Newton and Leibnitz, working independently, contributed most to the development of the differential calculus. Originally a derivative was called "fluxia" by Newton, and the term "derivative" was introduced by Arbogastus at the end of the eighteenth century. The term "differential" was introduced by Leibnitz.

Newton, Isaac (1642–1727). English physicist and mathematician who is probably the single most important contributor to the development of modern science. He was born in a poor farming family but, luckily for humanity, was a bad farmer and was sent to Cambridge to study to become a preacher. His genius and devotion to science resulted in discovering fundamental laws of motion (Newton's laws), gravity and motion of planets, classical theory of mechanics and optics, a system of chemistry, theory of sound, differential and integral calculus, etc. Actually he mathematized all of the physical sciences, reducing their study to rigorous, universal, and rational procedures.

Leibnitz, Gottfried (1646–1716). German philosopher, physicist, and mathematician. He made contributions to the study of forces, weights, and theory of differential equations. Independently of Newton, he also developed differential and integral calculus, and introduced the integral sign and derivative notation.

3.1.2 Velocity

For uniform linear motion, velocity is defined as change $\Delta d = d_2 - d_1$ in distance d_1 in the moment t_1, and d_2 in the moment t_2, with respect to change in time $\Delta t = t_2 - t_1$. Thus, velocity V is a quotient:

$$V = \frac{\Delta d}{\Delta t}$$

that gives an *average* velocity on the interval $\Delta t = (t_1, t_2)$.

For nonuniform motion average velocity is not accurate enough to describe the velocity in moment t_2. However, the smaller the interval Δt, the more accurately the velocity is described. Thus, the best measure to describe velocity V in the moment t_2 is the limit:

$$V = \lim_{\Delta t \to 0} \frac{\Delta d}{\Delta t}$$

Actually, the function $V(t)$ depends on the function $d(t)$, or in other terms the function $d(t)$ "derives" function $V(t)$. This is why these types of functions were termed "derivatives."

3.1.3 Definition of a Derivative

Let us consider $y = f(x)$, which is a continuous function of an argument x defined at an interval (a, b). Let x be a point within this interval. For a difference Δx (positive or negative), the function $y = f(x)$ will get a difference Δy:

$$\Delta y = f(x + \Delta x) - f(x)$$

An infinitesimal difference Δx results in an infinitesimal difference Δy. A *derivative* of a function $y = f(x)$ is a limit

$$\frac{d}{dx} f(x) = \dot{y} = \dot{f}(x) = \lim_{\Delta x \to 0} \frac{f(x + \Delta x) - f(x)}{\Delta x}$$

Note. When calculating the limit above, argument x is considered to be a constant value.

♦ <u>Example 3.1</u> Find the derivative for the function $y = x^2$ when $x = 3$.

Solution. $x + \Delta x = 3 + \Delta x$; $\Delta y = (3 + \Delta x)^2 - 3^2 = 6\Delta x + \Delta x^2$;

$$\frac{\Delta y}{\Delta x} = \frac{6\Delta x + \Delta x^2}{\Delta x} = 6 + \Delta x; \quad \lim_{\Delta x \to 0} \frac{\Delta y}{\Delta x} = \lim_{\Delta x \to 0} 6 + \Delta x = 6.$$

♦ Example 3.2 Find the derivative for the function $y = x^2$ for an arbitrary x.

Solution. $\Delta y = (x + \Delta x)^2 - x^2 = 2x\Delta x + \Delta x^2$

$$\frac{dy}{dx} = \lim_{\Delta x \to 0} \frac{\Delta y}{\Delta x} = \lim_{\Delta x \to 0} \frac{2x\Delta x + \Delta x^2}{\Delta x} = \lim_{\Delta x \to 0} (2x + \Delta x) = 2x.$$

3.1.4 Properties of the Derivative

- Constant factor c can be moved out of differentiation operation:

$$\frac{d}{dx}[c \cdot f(x)] = c \cdot \frac{d}{dx} f(x)$$

- The derivative of the algebraic sum of the functions is equal to the algebraic sum of the derivatives:

$$\frac{d}{dx}[f_1(x) + f_2(x) - f_3(x)] = \frac{d}{dx} f_1(x) + \frac{d}{dx} f_2(x) - \frac{d}{dx} f_3(x)$$

♦ Example 3.3

$$\frac{d}{dx}[4x^2] = 4 \cdot \frac{d}{dx} x^2 = 4 \cdot 2x = 8x; \qquad \frac{d}{dx}(3x^2 - x) = \frac{d}{dx} 3x^2 - \frac{d}{dx} x = 6x - 1$$

3.1.5 Differential

Let us consider that the difference Δy of a function $y = f(x)$ is represented as the sum of two factors:

$$\Delta y = A \cdot \Delta x + B$$

where A does not depend on Δx and B is the parameter of the higher order with respect to Δx when $x \to 0$.

Then the first factor is termed a *differential* of a function $f(x)$ and typically denoted as dy or $df(x)$.

Basic properties of the differential:

• A factor A is equal to the derivative of a function or, in other words, a differential of the function is equal to the product of the derivative and argument difference:

$$dy = \dot{y} \cdot \Delta x, \ \text{ or } df(x) = \dot{f}(x) \cdot \Delta x$$

• If a derivative is not equal to zero, then the differential of the function and its difference Δy are equivalent. If a derivative is equal to zero, then dy and Δy are not equivalent.

♦ Example 3.4 Find the differential of the function $y = \dfrac{1}{x}$ [1].

Solution. $\dot{y} = -\dfrac{1}{x^2}$. Thus, $dy = \dot{y} \cdot \Delta x = -\dfrac{\Delta x}{x^2}$

3.1.6 A Tangent

Let us consider that a line L represents a graph of the function $y = f(x)$. Then,

$$\Delta x = AB, \ \Delta y = BD$$

A *tangent* at a point A on a curve L is a line AC with which a secant AD tries to be coplaced when the point D moves to be coplaced with the point A staying at curve L (Figure 3.1).

A tangent AC splits BD into two sections: BC and CD. The line BC is proportional to AB and is equal to a differential dy:

$$BC = AB \cdot \tan \varphi$$

That is equivalent to:

$$dy = \Delta x \cdot \dot{f}(x)$$

Thus, on a graph, the differential of a function is represented by the difference of a tangent ordinate.

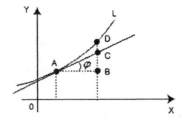

Figure 3.1 A tangent.

3.1.7 Properties of the Differential

• Constant factor C can be moved out of differential operation:

$$d[C \cdot f(x)] = C \cdot df(x)$$

• The differential of the algebraic sum of the functions is equal to the algebraic sum of differentials:

$$d[f_1(x) + f_2(x) - f_3(x)] = df_1(x) + df_2(x) - df_3(x)$$

• The differential of the function is equal to the product of a derivative and differential of the argument:

$$df(x) = \dot{f}(x) \cdot dx$$

3.1.8 The Differentials of Some Basic Functions

• The differential of constant function is equal to zero: $dc = 0$.
• The differential of the independent variable is equal to its difference: $dx = \Delta x$.
• The differential of a linear function is: $d(ax + b) = \Delta(ax + b) = a \cdot \Delta x$.
• The differential of the function x^n is: $dx^n = n \cdot x^{n-1} \Delta x$.

3.1.9 Differentiable Function

A continuous function that has a differential at a given point is called a *differentiable* function. If a function is differentiable, the limit

$$\lim_{\Delta x \to 0} \frac{f(a + \Delta x) - f(a)}{\Delta x}$$

exists, and at $x = a$ the graph of $f(x)$ has a unique, nonvertical, well-defined tangent.

◆ Example 3.5 The function $y = x^2$ is differentiable at point $x = 3$ since the

$$\text{limit } \frac{\Delta y}{\Delta x} = \lim_{\Delta x \to 0} \frac{6\Delta x + x^2}{\Delta x} = \lim_{\Delta x \to 0} (6 + \Delta x) = 6 \text{ exists [1].}$$

◆ Example 3.6 The function $y = \sqrt[3]{x}$ is not differentiable at point $x = 0$ since

$$\text{the limit } \frac{\Delta y}{\Delta x} = \lim_{\Delta x \to 0} \frac{\sqrt[3]{0 + \Delta x} - \sqrt[3]{0}}{\Delta x} = \infty \text{ (equal to infinity) [1].}$$

3.1.10 Differential of a Composite Function

A *composite function* is a function from another function of the following type:

$$y = f(u), \ u = \varphi(x) \text{ or } y = f[\varphi(x)]$$

◆ Example 3.7 If $y = u^2$, $u = 1 + 3x$, then a composite function is $y = (1 + 3x)^2$.

If $y = f(u)$ and $u = \varphi(x)$ are continuous functions, then the function $y = f[\varphi(x)]$ is also a continuous one.

The formula $df(x) = f'(x)dx$ is valid when x is a function of some other argument [e.g., $x = \varphi(t)$]. Thus, the differential of a composite function can be found by applying the above formula step by step to each function.

◆ Example 3.8 Find a differential of the function $y = (1 + x^2)^3$ [1].
Solution. We can write the complex function as: $y = u^3$, $u = 1 + x^2$.
Thus, $dy = 3u^2 du$, $du = 2x\,dx$, which results in

$$dy = 3(1 + x^2)^2 \cdot 2x\,dx = (6x + 12x^3 + 6x^5)dx$$

3.1.11 Derivative of a Composite Function

A derivative of a composite function is equal to the derivative of the function with respect to the auxiliary variable, multiplied by the derivative of this variable with respect to the argument. For the function $y = f[\varphi(x)]$, $y = f(u)$, $u = \varphi(x)$

$$\frac{dy}{dx} = \frac{dy}{du} \cdot \frac{du}{dx}$$

Analogously, if $y = f(u)$, $u = \varphi(v)$, $v = \phi(x)$

$$\frac{dy}{dx} = \frac{dy}{du} \cdot \frac{du}{dv} \cdot \frac{dv}{dx}$$

and so forth for greater numbers of factors.

♦ <u>Example 3.9</u> Find the derivative of the function $y = \sqrt{a^2 - x^2}$ [1].

Solution. Denoting $y = u^{1/2}$, $u = a^2 - x^2$,

$$\frac{dy}{du} = \frac{1}{2} u^{-1/2} = \frac{1}{2\sqrt{a^2 - x^2}}, \quad \frac{du}{dx} = -2x;$$

$$\frac{dy}{dx} = \frac{1}{2\sqrt{a^2 - x^2}} \cdot (-2x) = -\frac{x}{\sqrt{a^2 - x^2}}$$

♦ <u>Example 3.10</u> Find the derivative of the function $y = \sin^2 2x$ [1].

Solution. Denoting $y = u^2$, $u = \sin v$, $v = 2x$;

$$\frac{dy}{dx} = 2u \cdot \cos v \cdot 2 = 2 \sin 2x \cos 2x \cdot 2 = 2 \sin 4x$$

3.1.12 Derivative of a Product

A differential of a product of two functions is equal to the sum of the products of each function and differential of another function:

$$d(uv) = udv + vdu$$

Analogously,

$$d(uvw) = vwdu + uwdv + uvdw$$

and so forth for greater numbers of functions.

A derivative of a product can be found following the same rule:

$$\frac{d(uv)}{dx} = u\frac{dv}{dx} + v\frac{du}{dx} \qquad \frac{d(uvw)}{dx} = vw\frac{du}{dx} + uw\frac{dv}{dx} + uv\frac{dw}{dx}$$

♦ **Example 3.11** Find the derivative of the product $x \cdot \sin\dfrac{1}{x}$ [1].

Solution.

$$\frac{d}{dx}\left(x\cdot\sin\frac{1}{x}\right) = x\cdot\frac{d\sin\dfrac{1}{x}}{dx} + \sin\frac{1}{x}\cdot\frac{dx}{dx} = x\cos\frac{1}{x}\cdot\frac{d}{dx}\left(\frac{1}{x}\right) + \sin\frac{1}{x}\cdot1$$

$$= x\cdot\cos\frac{1}{x}\cdot\left(-\frac{1}{x^2}\right) + \sin\frac{1}{x} = -\frac{1}{x}\cdot\cos\frac{1}{x} + \sin\frac{1}{x}$$

3.1.13 Derivative of a Quotient (Fraction)

A differential of a fraction is equal to the product of a denominator and differential of a numerator minus the product of a numerator and differential of a denominator, all divided by a square of the denominator:

$$d\frac{u}{v} = \frac{vdu - udv}{v^2}$$

The same rule is applicable to a derivative:

$$\frac{d}{dx}\left(\frac{u}{v}\right) = \frac{v\dfrac{du}{dx} - u\dfrac{dv}{dx}}{v^2}$$

♦ **Example 3.12** Find the derivative of a function $y = (2x+1)/(x^2+1)$ [1].

Solution.

$$\frac{dy}{dx} = \frac{(x^2+1)\dfrac{d}{dx}(2x+1) - (2x+1)\dfrac{d}{dx}(x^2+1)}{(x^2+1)^2} = \frac{(x^2+1)\cdot2 - (2x+1)\cdot2x}{(x^2+1)^2}$$

$$= \frac{2(-x^2-x+1)}{(x^2+1)^2}$$

3.1.14 Derivative of a Reciprocal Function

If a function $y = f(x)$ results in a function $x = \varphi(y)$, then the function $\varphi(y)$ is termed a *reciprocal* one with respect to function $f(x)$.

♦ Example 3.13 For the function $y = x^2$, the reciprocal function is $x = \sqrt{y}$; for the function $y = 10^x$, the reciprocal function is $y = \log_{10} x$.

The derivative of a reciprocal function is equal to unity divided by the derivative of the underlying function:

$$\frac{dx}{dy} = 1 \bigg/ \frac{dy}{dx}$$

♦ Example 3.14 For a function $y = x^2$, $x \geq 0$, the reciprocal function is $x = \sqrt{y}$. Thus,

$$\frac{dy}{dx} = 2x, \quad \frac{dx}{dy} = \frac{1}{2x} = \frac{1}{2}\frac{1}{\sqrt{y}}$$

3.1.15 Derivative of Logarithmic Functions

The simplest is the rule of differentiation for a natural logarithm $\ln(x)$ with a base:

$$e = \lim_{\Delta x \to \infty}\left(1 + \frac{1}{x}\right)^x \approx 2.71828$$

In this case:

$$d \ln x = \frac{dx}{x} \qquad \frac{d}{dx}\ln x = \frac{1}{x}$$

For any arbitrary base of the logarithm

$$d \log_a x = \log_a e \cdot \frac{dx}{x} \qquad \frac{d}{dx}\log_a x = \log_a e \cdot \frac{1}{x}$$

In particular, for a logarithm with base $a = 10$ $\left(\log_{10} x = \lg x\right)$

$$d \lg x = \frac{C dx}{x} \qquad \frac{d}{dx} \lg x = C \cdot \frac{1}{x}, \quad C \approx 0.4343$$

3.1.16 How to Use a Differential for Approximate Calculations

If the function and its derivative can be easily found for a point x_0 but not for other points close to x_0, the approximate formula can be used:

$$f(x) = f(x_0 + \Delta x) = f(x_0) + \dot{f}(x_0) \cdot \Delta x$$

♦ Example 3.15 Find a square root of 10.

Solution. $f(x) = \sqrt{x}; x_0 = 9, \Delta x = 1, \dot{f}(x) = \frac{1}{2\sqrt{x}}$. Thus,

$$\sqrt{10} \approx \sqrt{x_0} + \frac{1}{2\sqrt{x_0}} \cdot \Delta x = 3 + \frac{1}{2 \cdot 3} \cdot 1 \approx 3 + 0.16 \approx 3.16.$$

There are several useful approximate formulas (x is a small quantity):

$$\sqrt{1 \pm x} \approx 1 \pm 0.5x; \qquad \frac{1}{\sqrt{1 \pm x}} \approx 1 \mp 0.5x; \qquad \frac{1}{1 \pm x} \approx 1 \mp x; \qquad \frac{1}{(1 \pm x)^2} \approx 1 \mp 2x;$$

$$\sqrt[3]{1 \pm x} \approx 1 \pm \frac{1}{3}x; \qquad \sin x \approx x; \qquad \cos x \approx 1 - \frac{1}{2}x^2; \qquad \tan x \approx x;$$

$$e^x \approx 1 + x; \qquad \ln(1 \pm x) \approx \pm x; \qquad 10^x \approx 1 + (1/C)x$$

3.1.17 Differentiation of Implicit Functions

If the function is expressed in an implicit form (e.g., $x^2 + y^2 = a$), the equation can be solved to find the explicit form $y = f(x)$, and then standard rules of differentiation are applied.

♦ Example 3.16 Find the derivative of the implicit function $x^2 + y^2 = 25$ at the point $x = 4$, $y = -3$ [1].
Solution. Solving the given equation (and taking into account that $y = -3$ is negative for $x = 4$), we obtain

$$y = -\sqrt{25 - x^2}. \text{ Thus, } \frac{dy}{dx} = \frac{x}{\sqrt{25 - x^2}} = \frac{4}{3}.$$

Generally, there is no need to find an explicit expression to find a derivative. It is sufficient to equate the differentials of both parts of the equation and then determine dy/dx, or differentiate both sides term by term with respect to an argument.

♦ Example 3.17. Find the derivative of an implicit function $y^3 = x^2$ [2].
Solution. Let us differentiate both sides:

$$\frac{dy^3}{dx} = \frac{dx^2}{dx}$$

Now y^3 is a composite function of x,

$$\frac{dy^3}{dx} = \frac{dy^3}{dy}\cdot\frac{dy}{dx} = 3y^2\frac{dy}{dx}.$$

Since $3y^2\dfrac{dy}{dx} = \dfrac{dx^2}{dx} = 2x$, we obtain $\dfrac{dy}{dx} = \dfrac{2x}{3y^2}.$

Since $y^2 = \dfrac{x^2}{y}$, $\dfrac{dy}{dx} = \dfrac{2x\cdot y}{3x^2} = \dfrac{2y}{3x}.$

3.1.18 Logarithmic Differentiation

Sometimes it is convenient to take a logarithm of a function before finding its derivative. Such a technique is described as *logarithmic differentiation.*

♦ Example 3.18 Find a derivative of the implicit function [2]:
$y = (\sin x)^x,\ 0 < x \le \pi$.

Solution. First we take logarithms of both parts of the equation:

$\ln y = x \ln(\sin x).$

Then implicit differentiation with respect to x as in the previous paragraph gives:

$$\frac{1}{y}\frac{dy}{dx} = \ln\sin x + x\frac{\cos x}{\sin x} = \ln\sin x + x\cot x$$

Thus, $\dfrac{dy}{dx} = y(\ln\sin x + x\cot x) = (\sin x)^x(\ln\sin x + x\cot x).$

3.1.19 Parametric Differentiation

Let us assume we have two functions of the argument t:

$$x = f(t), \quad y = \varphi(t)$$

Thus, one of them (e.g., function y) is a function of another one (e.g., x). The equation above describes a function specified *parametrically*, and the auxiliary variable t is called a *parameter*. The derivative can be found as:

$$\frac{dy}{dx} = \frac{dy/dt}{dx/dt} = \frac{\dot{y}}{\dot{x}}$$

♦ Example 3.19 Find the derivative of the parametric function [1]:

$$x = R \cdot \cos \varphi, \quad y = R \cdot \sin \varphi.$$

Solution. Parameter $t = \varphi$.

$$\frac{dy}{d\varphi} = R \cos \varphi; \quad \frac{dx}{d\varphi} = -R \sin \varphi;$$

$$\frac{dy}{dx} = \frac{R \cos \varphi}{-R \sin \varphi} = -\cot \varphi.$$

Actually, the parametric function above is the equation of a circle with radius R and the derivative is the angular coefficient of the tangent AB (Figure 3.2).

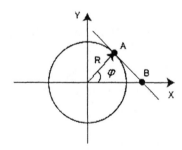

Figure 3.2 Derivative of a parametric function.

3.1.20 Derivatives of Higher Orders

If $\dfrac{dy}{dx}$ is a derivative of a function $y = f(x)$, then a derivative $\dfrac{d}{dx}\left(\dfrac{dy}{dx}\right) = \dfrac{d^2 y}{dx^2}$

is called the *second derivative,* and typically is written as:

$$\frac{d^2 y}{dx^2},\ \text{or}\ \frac{d^2 f}{dx^2},\ \text{or}\ f''(x),\ \text{or}\ f^{(2)}(x).$$

Analogously, the derivative with respect to the second derivative is called the *third derivative*

$$\frac{d^3 y}{dx^3},\ \text{or}\ \frac{d^3 f}{dx^3},\ \text{or}\ f'''(x),\ \text{or}\ f^{(3)}(x)\ \text{and so forth.}$$

♦ <u>Example 3.20</u> Find higher derivatives of the function $y = x^4$.
Solution.

$$f^{(1)}(x) = 4x^3;\ \ f^{(2)}(x) = \frac{d}{dx}\left(4x^3\right) = 12x^2;\ \ f^{(3)}(x) = \frac{d}{dx}\left(12x^2\right) = 24x;$$

$$f^{(4)}(x) = \frac{d}{dx}(24x) = 24;\ \ f^{(5)}(x) = 0.$$

3.1.21 Differentials of Higher Orders

Let us consider a series of argument values with uniform increment Δx:

$$x,\ x + \Delta x,\ x + 2\Delta x,\ x + 3\Delta x, \ldots$$

and corresponding values of the functions:

$$y = f(x),\ y_1 = f(x + \Delta x),\ y_2 = f(x + 2\Delta x), \ldots$$

Let us denote:

$$\Delta y = f(x + \Delta x) - f(x)$$
$$\Delta y_1 = f(x + 2\Delta x) - f(x + \Delta x)$$
$$\Delta y_2 = f(x + 3\Delta x) - f(x + 2\Delta x),\ \text{etc.}$$

Values $\Delta y, \Delta y_1, \Delta y_2 ...$, are called the *first differences.* Values $\Delta y_1 - \Delta y, \Delta y_2 - \Delta y_1,...$ are called the *second differences* and are written as:

$$\Delta^2 y = \Delta y_1 - \Delta y ,$$
$$\Delta^2 y_1 = \Delta y_2 - \Delta y_1 ,...$$

Analogously, the third differences are introduced:

$$\Delta^3 y = \Delta^2 y_1 - \Delta^2 y ,$$
$$\Delta^3 y_1 = \Delta^2 y_2 - \Delta^2 y_1 ,...$$

The first factor of the first difference is called the *first differential,* the first factor of the second difference is called the *second differential,* the first factor of the third difference is called the *third differential,* and so forth.

♦ Example 3.21   Find the first three differentials of the function $y = x^3$ for $x = 2$ [1].
Solution. First differences are:

$$\Delta y = (2 + \Delta x)^3 - 2^3 = 12\Delta x + 6\Delta x^2 + \Delta x^3$$
$$\Delta y_1 = (2 + 2\Delta x)^3 - (2 + \Delta x)^3 = 12\Delta x + 18\Delta x^2 + 7\Delta x^3$$
$$\Delta y_2 = (2 + 3\Delta x)^3 - (2 + 2\Delta x)^3 = 12\Delta x + 30\Delta x^2 + 19\Delta x^3$$
....

Second differences are:

$$\Delta^2 y = \Delta y_1 - \Delta y = 12\Delta x^2 + 6\Delta x^3$$
$$\Delta^2 y_1 = \Delta y_2 - \Delta y_1 = 12\Delta x^2 + 12\Delta x^3$$
....

Third differences are:

$$\Delta^3 y = \Delta^2 y_1 - \Delta^2 y = 6\Delta x^3$$
....

Thus the first differential is $12\Delta x$, the second one is $12\Delta x^2$, and the third one is $6\Delta x^3$.

In general, the differential of nth order can be found as:

$$d^n f(x) = f^{(n)}(x) \cdot \Delta x^n$$

For the independent argument x we can write: $\Delta x = dx$. Thus,

$$d^n f(x) = f^{(n)}(x) \cdot dx^n$$

Higher derivatives can be expressed via differentials as:

$$f^{(n)}(x) = \frac{d^n f(x)}{dx^n}$$

For the derivative of a second order:

$$f^2(x) = \frac{dx \, d^2 y - dy \, dx^2}{dx^3}$$

It is valid for the arbitrary choice of the argument. If argument is x, then $dx^2 = 0$ and:

$$f^2(x) = \frac{d^2 y}{dx^2}$$

3.1.22 How to Solve Indeterminacies

If a function is indeterminate at a point a, but has a limit when $x \to a$, the procedure for finding this limit is called *solving indeterminacy*. Two basic indeterminacies are of type $\dfrac{0}{0}$ or $\dfrac{\infty}{\infty}$.

The general procedure is set by a *rule of L'Hospital*: if two functions $f(x)$ and $\varphi(x)$ are infinitesimal or nonterminating when $x \to a$ (or $x \to \infty$), then

$$\lim_{x \to a} \frac{f(x)}{\varphi(x)} = \lim_{x \to a} \frac{\dot{f}(x)}{\dot{\varphi}(x)}$$

L'Hospital, Guillaume (1661–1704). French mathematician. At age 15 solved a difficult problem about cycloids posed by Pascal. Actually the rule known today as the *rule of L'Hospital* was first introduced by Ivan Bernoulli.

♦ <u>Example 3.22</u> Find the limit of the function $\dfrac{x^2-1}{x^3-1}$ for $x \to 1$ [1].

Solution. This is an indeterminacy of $\dfrac{0}{0}$ type.

$$\lim_{x \to 1}\frac{x^2-1}{x^3-1}=\lim_{x \to 1}\frac{(x-1)(x+1)}{(x-1)(x^2+x+1)}=\lim_{x \to 1}\frac{x+1}{x^2+x+1}=\frac{2}{3}$$

♦ <u>Example 3.23</u> Find $\lim\limits_{x \to \infty}\dfrac{\ln x}{x^2}$ [1].

Solution. This is an indeterminacy of $\dfrac{\infty}{\infty}$ type. Although both functions $f(x)=\ln x$ and $\varphi(x)=x^2$ are nonterminating when $x \to \infty$, the limit of their ratio is:

$$\lim_{x \to \infty}\frac{\ln x}{x^2}=\lim_{x \to \infty}\frac{\frac{1}{x}}{2x}=0$$

If not only functions $f(x)$ and $\varphi(x)$ but also its derivatives are infinitesimal or nonterminating, the rule of L'Hospital can be applied to the derivatives of derivatives. For example,

$$\lim_{x \to a}\frac{f^{(1)}(x)}{\varphi^{(1)}(x)}=\lim_{x \to a}\frac{f^{(2)}(x)}{\varphi^{(2)}(x)}$$

Other indeterminacies, like $0 \cdot \infty$, can be solved by converting them into indeterminacies $\dfrac{0}{0}$ or $\dfrac{\infty}{\infty}$.

♦ <u>Example 3.24</u> Find $\lim\limits_{x \to 0} x \cdot \cot \dfrac{x}{2}$ [1].

Solution. $\lim\limits_{x \to 0} x \cdot \cot \dfrac{x}{2}=\lim\limits_{x \to 0}\dfrac{x}{\tan \dfrac{x}{2}}=\lim\limits_{x \to 0}\left[1:\dfrac{1}{2\cos^2 \dfrac{x}{2}}\right]=2.$

3.1.23 Taylor's Formula

If a function $f(x)$ at the interval (a, b) has the derivatives up to the $(n+1)$st order, then it can be expanded as *Taylor's series*:

$$f(b) = f(a) + \frac{f^{(1)}(a)}{1!}(b-a) + \frac{f^{(2)}(a)}{2!}(b-a)^2 + \dots \frac{f^{(n)}(a)}{n!}(b-a)^n$$

$$+ \frac{f^{(n+1)}(\xi)}{(n+1)!}(b-a)^{(n+1)}$$

where ξ is a value between a and b.

In practical engineering applications, the factors of higher order are often neglected, and with acceptable accuracy the Taylor's series for a function $f(x)$ at a point a can be written as ($b = x$):

$$f(x) = f(a) + \frac{f^{(1)}(a)}{1!}(x-a) + \frac{f^{(2)}(a)}{2!}(x-a)^2 + \dots$$

When $a = 0$, the Taylor's series is called a *Maclaurin series*:

$$f(x) = f(0) + f^{(1)}(0) \cdot x + \frac{f^{(2)}(0)}{2!} \cdot x^2 + \dots$$

◆ <u>Example 3.25</u> Find the Maclaurin series for the function $f(x) = e^x$ and calculate e^x for $x = 0.1$.

Taylor, Brook (1685–1731). English mathematician. Contributed to the mathematical theory of perspective, and was one of the first to recognize the existence of singular solutions to differential equations. Best known for invention of the method for expanding functions in terms of polynomials about an arbitrary point known as *Taylor's series* (published in 1715).

Maclaurin, Colin (1698–1746). Scottish mathematician. He became a disciple of Newton and published the first systematic formulation of Newton's methods in 1742. Best known for developing a method for expanding functions about the origin in terms of series now known as *Maclaurin series* which was generalized to expansion about an arbitrary point by Taylor.

Solution. $e^x = e^0 = 1;$ $f^{(1)}(0) = ... f^{(n)}(0) = 1;$

Thus, $e^x = 1 + x + \dfrac{x^2}{2} + ... \dfrac{x^n}{n!} + ...$

With accuracy up to the first three factors, when $x = 0.1$:

$$e^x \approx 1 + 0.1 + \frac{0.1^2}{2} = 1.105$$

3.1.24 How to Find Maxima and Minima of the Function

A *necessary* condition for a function $y = f(x)$ to have a maximum or minimum at the point a is that $f'(a) = 0$ (or infinity, or is indeterminate). This is not a *sufficient* condition, since there are some functions that have horizontal tangents at nonmaximal/minimal values. But in most tasks a condition

$$f'(a) = 0$$

might be used to find maximum/minimum of many practical functions.

An indication, whether it is maximum or minimum, is in variation of a derivative sign in the vicinity of a. When moving from $x < a$ to $x > a$, if the sign of $f'(x)$ changes from positive to negative, it is an indication of *maximum*. If vice versa, it is an indication of *minimum*. If it is hard to judge from a sign, the following rule can be used: if $f'(a) = 0$ and $f''(a) < 0$, then $f(x)$ has *maximum* at the point a; if $f'(a) = 0$ and $f''(a) > 0$, then function $f(x)$ has a *minimum* at the point a.

♦ Example 3.26 Find maxima and minima for the function [1]:

$$f(x) = \frac{1}{2}x^4 - x^2 + 1$$

Solution. Solving the equation

$$f'(x) = 2x^3 - 2x = 0$$

obtain its roots as $x_1 = -1;$ $x_2 = 0;$ $x_3 = 1$. For the second derivative:

$$f''(x) = 6x^2 - 2 = 2(3x^2 - 1)$$

we find that $f''(-1) > 0;$ $f''(0) < 0;$ $f''(1) > 0$.

Thus, $x = -1$ and $x = 1$ are the points where function $f(x)$ has minima, and $x = 0$ is the point of maximum.

If the second derivative at the point a is also equal to zero, the rule is as follows: if the first nonzero derivative $f^{2k}(a)$ has the even order, then the function $f(x)$ has at this point maximum when $f^{2k}(a) < 0$ and minimum when $f^{2k}(a) \succ 0$. If the first nonzero derivative $f^{2k+1}(a)$ at the point a has the odd order, the function does not have maximum or minimum values at this point. The function increases when $f^{2k+1}(a) \succ 0$ and decreases when $f^{2k+1}(a) < 0$.

3.1.25 Common Derivatives

Nowadays derivatives of arbitrary functions can be found with the aid of a computer in the symbolic form (see Chapter 9). The most common derivatives worth remembering are given in Appendix 9.

3.2 Fundamentals of Integration

3.2.1 Historical Background

Integral calculus evolved from the tasks of finding areas, volumes, and centers of gravity. The basic ideas can be traced back to Archimedes. A systematic view of integration methods was formed in the seventeenth century in the works of Newton, Leibniz, Pascal, Fermat, and other mathematicians. The term "integral" was introduced by Leibniz (who derived it from the Latin word *integralis* which means "a whole") with respect to expression

$$\int y dx$$

Fourier expanded it to

$$\int_a^b y dx$$

Later, Euler and Chebyshev made a considerable contribution to the development of the contemporary methods of integration techniques. In general, the basic task of integral calculus is to find the original function when its differential is known.

3.2.2 The Original Function

If $f(x)$ is a derivative of the function $F(x)$ [i.e., $f(x)dx = dF(x)$], the function $F(x)$ is called an *original* one for the function $f(x)$.

♦ Example 3.27 $3x^2$ is a derivative of the function x^3, thus $3x^2 dx = d(x^3)$ and function x^3 is the original function for the function $3x^2$. The function $x^3 + 4$ is also an original function for $3x^2$, since $3x^2 dx = d(x^3 + 4)$

Any continuous function $f(x)$ has an infinite number of original functions $F(x) + C$, where C is a *constant of integration*.

3.2.3 An Indefinite Integral

The generic original function including a constant value C is called an *indefinite integral* of the expression $f(x)dx$:

$$\int f(x)dx = F(x) + C$$

The function $f(x)$ is called the *integrand*.

Fermat, Pierre (1601–1665). French lawyer who enjoyed doing math in his spare time. Although he pursued mathematics as an amateur, his work in number theory was of such exceptional quality and erudition that he is generally regarded as one of the greatest mathematicians of all times. He had the habit of scribbling notes in the margins of books and letters rather than publishing them. He is most famous for writing a note in the margin of a book that he had discovered a proof that the equation $x^n + y^n = z^n$ has no integer solutions for $n > 2$. He stated "I have discovered a truly marvelous proof of this, which, however, the margin is not large enough to contain." The proposition, which came to be known as *Fermat's last theorem*, for centuries baffled all attempts to prove it (and it was believed to have sent some unlucky attempters to lunatic asylums) until A. Wiles succeeded in 1995.

Chebyshev, Pafnutiy (1821–1894). Russian mathematician who made considerable contributions to number theory, algebra, analysis, probability, and applied mathematics.

Pascal, Blaise (1623–1662). French mathematician and philosopher. He contributed to integral calculus and research of barometric theory. Best known for *Pascal's principle*, which states that the pressure is constant throughout a static fluid.

♦ Underline: Example 3.28 From the previous example, we see that $3x^2 dx = d(x^3)$. Thus,

$$\int 3x^2 dx = x^3 + C$$

Only one original function from a set of all original functions can take a value b for a specified value of the argument $x = a$. Thus, for a known indefinite integral

$$\int f(x)dx = F(x) + C$$

a constant C can be found from the equation

$$b = F(a) + C$$

♦ Example 3.29 Find the original function for the function $\frac{1}{2}x$, which takes value 3 for $x = 2$.

Solution. With the aid of a computer (see Example 9.7, Chapter 9), we can find

$$\int \frac{1}{2}x\,dx = \frac{1}{4}x^2 + C$$

Since

$$\frac{1}{4}2^2 + C = 3,$$

$C = 2$ and the original function is $y = \frac{1}{4}x^2 + 2$.

3.2.4 Properties of the Indefinite Integral

• A sign of a differential in front of a sign of an integral cancels the latter:

$$d\int f(x)dx = f(x)dx$$

In other words, the derivative of the indefinite integral is equal to the integrand:

$$\frac{d}{dx}\int f(x)dx = f(x)$$

- A sign of an integral in front of a sign of a differential cancels the latter, but a constant factor is introduced, for example $\int d\sin x = \sin x + C.$
- A constant factor can be moved out of the sign of the integral:

$$\int C\, f(x)dx = C\int f(x)dx$$

- An integral of an algebraic sum is equal to the sum of integrals of each integrand

$$\int [f_1(x)+f_2(x)-f_3(x)]dx = \int f_1(x)dx + \int f_2(x)dx - \int f_3(x)dx$$

3.2.5 Integration by Substitution

Sometimes the integral that cannot be found analytically can be converted into one by introducing an auxiliary variable z related to an argument x via some function.

♦ Example 3.30 Evaluate the integral $\int \sqrt{2x-1}\, dx$ [1].

Solution. Let us denote $z = 2x - 1$. Differentiating, we obtain $\dfrac{dz}{dx} = 2$, thus

$$dx = \frac{dz}{2}.$$

Then the original integral can be converted into a standard $\int x^n dx$ type:

$$\int \sqrt{2x-1}\, dx = \int \sqrt{z}\,\frac{dz}{2} = \frac{1}{2}\int z^{1/2} dz = \frac{1}{2}\cdot\frac{z^{3/2}}{3/2}+C = \frac{z^{3/2}}{3}+C = \frac{1}{3}(2x-1)^{3/2}+C$$

3.2.6 Integration by Parts

In this case, an integral of $\int u\, dv$ type is converted into the integral $\int v\, du$ type by applying formula:

$$\int u\, dv = uv - \int v\, du$$

Certainly, it makes sense to apply this technique if $\int v\,du$ can be evaluated easier than $\int u\,dv$.

♦ Example 3.31 Evaluate the integral $\int x \cdot e^x dx$ [1].

Solution. $u = x; v = e^x; \quad x(e^x dx) = x d(e^x)$

Applying the formula above

$$\int x d e^x = x e^x - \int e^x dx$$

we find

$$\int x e^x dx = \int x(d e^x) = x e^x - \int e^x dx = x e^x - e^x + C$$

3.2.7 Definite Integral

Let us consider a function $f(x)$, which is a continuous one within the interval (a, b). Within (a, b), we take a sequence of n points $x_1, x_2, ..., x_n$ (Figure 3.3). Let us denote $a = x_0$, $b = x_{n+1}$ and specify the points $\xi_1, \xi_2, ...$ within each of the smaller intervals $(x_0, x_1), (x_1, x_2), ...,$ and so forth.

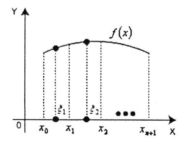

Figure 3.3 A concept of a definite integral.

Let us write the sum

$$\sum_{n} f(\xi_1)(x_1 - x_0) + f(\xi_2)(x_2 - x_1) + \ldots + f(\xi_{n+1})(x_{n+1} - x_n)$$

A *definite integral* is a limit of the sum $\sum_{n}$ when the longest of all the intervals $\Delta x_i = (x_i, x_{i+1})$ tends to zero

$$\lim_{\Delta x_i \to 0} \sum_{n} = \int_{a}^{b} f(x)dx$$

The ends of the interval are called *limits of integration*. In the case $a = b$:

$$\int_{a}^{b} f(x)dx = \int_{a}^{a} f(x)dx = 0$$

3.2.8 Properties of a Definite Integral

• Sign of a definite integral changes when limits of integration are interchanged:

$$\int_{a}^{b} f(x)dx = -\int_{b}^{a} f(x)dx$$

• If c is a point within interval (a, b):

$$\int_{a}^{b} f(x)dx = \int_{a}^{c} f(x)dx + \int_{c}^{b} f(x)dx$$

• Integral of an algebraic sum of the factors is equal to the sum of integrals of each factor:

$$\int_{a}^{b} [f_1(x) + f_2(x) - f_3(x)]dx = \int_{a}^{b} f_1(x)dx + \int_{a}^{b} f_2(x)dx - \int_{a}^{b} f_3(x)dx$$

• A constant coefficient C can be moved out of the integral sign:

$$\int_a^b C f(x)dx = C \int_a^b f(x)dx$$

3.2.9 How to Evaluate a Definite Integral

If an indefinite integral can be evaluated as:

$$\int f(x)dx = F(x) + C$$

then the corresponding definite integral is:

$$\int_a^b f(x)dx = F(b) - F(a)$$

Note. Constant of integration C can be omitted when evaluating an indefinite integral since it is cancelled during subtraction.

♦ <u>Example 3.32</u> Evaluate $\int_{-2}^{3} 3x^2 dx$.

Solution. $\int 3x^2 dx = x^3$; Thus,

$$\int_{-2}^{3} 3x^2 dx = x^3 \Big|_{x=3} - x^3 \Big|_{x=-2} = 27 - (-8) = 35$$

The same techniques (integration by substitution, integration by parts, and so forth) discussed earlier can be used to evaluate definite integrals.

♦ <u>Example 3.33</u> Evaluate $\int_{5}^{13} \sqrt{2x-1}\, dx$ [2].

Solution. In Example 3.30, it was evaluated that:

$$\int \sqrt{2x-1}\, dx = \frac{1}{3}(2x-1)^{3/2}$$

Since a = 5, b = 13 (compare to Example 9.11, Chapter 9):

$$\int\limits_5^{13} \sqrt{2x-1}\, dx = \frac{1}{3}(2\cdot 13-1)^{3/2} - \frac{1}{3}(2\cdot 5-1)^{3/2}$$

$$= \frac{1}{3}\left(25^{3/2} - 9^{3/2}\right) = \frac{1}{3}\cdot 25^{3/2} - \frac{1}{3}\cdot 9^{3/2} = 32\frac{2}{3}$$

♦ <u>Example 3.34</u> Evaluate $\int\limits_0^1 xe^x dx$.

Solution. In Example 3.31, it was found that: $\int xe^x dx = xe^x - e^x$.

Thus, $\int\limits_0^1 xe^x dx = x\cdot e^x\Big|_0^1 - e^x\Big|_0^1 = (e-0)-(e-1) = 1$.

3.2.10 How to Evaluate a Definite Integral Approximately

In order to obtain quick estimates in engineering applications, it is often convenient to evaluate the definite integral by presenting it in discrete form. In this case, the interval $(a,\ b)$ of the function $y = f(x)$ is split into K equal intervals, and the function is evaluated in $K+1$ points ($x_0 = a$, $x_1 = x_0 + \Delta h$, $x_2 = x_0 + 2\Delta h$, ..., $x_n = b = x_0 + K\Delta h$) taken with the increment $\Delta h = \dfrac{b-a}{K}$.

Thus,

$$y_0 = f(x_0) = f(a);\quad y_1 = f(x_1) = f(a+\Delta h);\ ...$$

$$y_k = f(x_k) = f(a+k\cdot\Delta h);\ ...\ y_K = f(x_K) = f(b)$$

In this case, the simplest formula to calculate the definite integral is given by the *trapezoid rule*:

$$\int\limits_a^b f(x)dx = \frac{\Delta h}{2}\cdot (y_0 + y_K + 2\cdot\sum_{k=1}^{K-1} y_k)$$

Obviously, the larger the number K is chosen, the more accurate calculation is. An example of how to calculate the definite integral based on the trapezoid rule is given in Chapter 11 (Problem 11.6).

3.2.11 Improper Integrals

The definition of a definite integral in Section 3.2.7 was introduced for a finite domain of integration (a, b) and a continuous function $f(x)$. In case either a, b, or both, are infinite, or $f(x)$ has a discontinuity, the integrals are called *improper integrals.*

In the former case leading to improper integrals (let us assume $b = \infty$), we define a limit:

$$\int_a^\infty f(x)\,dx = \lim_{X \to \infty} \int_a^X f(x)\,dx$$

if this limit exists.

Otherwise $\int_a^\infty f(x)\,dx$ has no meaning.

♦ Example 3.35 Evaluate $\int_0^\infty 2^{-x}\,dx$ [1].

Solution. $\lim_{X \to \infty} \int_0^X 2^{-x}\,dx = \dfrac{1}{\ln 2}\left(-2^{-x}\right)\Big|_0^X = \dfrac{1}{\ln 2}\left(1 - \dfrac{1}{2^X}\right)$

$\lim_{X \to \infty} \dfrac{1}{\ln 2}\left(1 - \dfrac{1}{2^X}\right) = \dfrac{1}{\ln 2}$.   Thus, $\int_0^\infty 2^{-x}\,dx = \dfrac{1}{\ln 2} \approx 1.4$.

When the integrand $f(x)$ has a discontinuity (let us say at point a), the same rule as above is applied:

$$\int_a^b f(x)\,dx = \lim_{X \to a} \int_X^b f(x)\,dx$$

if this limit exists. Otherwise, the improper integral has no meaning.

♦ Example 3.36 Evaluate $\displaystyle\int_0^a \dfrac{a^2\,dx}{\sqrt{a^2 - x^2}}$ [1].

Solution. This is an improper integral since $f(x) = \infty$ (the integral is a discontinuous one) for $x = a$. But it can be evaluated because the function

$$\int_0^X \frac{a^2\,dx}{\sqrt{a^2 - x^2}} = a^2 a \sin \frac{X}{a}$$

tends to the value $\dfrac{\pi a^2}{2}$ when $X \to a$.

Thus, $\displaystyle\int_0^a \frac{a^2\,dx}{\sqrt{a^2 - x^2}} = \frac{\pi a^2}{2}$.

3.2.12 How to Evaluate Areas and Volumes

• If $f(x) \geq 0$ is a continuous function on the interval (a, b) [Figure 3.4(a)], then

area A of the *curved trapezoid ABCD* is $A = \displaystyle\int_a^b f(x)\,dx$.

• If $f(x) \geq 0$ and $h(x) \geq 0$ are continuous functions on the interval (a, b), the area A of the *plane region* bounded by these functions [Figure 3.4(b)] is:

$$A = \int_a^b f(x)\,dx - \int_a^b h(x)\,dx$$

• If $\rho = f(\varphi)$ is a function in the polar coordinates, area A of the *sector* within two radiuses OA and OB, and a function $f(\varphi)$ [Figure 3.5(a)] is:

$$A = \frac{1}{2} \int_{\varphi_1}^{\varphi_2} f^2(\varphi)\,d\varphi$$

• If $f(x) \geq 0$ is a continuous function on the interval (a, b), the length of a curve l (arc length) and surface area S generated by this curve when it is rotated through radius 2π about the X axis [Figure 3.5(b)] are:

$$l = \int_a^b \sqrt{1 + [f'(x)]^2}\,dx; \qquad S = 2\pi \int_a^b f(x)\sqrt{1 + [f'(x)]^2}\,dx$$

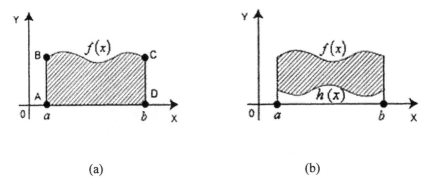

(a)   (b)

Figure 3.4 (a) Area of a curved trapezoid and (b) plane region.

- If $f(x) \geq 0$ is a continuous function on the interval (a, b), the volume V of a *solid revolution* with X as an axis of symmetry [Figure 3.5(b)] is:

$$V = \pi \int_a^b f^2(x)\,dx$$

◆ <u>Example 3.37</u> Find the area of a circle with radius r [Figure 3.6(a)].

Solution. $A = \dfrac{1}{2} \int\limits_0^{2\pi} r^2\, d\varphi = \dfrac{1}{2} \cdot r^2 \int\limits_0^{2\pi} d\varphi = \dfrac{1}{2} r^2 \cdot \varphi \Big|_0^{2\pi} = \dfrac{1}{2} r^2 \cdot 2\pi = \pi r^2$

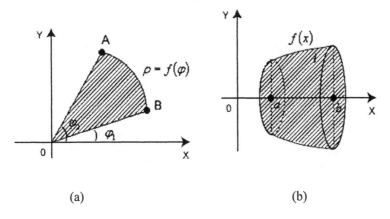

(a)   (b)

Figure 3.5 (a) Area bounded by a curve in polar coordinates and (b) area and volume generated by a rotating curve.

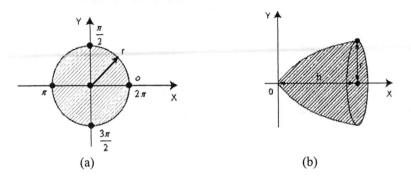

(a)   (b)

Figure 3.6 (a) Area of the circle and (b) volume of paraboloid.

♦ Example 3.38 Find the volume of a parabolic reflector (paraboloid) with

radius r, height h, and the equation of generating parabola $y^2 = \dfrac{r^2 x}{h}$ [Figure

3.6(b)] [1].
Solution.

$$V = \pi \int_0^h \frac{r^2 x}{h}\, dx = \frac{\pi r^2}{h} \int_0^h x\, dx = \frac{\pi r^2}{2h} x^2 \Big|_0^h = \frac{\pi r^2 h}{2}$$

3.2.13 How to Evaluate Mean Values

The *mean value* of a function $y = f(x)$ [Figure 3.7(a)] can be evaluated as:

$$y_0 = \frac{1}{b-a} \int_a^b f(x)\, dx$$

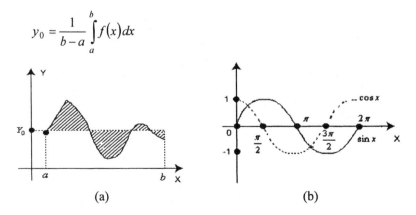

(a)   (b)

Figure 3.7 (a) The mean value and (b) functions $\sin x$ and $\cos x$.

♦ Example 3.39   Find the mean value of the function $y = \sin x$ at intervals $(0, 2\pi)$ and $(0, \pi)$.

Solution. For interval $(0, 2\pi)$

$$y_0 = \frac{1}{2} \int\limits_0^{2\pi} \sin x \, dx = \frac{1}{2\pi} \cdot \left(-\cos x\right)\Big|_0^{2\pi} = \frac{1}{2\pi} \cdot (1 - 1) = 0$$

For interval $(0, \pi)$

$$y_0 = \frac{1}{\pi} \int\limits_0^{\pi} \sin x \, dx = \frac{1}{\pi} \cdot \left(-\cos x\right)\Big|_0^{\pi} = \frac{1}{\pi} \cdot [-(-1-1)] = \frac{2}{\pi} \approx 0.637$$

Judgments made on the mean values sometimes are deceptive. The mean value of the alternating current is zero, but it still might kill you.

That is why another important parameter often used in electrical engineering is the *root-mean-square* value:

$$\text{rms} = \sqrt{\frac{1}{b-a} \int\limits_a^b [f(x)]^2 \, dx}$$

♦ Example 3.40   Find the rms of the function $y = \sin x$ at $(0, 2\pi)$.

Solution. For interval $(0, 2\pi)$

$$(\text{rms})^2 = \frac{1}{2\pi} \int\limits_0^{2\pi} \sin^2 x \, dx = \frac{1}{2\pi} \left(\frac{x}{2} - \frac{\sin x \cos x}{2} \right)$$

$$= \frac{1}{2\pi} \cdot \frac{x}{2}\Big|_0^{2\pi} - \frac{1}{2\pi} \cdot \frac{\sin x \cos x}{2}\Big|_0^{2\pi} = (\frac{2\pi}{2\pi \cdot 2} - 0) - (0 - 0) = \frac{1}{2}$$

Thus, $\text{rms} = \dfrac{1}{\sqrt{2}}$

3.2.14 Common Integrals

Integrals of arbitrary functions can be found with the aid of a computer in the symbolic form (see Chapter 9). The most common integrals which are worth remembering are given in Appendix 10.

3.3 Differentiation and Integration of the Functions with Several Variables

3.3.1 Functions of Several Variables

If there is a correspondence between several values of the arguments and a single (or multiple) value of the functions, such a function is called *a function of several variables*: $z = f(x, y)$ is a function of two variables; $u = f(x, y, z)$ is a function of three variables, and so forth.

The function can be explicit or implicit.

♦ Example 3.41 The height h of a point on the Earth's surface is the explicit function of its latitude φ and longitude θ :

$$h = f(\varphi, \theta), \quad -180^0 \leq \varphi \leq 180^0, \quad -90^0 \leq \theta \leq 90^0$$

♦ Example 3.42 The volume of one kilogram of air V is the implicit function of air pressure p and temperature T : $pV = C(273.2 + T)$, where C is a constant.

3.3.2 Partial Derivatives

The *partial derivative* of a function $u = f(x, y, z, ..., v)$ with respect to an argument x is:

$$u'_x = \frac{\partial u}{\partial x} = \lim_{\Delta x \to 0} \frac{f(x + \Delta x, y, z, ..., v)}{\Delta x}$$

Analogously, the derivatives with respect to other arguments can be defined:

$$u'_y = \frac{\partial u}{\partial y} = \lim_{\Delta y \to 0} \frac{f(x, y + \Delta y, z, ..., v)}{\Delta y} \quad \text{and so forth.}$$

Hereinafter, we will consider the function of three arguments $u = f(x, y, z)$ as an example of the function of several variables.

♦ Example 3.43 Find the partial derivative with respect to argument x at

the point $P = (0,0,1)$ of the function

$$u = f(x, y, z) = 2x^2 + y^2 - 3z^2 - 3xy - 2xz$$

Solution. For the purpose of finding the derivative with respect to x we can consider this function only as the function of x and disregard other arguments. Thus:

$$\frac{\partial u}{\partial x} = 4x - 3y - 2z\big|_{x=0, y=0, z=1} = -2$$

3.3.3 Partial Differential

If $\Delta x, \Delta y, \Delta z$ are differences for the function $u = f(x, y, z)$ at the points x_0, y_0, z_0, then the *complete difference* of the function is:

$$\Delta_x u = \Delta_x f(x, y, z) = f(x_0 + \Delta x, y_0 + \Delta y, z_0 + \Delta z) - f(x_0, y_0, z_0)$$

If some differences are equal to zero (e.g., $\Delta y = \Delta z = 0$), then the function has a *partial differential*:

$$\Delta u = \Delta f(x, y, z) = f(x_0 + \Delta x, y_0, z_0) - f(x_0, y_0, z_0)$$

If a partial difference can be represented as the sum of two factors:

$$\Delta_x u = A \cdot \Delta x + B$$

where A does not depend on Δx, and B is of higher order with respect to Δx when $x \to 0$, then the first factor is termed a *partial differential* of the function $f(x, y, z)$ with respect to the argument x:

$$d_x u = d_x f(x, y, z) = A \cdot \Delta x$$

The partial differentials with respect to y, z are defined analogously. In other words, a partial differential with respect to x is the differential of the function $f(x, y, z)$ evaluated with assumption that y and z do not vary $(\Delta y = \Delta z = 0)$. The coefficient A is equal to the partial derivative with respect to x:

$$d_x u = u'_x \cdot dx$$

Analogously,

$$d_y u = u'_y \cdot d_y; \quad d_z u = u'_z \cdot dz$$

♦ Example 3.44 Find the partial differentials of the function

$$u = x^2 y + y^2 x$$

Solution. The differentials can be found if we assume first that y = const, and then x = const:

$$d_x u = \left(2xy + y^2\right) dx$$
$$d_y u = \left(x^2 + 2xy\right) dy$$

Thus, a partial derivative u'_x of the function $u = f(x, y, z)$ is the ratio of the partial differential $d_x u$ over the differential of the argument dx :

$$\frac{\partial u}{\partial x} = u'_x = \frac{d_x u}{dx}$$

3.3.4 Total Differential

If a total difference $\Delta f(x, y, z)$ of the function $f(x, y, z)$ can be represented as the sum of two factors:

$$\Delta f(x, y, z) = \left(A\Delta x + B\Delta y + C\Delta z\right) + \varepsilon$$

where coefficients A, B, C do not depend on $\Delta x, \Delta y, \Delta z$, and ε is of higher order with respect to $\rho = \sqrt{\Delta x^2 + \Delta y^2 + \Delta z^2}$, then the first factor

$$A\Delta x + B\Delta y + C\Delta z$$

is called a *complete differential* (or just a differential) and denoted as $df(x, y, z)$.

The coefficients A, B, C are equal to:

$$A = \frac{\partial u}{\partial x}, \quad B = \frac{\partial u}{\partial y}, \quad C = \frac{\partial u}{\partial z}$$

and the complete differential is equal to the sum of the partial differentials:

$$du = df(x, y, z) = d_x u + d_y u + d_z u = \frac{\partial u}{\partial x} dx + \frac{\partial u}{\partial y} dy + \frac{\partial u}{\partial z} dz$$

3.3.5 How to Evaluate a Partial Derivative

In order to evaluate a partial derivative, it is often more convenient to find a complete differential first. The latter one is evaluated based on the same rules as for the differential of the function of a single argument (see Section 3.1).

♦ Example 3.45 Find the partial derivatives for the function $u = a \tan\left(\dfrac{y}{x}\right)$ [1].

Solution. First we evaluate the complete differential.

$$du = \frac{d\left(\dfrac{y}{x}\right)}{1+\left(\dfrac{y}{x}\right)^2} = \frac{x \cdot dy - y \cdot dx}{x^2 + y^2} = \frac{x}{x^2+y^2} \cdot dy - \frac{y}{x^2+y^2} \cdot dx = u'_x \cdot dx + u'_y \cdot dy$$

Thus,

$$u'_x = -\frac{y}{x^2+y^2}; \quad u'_y = \frac{x}{x^2+y^2}$$

3.3.6 How to Differentiate a Composite Function

The rules for differentiation of a composite function can be carried over to partial differentiation. If z is the function of x and y, and x, y are the functions of u and v, for example:

$$z = F[x(u,v), y(u,v)]$$

the rule to find partial derivatives is as follows:

$$\frac{\partial z}{\partial u} = \frac{\partial z}{\partial x} \cdot \frac{\partial x}{\partial u} + \frac{\partial z}{\partial y} \cdot \frac{\partial y}{\partial u}$$

$$\frac{\partial z}{\partial v} = \frac{\partial z}{\partial x} \cdot \frac{\partial x}{\partial v} + \frac{\partial z}{\partial y} \cdot \frac{\partial y}{\partial v}$$

This rule stays the same for any number of variables. For any function

$$z = F[x(u,v,...,t), y(u,v,...,t),..., w(u,v,...,t)]$$

$$\frac{\partial z}{\partial u} = \frac{\partial z}{\partial x} \cdot \frac{\partial x}{\partial u} + \frac{\partial z}{\partial y} \cdot \frac{\partial y}{\partial u} + ... + \frac{\partial z}{\partial w} \cdot \frac{\partial w}{\partial u}$$

$$\frac{\partial z}{\partial v} = \frac{\partial z}{\partial x} \cdot \frac{\partial x}{\partial v} + \frac{\partial z}{\partial y} \cdot \frac{\partial y}{\partial v} + ... + \frac{\partial z}{\partial w} \cdot \frac{\partial w}{\partial v}$$

....

$$\frac{\partial z}{\partial t} = \frac{\partial z}{\partial x} \cdot \frac{\partial x}{\partial t} + \frac{\partial z}{\partial y} \cdot \frac{\partial y}{\partial t} + ... + \frac{\partial z}{\partial w} \cdot \frac{\partial w}{\partial t}$$

♦ Example 3.46 Find $\dfrac{dz}{dt}$ when $z(t) = \sin(3x - y)$ and $x = 2t^2 - 3$,

$y = 0.5t^2 - 5t + 1$ [2].
Solution.

$$\frac{\partial z}{\partial t} = \frac{\partial z}{\partial x} \cdot \frac{\partial x}{\partial t} + \frac{\partial z}{\partial y} \cdot \frac{\partial y}{\partial t}$$

Thus,

$$\frac{dz}{dt} = 3[\cos(3x - y)] \cdot 4t - [\cos(3x - y)](t - 5) = (11t + 5)\cos(3x - y) =$$

$$(11t + 5)\cos\left(\frac{11}{2}t^2 + 5t - 10\right)$$

3.3.7 How to Differentiate an Implicit Function

Let us consider the equation

$$F(x, y, z) = 0$$

that typically defines z as the implicit function of arguments x and y. The rule to differentiate such a function is as follows:

- The total differential of the function has to be found (see Section 3.3.4);
- Coefficients for dx, dy provide the corresponding partial derivatives of the function.

♦ Example 3.47 Find the total differential and partial derivatives for the implicit function $z = f(x,y)$ given by equation [1]:

$x^2 + y^2 + z^2 = 9$ for $x = 1, y = -2, z = -2$
Solution. Differentiating the given equation, we find

$$2x \cdot dx + 2y \cdot dy + 2z \cdot dz = 0$$

Thus, the total differential of the function $z = f(x,y)$:

$$dz = -\frac{x}{z}dx - \frac{y}{z}dy$$

For the point (1, −2, −2): $dz = \frac{1}{2}dx - dy$

Partial derivatives are: $\dfrac{\partial z}{\partial x} = \dfrac{1}{2};\ \dfrac{\partial z}{\partial y} = -1$

3.3.8 Partial Derivatives of Higher Orders

The second derivatives for the function $z = f(x,y)$ are the partial derivatives for the functions:

$$\frac{\partial z}{\partial x} = f_x'(x, y) \quad \frac{\partial z}{\partial y} = f_y'(x, y)$$

There are four second derivatives:

$$\frac{\partial}{\partial x}\left(\frac{\partial z}{\partial x}\right) = \frac{\partial^2 z}{\partial x^2} = f_{xx}''(x, y) \tag{3.1}$$

$$\frac{\partial}{\partial y}\left(\frac{\partial z}{\partial x}\right) = \frac{\partial^2 z}{\partial y\,\partial x} = f_{yx}''(x, y) \tag{3.2}$$

$$\frac{\partial}{\partial x}\left(\frac{\partial z}{\partial y}\right) = \frac{\partial^2 z}{\partial x\,\partial y} = f_{xy}''(x, y) \tag{3.3}$$

$$\frac{\partial}{\partial y}\left(\frac{\partial z}{\partial y}\right) = \frac{\partial^2 z}{\partial y^2} = f_{yy}''(x, y) \tag{3.4}$$

The mixed derivatives (3.2) and (3.3) for functions that are continuous at the given point are equal:

$$f_{yx}''(x, y) = f_{xy}''(x, y)$$

The higher-order partial derivatives $\dfrac{\partial^{m+n} f}{\partial x^m \partial y^n}$ are defined in a similar manner.

♦ Example 3.48 Find the second partial derivative for the function [1]:

$$z = x^3 y^2 + 2x^2 y - 6$$

Solution.

$$\frac{\partial z}{\partial x} = 3x^2 y^2 + 4xy; \quad \frac{\partial z}{\partial y} = 2x^3 y + 2x^2$$

Thus, $\dfrac{\partial^2 z}{\partial x^2} = 6xy^2 + 4y; \quad \dfrac{\partial^2 z}{\partial y^2} = 2x^3; \quad \dfrac{\partial^2 z}{\partial x \, \partial y} = \dfrac{\partial^2 z}{\partial y \, \partial x} = 6x^2 y + 4x$

3.3.9 Taylor's Formula

For the function of several arguments the *Taylor's formula* is similar to the one for a function of a single argument (see Section 3.1.23). But total differentials are used. Thus, for the function of two arguments $z = f(x,y)$:

$$f(x + \Delta x, y + \Delta y) = f(x, y) + \frac{1}{1!}\left[f_x'(x, y)\Delta x + f_y'(x, y)\Delta y \right]$$

$$+ \frac{1}{2!}\left[f_{xx}''(x, y)\Delta x^2 + 2f_{xy}''(x, y)\Delta x \Delta y + f_{yy}''(x, y)\Delta y^2 \right]$$

$$+ \frac{1}{3!}\left[f_{xxx}'''(x, y)\Delta x^3 + 3f_{xxy}'''(x, y)\Delta x^2 \Delta y + 3f_{xyy}'''(x, y)\Delta x \Delta y^2 + f_{yyy}'''(x, y)\Delta y^3 \right] + \ldots$$

♦ <u>Example 3.49</u> Obtain the Taylor series for the function $z = \sin xy$ about the point *(a, b)* for $a = 1, b = \dfrac{\pi}{3}$ neglecting terms of degree two and higher [2].

Solution.

$$f(x, y) = f\left(1, \frac{1}{3}\pi\right) + \frac{1}{1!}\left[(x-1)\frac{\partial f}{\partial x}\Bigg|_{1,\frac{\pi}{3}} + \left(y - \frac{\pi}{3}\right)\frac{\partial f}{\partial y}\Bigg|_{1,\frac{\pi}{3}} \right]$$

$$f\left(1, \frac{1}{3}\pi\right) = \sin\frac{\pi}{3} = \frac{\sqrt{3}}{2}$$

$$\frac{\partial f}{\partial x}\Bigg|_{1,\frac{\pi}{3}} = y\cos xy = \frac{\pi}{3}\cos\frac{\pi}{3} = \frac{\pi}{3}\cdot\frac{1}{2} = \frac{\pi}{6}$$

$$\frac{\partial f}{\partial y}\Bigg|_{1,\frac{\pi}{3}} = x\cos xy = \cos\frac{\pi}{3} = \frac{1}{2}$$

Thus,

$$\sin xy = \frac{\sqrt{3}}{2} + \frac{\pi}{6}(x-1) + \frac{1}{2}\left(y - \frac{\pi}{3}\right)$$

3.3.10 Double Integral

Let us assume that the function of two arguments $f(x,y)$ is a continuous one within a domain D with area A. If we split this domain into n subdomains $D_1,...,D_n$ with areas $\Delta A_1,...,\Delta A_n$ and choose the one with the largest diameter d_k, then a *double integral* of the function $f(x,y)$ in domain D is:

$$\iint\limits_D f(x,y)\,dA = \iint\limits_D f(x,y)\,dx\,dy = \lim_{d_k \to 0}\left[f(x_1,y_1)\Delta A_1 + ...f(x_n,y_n)\Delta A_n\right]$$

Geometrically, the double integral is equal to the volume of the cylindrical body built on the basis D, and limited on top by the corresponding surface $z = f(x,y)$ (Figure 3.8).

3.3.11 Properties of the Double Integral

• If a domain D is split into D_1 and D_2, then

$$\iint\limits_D f(x,y)\,dA = \iint\limits_{D_1} f(x,y)\,dA + \iint\limits_{D_2} f(x,y)\,dA$$

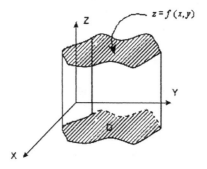

Figure 3.8 A concept of a double integral.

The same rule applies if $D = D_1 + D_2 + ... + D_n$.

- The double integral of an algebraic sum of functions is equal to the algebraic sum of double integrals for each function. For three functions:

$$\iint_D [f(x, y) + \varphi(x, y) + \psi(x, y)] dA = \iint_D f(x, y) dA + \iint_D \varphi(x, y) dA + \iint_D \psi(x, y) dA$$

- Constant multiplier C can be moved out of the integral sign

$$\iint_D C f(x, y) dA = C \iint_D f(x, y) dA$$

- If α is the minimum value of the function $f(x,y)$, β is the maximum value of the function $f(x,y)$ and A is the area of domain D, then

$$\alpha \cdot A \le \iint_D f(x, y) dA \le \beta \cdot A$$

3.3.12 How to Evaluate a Double Integral

Rule #1. If domain D is given by the inequalities

$$a \le x \le b; \ c \le y \le d$$

that is, it is rectangular [Figure 3.9(a)], then

$$\iint_D f(x, y) dx\, dy = \int_a^b \int_c^d f(x, y) dx\, dy = \int_c^d dy \int_a^b f(x, y) dx \qquad (3.5)$$

or

$$\iint_D f(x, y) dx\, dy = \int_a^b \int_c^d f(x, y) dy\, dx = \int_a^b dx \int_c^d f(x, y) dy \qquad (3.6)$$

The last integral in (3.5) is evaluated assuming that y is a constant. Then the first integral is evaluated. The same is for integral (3.6), only x is assumed constant.

If $f(x, y) = \varphi(x) \cdot \psi(y)$, then

$$\iint_D f(x,y)\,dx\,dy = \int_a^b \int_c^d f(x,y)\,dx\,dy = \int_a^b \varphi(x)\,dx \int_c^d \psi(y)\,dy$$

♦ <u>Example 3.50</u> Evaluate $I = \int_1^3 \int_2^5 \left(5x^2 y - 2y^3\right)dx\,dy$ [1].

Solution.

$$I = \int_1^3 dy \int_2^5 \left(5x^2 y - 2y^3\right)dx$$

$$\int_2^5 \left(5x^2 y - 2y^3\right)dx = 5y\int_2^5 x^2\,dx - 2y^3\int_2^5 dx = 5y\cdot\frac{x^3}{3}\bigg|_2^5 - 2y^3 x\bigg|_2^5 = 195y - 6y^3$$

$$\int_1^3 \left(195y - 6y^3\right)dy = 195\cdot\frac{y^2}{2}\bigg|_1^3 - 6\cdot\frac{y^4}{4}\bigg|_1^3 = 780 - 120 = 660$$

<u>*Rule #2*</u>. If for the point P the function is given in polar coordinates $f(P) = f(r,\varphi)$, then

$$\iint_D f(P)\,dA = \iint_D f(r,\varphi)r\,dr\,d\varphi$$

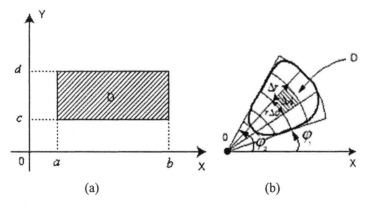

(a)   (b)

Figure 3.9 (a) A rectangular domain and (b) domain in polar coordinates.

where $dA = r\,dr\,d\varphi$ is the element of the area in the polar coordinates. If domain D is such that the pole O is located out of it, and every polar ray intersects the domain boundaries not more than twice [Figure 3.9(b)], then

$$\iint_D f(r,\varphi)r\,dr\,d\varphi = \int_{\varphi_1}^{\varphi_2} d\varphi \int_{r_1}^{r_2} f(r,\varphi)r\,dr \qquad (3.7)$$

If domain D is such that the pole O is located within its boundaries and every polar ray intersects it only once [Figure 3.10(a)], then $r_1 = 0$, $\varphi_1 = 0$, and $\varphi_2 = 2\pi$.

♦ Example 3.51 Evaluate $I = \iint_D r\sin\varphi\,dA$ if D is a half-circle with diameter d [Figure 3.10(b)] [1].

Solution. For the given half-circle $r = d\cos\varphi$. In (3.7), limits are:

$$r_1 = 0, r_2 = d\cos\varphi, \ \varphi_1 = 0, \varphi_2 = \frac{\pi}{2}$$

Thus,

$$\iint_D r\sin\varphi\,dA = \int_0^{\pi/2} d\varphi \int_0^{d\cos\varphi} r^2\sin\varphi\,dr = \int_0^{\pi/2}\sin\varphi\,d\varphi \int_0^{d\cos\varphi} r^2\,dr =$$

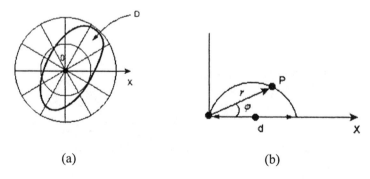

(a)   (b)

Figure 3.10 (a) Domain with the pole within its boundaries and (b) half-circle.

$$\int\limits_{0}^{\pi/2} \sin\varphi \frac{d^3\cos^3\varphi}{3} d\varphi = \frac{d^3}{3} \int\limits_{0}^{\pi/2} \sin\varphi\cos^3\varphi\, d\varphi$$

With the aid of a computer (see Example 9.9, Chapter 9), we find

$$I = \int \sin\varphi\cos^3\varphi\; d\varphi = -\cos^4\varphi\big/4$$

Thus,

$$\iint\limits_{D} r\sin\varphi\, dA = \frac{d^3}{3}\cdot\left(-\frac{\cos^4\varphi}{4}\right)\Bigg|_{0}^{\pi/2} = \frac{d^3}{3}\cdot\left(\frac{0}{4}-\frac{1}{4}\right) = \frac{d^3}{12}$$

Rule #3. If domain D intersects with any vertical line not more than in two points (P_1, P_2), then D is given by inequalities [Figure 3.11(a)]

$$a \le x \le b, \quad \varphi_1(x) \le y \le \varphi_2(x)$$

and the double integral is evaluated as:

$$\iint\limits_{D} f(x,y)\Delta A = \int\limits_{a}^{b} dx \int\limits_{\varphi_1(x)}^{\varphi_2(x)} f(x,y)dy$$

Analogously, if it intersects with any horizontal line not more than in two points [Figure 3.11(b)], then

$$c \le y \le d, \quad \varphi_1(y) \le x \le \varphi_2(y)$$

$$\iint\limits_{D} f(x,y)dx\,dy = \int\limits_{c}^{d} dy \int\limits_{\varphi_1(y)}^{\varphi_2(y)} f(x,y)dx$$

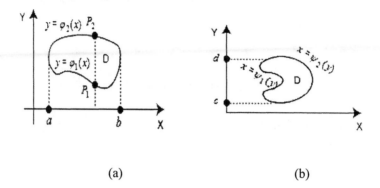

(a)   (b)

Figure 3.11 (a) Domain intersection with vertical lines and (b) horizontal lines.

If the domain D does not meet any of those requirements, then it is split in several parts [Figure 3.12(a)] in a way that previous formulas are applicable to each part.

♦ **Example 3.52** Evaluate $I = \iint\limits_{D}\left(y^2 + x\right)dx\,dy$ if D is bounded by the

parabola $y = x^2$ and $y^2 = x$ [Figure 3.12(b)] [1].

Solution. $I = \int\limits_{0}^{1} dx \int\limits_{y=x^2}^{y=\sqrt{x}}\left(y^2 + x\right)dx$

Assuming $x = $ const,

$$\int\limits_{x^2}^{\sqrt{x}}\left(y^2 + x\right)dy = \left[\frac{y^3}{3} + xy\right]\Bigg|_{x^2}^{\sqrt{x}} = \frac{\left(\sqrt{x}\right)^3 - x^6}{3} + x\sqrt{x} - x^3$$

$$= \left(\frac{x^{3/2}}{3} + x^{3/2}\right) - \left(\frac{x^6}{3} + x^3\right) = \frac{4}{3}x^{3/2} - \frac{x^6}{3} - x^3 = f(x)$$

Thus,

$$I = \int\limits_{0}^{1} f(x)\,dx = \frac{4}{3}\cdot\frac{2}{5}\cdot x^{5/2}\Bigg|_{0}^{1} - \frac{1}{3}\cdot\frac{x^7}{7}\Bigg|_{0}^{1} - \frac{x^4}{4}\Bigg|_{0}^{1} = \frac{33}{140} = 0.236$$

3.3.13 A Triple Integral

The definition of the triple integral is similar to the definition of a double integral (see Section 3.3.10). If the function of three arguments $f(x,y,z)$ is continuous within the three-dimensional domain D and we split D into n pieces with volumes $\Delta V_1, \ldots, \Delta V_n$, then when the largest of the diameters of the partial domains d_k tends to zero, the *triple integral* is the limit of the sum:

$$\iiint_D f(x,y,z)\,dV = \iiint_D f(x,y,z)\,dx\,dy\,dz =$$

$$\lim_{d_k \to 0}\left[f(x_1,y_1,z_1)\Delta V_1 + \ldots f(x_n,y_n,z_n)\Delta V_n\right]$$

The properties of the triple integral are the same as the properties of the double integral (see Section 3.3.11).

3.3.14 How to Evaluate a Triple Integral

Rule #1. If domain D is given by inequalities:

$$a \le x \le b;\ c \le y \le d;\ e \le z \le f,$$

that is a parallelepiped with edges parallel to OX, OY, OZ (Figure 3.13) then the triple integral is evaluated as:

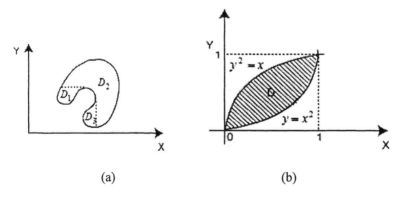

(a)   (b)

Figure 3.12 (a) Split domain and (b) domain bounded by the parabolas.

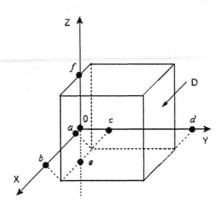

Figure 3.13 Parallelepiped domain.

$$\iiint_D f(x,y,z)\,dV = \int_e^f \int_c^d \int_a^b f(x,y,z)\,dx\,dy\,dz = \int_e^f dz \int_c^d dy \int_a^b f(x,y,z)\,dx$$

or interchanging the succession of the arguments:

$$\int_c^d dy \int_a^b dx \int_e^f f(x,y,z)\,dz, \quad \int_e^f dz \int_a^b dx \int_c^d f(x,y,z)\,dy$$

Rule #2. If for the point P the function $f(P) = f(r,\varphi,\theta)$ is given in spherical coordinates (see Appendix 6), then:

$$\iiint_D f(r,\varphi,\theta)\,dV = \iiint_D f(r,\varphi,\theta)r^2\,dr\,\sin\theta\,d\theta\,d\varphi$$

♦ Example 3.53 Evaluate the integral $I = \iiint_D \rho^2\,dV$ of the function [1]:

$$f(P) = \rho^2, \text{ if } \rho = r\sin\theta, 0 < \varphi \le 2\pi, 0 \le \theta \le \frac{\pi}{4}, 0 \le r \le 2R\cos\theta$$

Solution.

$$I = \iiint r^4 \sin^3 \theta \, dr \, d\theta \, d\varphi = \int\limits_0^{2\pi} d\varphi \int\limits_0^{2R\cos\theta} r^4 \, dr \int\limits_0^{\pi/4} \sin^3 \theta \, d\theta$$

$$= 2\pi \cdot \frac{(2R\cos\theta)^5}{5} \int\limits_0^{\pi/4} \sin^3 \theta \, d\theta = 2\pi \cdot \frac{32R^5}{5} \int\limits_0^{\pi/4} \cos^5 \theta \sin^3 \theta \, d\theta$$

$$= \frac{64\pi R^5}{5} \left(-\frac{1}{8} \sin^2 \theta \cos^6 \theta - \frac{1}{24} \cos^6 \theta \right) \Bigg|_0^{\pi/4}$$

$$= \frac{64\pi R^5}{5} \left(-\frac{1}{8} \sin^2 \frac{\pi}{4} \cos^6 \frac{\pi}{4} - \frac{1}{24} \cos^6 \frac{\pi}{4} + \frac{1}{24} \right)$$

$$= \frac{64\pi R^5}{5} \left(-\frac{1}{8} \cdot \frac{1}{2} \cdot \frac{1}{8} - \frac{1}{24} \cdot \frac{1}{8} + \frac{1}{24} \right) = \frac{64\pi R^5}{5} \cdot \frac{33}{1152} = \frac{11}{30} \pi R^5$$

Note. The integral $\int \cos^5 \theta \sin^3 \theta \, d\theta$ can be found with the aid of a computer (see Example 9.10, Chapter 9).

Rule #3. If for the point P the function $f(P) = f(r, \varphi, z)$ is given in the cylindrical coordinates (see <u>Appendix 6</u>), then

$$\iiint\limits_D f(r, \varphi, \theta) \, dV = \iiint\limits_D f(r, \varphi, \theta) r \, dr \, d\varphi \, dz$$

♦ <u>Example 3.54</u> Find integral $I = \iiint\limits_D z \, dV$ when domain D is a half-sphere with radius R [1].

Solution. The equation of the domain is the circle $x^2 + y^2 + z^2 = R^2$. Thus,

$$I = \int\limits_{0}^{\sqrt{R^2-z^2}} \int\limits_{0}^{2\pi} \int\limits_{0}^{R} z\, r\, dr\, d\varphi\, dz = \int\limits_{0}^{2\pi} d\varphi \int\limits_{0}^{R} z\, dz \int\limits_{0}^{\sqrt{R^2-z^2}} r\, dr = 2\pi \int\limits_{0}^{R} z\, dz \cdot \left.\frac{r^2}{2}\right|_{0}^{\sqrt{R^2-z^2}}$$

$$= 2\pi \int\limits_{0}^{R} \frac{R^2 - z^2}{2} \cdot z\, dz = \pi \left(R^2 \cdot \left.\frac{z^2}{2}\right|_{0}^{R} - \left.\frac{z^4}{4}\right|_{0}^{R} \right) = \pi \left(\frac{R^4}{2} - \frac{R^4}{4} \right) = \frac{\pi R^4}{4}$$

3.3.15 A Curved Integral

Let us consider the function $P(x, y)$, which is continuous in some domain on the plane XOY. If we take on this plane an arbitrary line AB [Figure 3.14(a)], split it in n sections, then if n tends to infinity and the largest of $|\Delta x_k|$ tends to zero, the *curved integral* of the function $P(x, y)$ along the path AB is:

$$\int\limits_{AB} P(x, y)\, dx = \lim_{|\Delta x_k| \to 0} \left[P(x_1, y_1)\Delta x_1 + \ldots + P(x_n, y_n)\Delta x_n \right]$$

Analogously, we can define the curved integral for the function $Q(x, y)$

$$\int\limits_{AB} Q(x, y)\, dy \quad \text{and}$$

$$\int\limits_{AB} P(x, y)\, dx + Q(x, y)\, dy; \quad \int\limits_{AB} P(x, y, z)\, dx + Q(x, y, z)\, dy + W(x, y, z)\, dz$$

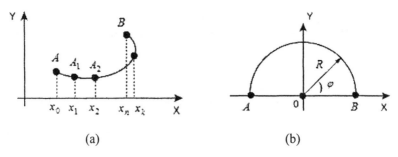

(a)   (b)

Figure 3.14 (a) A concept of the curved integral and (b) half-circle.

The sign of the curved integral depends on the direction of the motion along the path [from A to B (i.e., AB), or from B to A, (i.e., BA)].

$$\int\limits_{BA} P\,dx + Q\,dy = -\int\limits_{AB} P\,dx + Q\,dy$$

$$\int\limits_{BA} P\,dx + Q\,dy + W dz = -\int\limits_{AB} P\,dx + Q\,dy + W dz$$

The curved integral has the same properties as an ordinary integral (see Section 3.2.4).

3.3.16 How to Evaluate a Curved Integral

In order to evaluate the curved integral

$$\int\limits_{AB} P(x, y)\,dx + Q(x, y)\,dy \qquad (3.8)$$

the line AB has to be represented in the parametric form

$$x = f(t), \quad y = \phi(t)$$

Then the ordinary integral:

$$\int\limits_{t_A}^{t_B} \{P[f(t), \varphi(t)] \cdot f'(t) + Q[f(t), \varphi(t)] \cdot \phi'(t)\}\,dt$$

is equal to the curved integral (3.8).

♦ Example 3.55 Evaluate the curved integral

$$I = \int\limits_{AB} -y\,dx + x\,dy$$

along the upper half of the circle $x^2 + y^2 = R^2$ in the direction AB and BA [Figure 3.14 (b)] [1].
Solution. The curve AB can be represented by the parametric equations:

$$x = R\cos\varphi, \; y = R\sin\varphi$$
Thus,

$$\varphi_A = \pi, \varphi_B = 0, P(x, y) = -y; Q(x, y) = x;$$

$$f(\varphi) = R\cos\varphi; f'(\varphi) = -R\sin\varphi; \phi(\varphi) = R\sin\varphi; \phi'(\varphi) = R\cos\varphi$$

Then the curved integral in the direction *AB*:

$$I = \int_{\pi}^{0}[x(\varphi)\cdot f'(\varphi) + y(\varphi)\cdot\phi'(\varphi)]d\varphi = \int_{\pi}^{0}[-R\sin\varphi\cdot-R\sin\varphi + R\cos\varphi\cdot R\cos\varphi]d\varphi$$

$$= R^2\int_{\pi}^{0}d\varphi = -\pi R^2$$

In the direction *BA* the curved integral is:

$$I = R^2\int_{0}^{\pi}d\varphi = \pi R^2$$

3.3.17 The Green's Formula

If D is a plane domain bounded by the contour C [Figure 3.15(a)], functions $P(x, y)$, $Q(x, y)$, and its partial derivatives $\partial Q/\partial x$, $\partial P/\partial y$ are continuous in the domain D, then the *Green's formula* is valid:

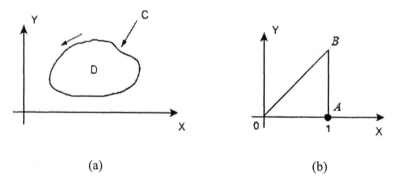

(a)   (b)

Figure 3.15 (a) Contour C and (b) the perimeter of the triangle.

$$\int_{+C} P(x,y)\,dx + Q(x,y)\,dy = \iint_D \left(\frac{\partial Q}{\partial x} - \frac{\partial P}{\partial y}\right) dx\,dy$$

where $+C$ means that we move along the contour C in the positive direction (counterclockwise for the standard orientation of the axes).

♦ Example 3.56 Evaluate the curved integral

$$I = \int \left(x - y^2\right) dx + 2xy\,dy$$

along the perimeter of the triangle *OAB* [Figure 3.15(b)] [1].
Solution.

$$P = x - y^2;\ Q = 2xy;\ \frac{\partial}{\partial x}Q = 2y;\ \frac{\partial}{\partial y}P = -2y$$

Thus,

$$I = \iint_D \left[2y - (-2y)\right] dx\,dy = \iint_D 4y\,dx\,dy$$

Because D is the triangle *OAB*:

$$I = \int_0^1 \int_0^x 4y\,dx\,dy = \int_0^1 dx \int_0^x 4y\,dy = 4\int_0^1 \left.\frac{y^2}{2}\right|_0^x dx = 2\int_0^1 x^2\,dx = \frac{2}{3}$$

3.4 Differential Equations

3.4.1 Definitions

A *differential equation* is an equation involving derivatives (or differentials) of the unknown functions.

If unknown functions depend only on a single argument, the equation is called an *ordinary differential equation,* and if unknown functions depend on several arguments, it is called a *partial differential equation.*

The *order* of the differential equation is the degree of the highest derivative that occurs in the equation.

Linear differential equations are those in which unknown variables (or a single variable) do not occur as nonlinear functions, products, or are raised to powers; otherwise, equations are called *nonlinear.*

The linear equation with a zero right-hand side is called a *homogeneous differential equation*; otherwise, it is a *nonhomogeneous equation.*

♦ Example 3.57 If $y = f(x)$, then

$\frac{dy}{dx} + 4y = 0$ is an ordinary homogeneous linear equation of the first order.

$\frac{dy}{dx} + 4y = \cos(x)$ is an ordinary nonhomogeneous linear equation of the first order.

$\frac{d^2 y}{dx^2} + \frac{1}{x} \cdot \frac{dy}{dx} - \frac{1}{x^2} y = 0$ is an ordinary homogeneous linear equation of the second order.

If $z = f(x, y)$, then

$e^z + \frac{\partial z}{\partial x} + \frac{\partial z}{\partial y} = 2x^2 + 6y$ is a nonlinear partial differential equation of the first order.

3.4.2 How to Solve Differential Equations: General Concept

The function $y = \varphi(x)$ is called a solution of the differential equation

$$\Phi\left[x, y, y', y'', \ldots, y^{(n)}\right] = 0$$

with a single unknown variable y if the latter becomes the identity when the function $y = \varphi(x)$ is substituted into the equation.

Note. The solution of a differential equation is not a single value (or set of values) but a function (or set of functions).

♦ Example 3.58 The function $y = \sin x$ is the solution of the equation

$$\frac{d^2 y}{dx^2} + y = 0$$

since $\left(\sin x\right)'' = -\sin x$. Thus $\left(\sin x\right)'' + \sin x = -\sin x + \sin x = 0$.

But it is not the only solution of this equation. It is easy to observe that functions $y = \frac{1}{2}\sin x$, $y = \cos x$, $y = 3\cos x$ are also the solutions for this equation.

The most general function that will satisfy the differential equation is called the *general solution*. Typically, it contains the number of arbitrary constants equal to the order of the differential equation. For a particular numerical value of the constant, we obtain a *particular solution*.

♦ <u>Example 3.59</u> Find the general solution of the equation of the first order [1]:

$$\frac{dy}{dx} = \cos x.$$

Solution. The generic function $y = \varphi(x)$ can be found as:

$$y = \int \cos x \, dx$$

Thus, $y = \varphi(x) = \sin x + C$ is the general solution of this equation. The functions $y = \sin x$, $y = \sin x + 1$, $y = \sin x - 3/4$, and so forth are the particular solutions.

Often the arbitrary constant in the general solution can be determined if *boundary conditions* are specified, that is, if the function $y = \varphi(x)$ takes some specified value y_0 when $x = x_0$. In the special case when all the boundary conditions are given for the same value of their independent variable (argument), they are called *initial conditions*.

♦ <u>Example 3.60</u> Find the solution of the differential equation $\frac{dy}{dx} = -3y$ for initial conditions $y_0 = 4$, when $x_0 = 0$ or $y(0) = 4$.

Solution. We can recall that if

$$y(x) = e^{-3x}$$

then

$$\frac{dy}{dx} = -3 \cdot e^{-3x} = -3y$$

and for any arbitrary C the function

$$y(x) = C \cdot e^{-3x}$$

turns the equation into the identity:

$$\frac{dy}{dx} = -3 \cdot Ce^{-3x} = -3y$$

Thus, $y(x) = C \cdot e^{-3x}$ is the general solution of the equation. But since $y(0) = 4$, then

$$y(0) = C \cdot e^{-3x}\Big|_{x=0} = C = 4$$

and $C = 4$.

The solution of the equation with initial conditions specified

$$y(x) = 4 \cdot e^{-3x}$$

If the solution can be obtained as the functional relationship between the dependent variable y and independent variable x (as in the example above) the differential equation has an *analytical solution*. Otherwise, it can be solved only by numerical methods with the aid of a computer that leads to a *numerical solution* (see Chapter 9).

3.4.3 How to Solve the First-Order Ordinary Differential Equations

Rule #1. The solution of some simple equations is obvious and based on inspection of the functions in the equation and relating them to known derivatives.

♦ Example 3.61 Find the solution of the differential equation

$$\frac{dy}{dx} = -4y$$

Solution. Obviously, if $y(x) = e^{-4x}$, then

$$\frac{dy}{dx} = -4 \cdot e^{-4x} = -4y$$

Thus, the function $y(x) = e^{-4x}$ is the solution of this differential equation.

Rule #2. If the first-order differential equation is the linear one, it has a general form:

$$\frac{dy}{dx} + P(x)y = Q(x)$$

The general analytical solution of this equation is:

$$y = \exp[-\int P(x)dx] \cdot \left\{ \int \exp[\int P(x)dx] \cdot Q(x)dx + C \right\}$$

If the equation is *homogeneous*, then $Q(x) = 0$; and the solution is:

$$y = C \cdot \exp[-\int P(x)dx]$$

♦ Example 3.62 Find the solution of the linear differential equation [2]:

$$\frac{dy}{dx} + \frac{1}{x} y = x$$

Solution. In this equation, $P(x) = \frac{1}{x}$, $Q(x) = x$

$$\exp[-\int P(x)dx] = \frac{1}{e^{\int \frac{1}{x}dx}} = \frac{1}{e^{\ln x}} = \frac{1}{x}$$

$$\int \exp\left[\int P(x)dx\right] \cdot Q(x)dx = \int x \cdot x \cdot dx = \int x^2 dx = \frac{x^3}{3} + C$$

Thus, the solution is $\frac{1}{x}\left(\frac{x^3}{3} + C\right) = \frac{x^2}{3} + \frac{C}{x}$.

♦ Example 3.63 Find the solution of the homogeneous linear differential equation [1]:

$$\frac{dy}{dx} - \frac{x}{1+x^2} y = 0$$

Solution. In this equation, $P(x) = -\frac{x}{1+x^2}$, $Q(x) = 0$. Thus,

$$y = C \cdot \exp\left[-\int -\frac{x\,dx}{1+x^2}\right] = C \cdot \exp\left[\frac{1}{2}\ln(1+x^2)\right] = C\sqrt{1+x^2}$$

(the integral in the previous formula can be found with the aid of a computer, see Chapter 9).

Rule #3. If the differential equation takes a *separable form*, that is, the equation

$$\frac{dy}{dx} = f(y, x)$$

can be manipulated by algebraic operations into the form

$$Q(y)\frac{dy}{dx} = -P(x)$$

where coefficient $Q(y)$ depends only on y and $P(x)$ only on x (i.e., the variables y and x are separable), or in the other representation

$$P(x)dx + Q(y)dy = 0$$

then the equation can be solved by integration that gives a general solution from

$$\int P(x)dx + \int Q(y)dy = C$$

or the particular solution from

$$\int_{x_0}^{x} P(x)dx + \int_{y_0}^{y} Q(y)dy = 0$$

if initial conditions $y_0(x_0)$ are known.

If the functions $P(x)$ and $Q(x)$ can be integrated analytically, then the analytical solution of the equation can be obtained.

♦ Example 3.64 Find a particular solution of the differential equation [1]:

$$\frac{1}{\sqrt{y}} \cdot \frac{dy}{dx} = -\sin x$$

for the initial conditions $y_0 = 3$ for $x_0 = \pi/2$.

Solution. The equation can be represented as

$$\frac{1}{\sqrt{y}} \cdot dy + \sin x \cdot dx = 0$$

After integration:

$$\int \frac{dy}{\sqrt{y}} + \int \sin x \cdot dx = C$$

we obtain

$$2\sqrt{y} - \cos x = C$$

For $x = \pi/2$, $y = 3$, we obtain $C = 2\sqrt{3}$. Thus,

$$2\sqrt{y} - \cos x = 2\sqrt{3}$$

and the solution is

$$y = \frac{\left(2\sqrt{3} + \cos x\right)^2}{4}$$

The same solution can be obtained from the equation:

$$\int_{3}^{y} \frac{dy}{\sqrt{y}} + \int_{\pi/2}^{x} \sin x \, dx = 0$$

Rule #4. If the equation takes the form or can be manipulated into the so-called *exact form* equivalent to

$$\frac{df(y,x)}{dx} = 0$$

then the solution is

$$f(y,x) = C.$$

♦ Example 3.65 Solve the equation [2]:

$$2yx \frac{dy}{dx} + y^2 - 2x = 0$$

Solution. Let us denote $f(y,x) = y^2 x - x^2$. Then

$$\frac{\partial f}{\partial y} = 2yx, \quad \frac{\partial f}{\partial x} = y^2 - 2x$$

Thus, the equation takes the form

$$\frac{\partial f(x,y)}{\partial y} \cdot \frac{dy}{dx} + \frac{\partial f(x,y)}{\partial x} = 0$$

which is equivalent to

$$\frac{df(y,x)}{dx} = \frac{d}{dx}\left(y^2 x - x^2\right) = 0$$

because the chain rule states that $\dfrac{\partial f(x,y)}{\partial x} = \dfrac{\partial f}{\partial y} \cdot \dfrac{dy}{dx} + \dfrac{\partial f}{\partial x}$.

The solution is obtained from the equation

$$y^2 x - x^2 = C$$

in the form (assuming $x > 0$ and $C > 0$):

$$y = \pm\sqrt{\frac{C}{x} + x}$$

Rule #5. If the differential equation does not take a separable form, that is, both P and Q in (3.9) are functions of x, y

$$P(x,y)dx + Q(x,y)dy = 0 \tag{3.9}$$

but the following condition is met:

$$\frac{\partial P}{\partial y} = \frac{\partial Q}{\partial x}$$

then the left-hand side of the (3.9) is the total differential of the function $F(x, y)$:

$$P(x,y)dx + Q(x,y)dy = d[F(x,y)]$$

Thus, $F(x,y) = C$.

♦ Example 3.66 Find a general solution of the differential equation [1]:

$$x^2 \cdot \frac{dy}{dx} = -2yx$$

Solution. The equation can be represented as:

$$2yx \cdot dx + x^2 \cdot dy = 0$$

$$\frac{\partial P}{\partial y} = \frac{\partial Q}{\partial x} = 2x \text{ , thus, the left-hand side is the total differential of the function}$$

$x^2 y$ and $2yx \cdot dx + x^2 \cdot dy = d\left(x^2 y\right),\ x^2 y = C.$

Thus, the general solution is:

$$y = \frac{C}{x^2}$$

Rule #6. If in the equation

$$P(x, y)dx + Q(x, y)dy = 0$$

the coefficients $P(x, y)$, $Q(x, y)$ do not meet the conditions $\dfrac{\partial P}{\partial y} = \dfrac{\partial Q}{\partial x}$

then the left-hand side does not compose total differential. Sometimes it is possible to find a multiplier $M(x, y)$, which converts

$$M(x, y)\left[P(x, y)dx + Q(x, y)dy\right]$$

into the total differential.

♦ Example 3.67 Find a general solution of the differential equation [1]:

$$x \cdot \frac{dy}{dx} = -2y$$

Solution. The equation can be represented as:

$$2y\, dx + x\, dy = 0, \quad \frac{\partial P}{\partial y} = 2, \quad \frac{\partial Q}{\partial x} = 1.$$

Thus, $\dfrac{\partial P}{\partial y} \neq \dfrac{\partial Q}{\partial x}$ and the left-hand side of the equation does not compose a total differential.

Multiplying the equation by $M(x, y) = x$, we can obtain the equivalent equation

$$x(2y\,dx + x\,dy) = 0 \quad \text{or} \quad 2yx\,dx + x^2 dy = d\left(x^2 y\right)$$

This is the equation from Example 3.66, and thus the general solution is

$$y = \frac{C}{x^2}$$

3.4.4 How to Solve the Second-Order Ordinary Differential Equations

The generic form for the second-order differential equation is

$$\Phi[x, y, y', y''] = 0$$

or with regard to y'':

$$y'' = f(x, y, y')$$

The general solution is the function

$$y = \varphi(x, C_1, C_2)$$

where C_1 and C_2 are the constants.

The general solution can be reduced to the particular solution when initial conditions $x = x_0$, $y = y_0$, $y' = y_0'$ are specified.

Rule #1. As in the first-order equations, there are obvious solutions for some simple equations based on integration of the equation functions.

♦ Example 3.68 Find the particular solution of the differential equation [1]:

$$y'' - x = 0$$

for initial conditions $x = 1$, $y = 1$, $y' = 2$.

Solution. The equation can be written as $y'' = x$, thus,

$$y' = \int x\,dx = \frac{x^2}{2} + C_1$$

and the general solution is:

$$y = \int y'dx = \int\left(\frac{x^2}{2} + C_1\right)dx = \frac{x^3}{6} + C_1 x + C_2$$

For specified initial conditions:

$$2 = \frac{1}{2} + C_1; \quad 1 = \frac{1}{6} + C_1 + C_2$$

that results in $C_1 = 3/2$, $C_2 = -2/3$. Thus, the solution is:

$$y = \frac{x^3}{6} + \frac{3}{2}x - \frac{2}{3}$$

Rule #2. If the second-order differential equation is the linear homogeneous constant-coefficient equation

$$ay'' + by' + cy = 0$$

where a, b, c are constants, the solution has the form

$$y = \exp(\mu x)$$

where μ satisfies the characteristic or auxiliary equation

$$a\mu^2 + b\mu + c = 0$$

Since a quadratic equation has two roots, μ_1 and μ_2, the following cases are possible.

A. $b^2 - 4ac > 0$; thus, μ_1 and μ_2 are two different real numbers. In this case, the general solution is:

$$y = C_1 \cdot e^{\mu_1 x} + C_2 \cdot e^{\mu_2 x}$$

B. $b^2 - 4ac = 0$; thus, $\mu_1 = \mu_2 = -b/2a$ (a repeated real root). The general solution is:

$$y = (C_1 + C_2 x) e^{-\frac{b}{2a}x}$$

C. $b^2 - 4ac < 0$; thus, μ_1 and μ_2 are a pair of complex-conjugate numbers:

$$\mu_{1,2} = -\frac{b}{2a} \pm j \cdot \beta$$

where

$$\beta = \frac{\sqrt{b^2 - 4ac}}{2a}$$

The general solution is:

$$y = e^{-\frac{b}{2a}x}(C_1 \cos \beta x + C_2 \sin \beta x)$$

◆ Example 3.69 Find the general solution of the equation [1]:

$$y'' + y' + y = 0$$

Solution. The characteristic equation $\mu^2 + \mu + 1 = 0$ has the roots

$$\mu_{1,2} = -\frac{1}{2} \pm j \cdot \frac{\sqrt{3}}{2}$$

Thus, the general solution is

$$y = e^{-\frac{x}{2}}\left(C_1 \cos \frac{\sqrt{3}}{2}x + C_2 \sin \frac{\sqrt{3}}{2}x\right)$$

◆ Example 3.70 Find the particular solution of the equation [1]:

$$8y'' + 2y' - 3y = 0$$

for initial conditions $x_0 = 0$, $y_0 = -6$, $y_0' = 7$.

Solution. The characteristic equation $8\mu^2 + 2\mu - 3 = 0$ has the roots $\mu_1 = 1/2$, $\mu_2 = -3/4$. Thus, the general solution is:

$$y = C_1 \cdot \exp\left(\frac{1}{2}x\right) + C_2 \cdot \exp\left(-\frac{3}{4}x\right)$$

To obtain the particular solution, we have to find the derivative:

$$y' = \frac{1}{2}C_1 \cdot \exp\left(\frac{1}{2}x\right) - \frac{3}{4}C_2 \cdot \exp\left(-\frac{3}{4}x\right)$$

For initial conditions specified:

$$-6 = C_1 + C_2; \quad 7 = \frac{1}{2}C_1 - \frac{3}{4}C_2$$

that results in $C_1 = 2$, $C_2 = -8$. The particular solution is:

$$y = 2\exp\left(\frac{1}{2}x\right) - 8 \cdot \exp\left(-\frac{3}{4}x\right)$$

Rule #3. If the second-order differential equation is the linear non-homogeneous constant-coefficient equation

$$ay'' + by' + cy = U(x) \tag{3.10}$$

its solution is based on the homogeneous equation

$$ay'' + by' + cy = 0 \tag{3.11}$$

and the following procedure:
Step 1. First, we rewrite the (3.10) as

$$y'' + P(x)y' + Q(x)y = R(x); \ P(x) = \frac{b}{a}; \quad Q(x) = \frac{c}{a}; \quad R(x) = \frac{U(x)}{a}$$

Step 2. Second, we solve (3.11) as per Rule #2 above and obtain the generic solution as

$$y = C_1 \cdot \varphi_1(x) + C_2 \cdot \varphi_2(x)$$

Step 3. Third, we find the constants C_1, C_2 from the set of equations $(C' = dC/dx)$:

$$\begin{cases} C_1' \cdot \varphi_1(x) + C_2' \cdot \varphi_2(x) = 0 \\ C_1' \cdot \varphi_1'(x) + C_2' \cdot \varphi_2'(x) = R(x) \end{cases}$$

♦ <u>Example 3.71</u> Find the solution of the equation [1]:

$$y'' + y = \tan x$$

Solution. The corresponding homogeneous equation is

$$y'' + y = 0$$

and the characteristic equation is

$$\mu^2 + 1 = 0$$

For this characteristic equation, $a = 1$, $b = 0$, $c = 1$; thus, the general solution is given by Rule #2C ($b^2 - 4ac < 0$). The roots of the characteristic equation are

$$\mu_{1,2} = -\frac{b}{2a} \pm j \cdot \beta = \pm j \quad (b = 0, \beta = 1)$$

and the general solution is:

$$y = C_1 \cos x + C_2 \sin x$$

The set of equations to find C_1, C_2 are

$$\begin{cases} C_1' \cdot \cos x + C_2' \cdot \sin x = 0 \\ -C_1' \cdot \sin x + C_2' \cdot \cos x = \tan x \end{cases}$$

The set has a solution

$$C_1' = -\tan x \sin x; \quad C_2' = \sin x$$

Thus,

$$C_1 = \int -\tan x \sin x \, dx + C_3 = \ln\left(\frac{\cos x}{1+\sin x}\right) + \sin x + C_3$$

$$C_2 = \int \sin x \, dx + C_4 = -\cos x + C_4$$

The general solution is:

$$y = [\ln\frac{\cos x}{1+\sin x} + \sin x + C_3]\cos x + \left(-\cos x + C_4\right)\cdot\sin x =$$

$$\cos x \ln\left(\frac{\cos x}{1+\sin x}\right) + C_3 \cos x + C_4 \sin x$$

3.4.5 How to Solve Linear Differential Equations of Any Order

The nth-order linear ordinary equation can be written as:

$$y^{(n)} + P_1(x)\cdot y^{(n-1)} + \ldots + P_n(x)y = R(x)$$

Rule #1. For an nth-order linear homogeneous constant-coefficient equation

$$y^{(n)} + P_1 \cdot y^{(n-1)} + \ldots + P_n y = 0$$

the solution can be obtained with the characteristic equation:

$$\mu^n + P_1 \cdot \mu^{n-1} + \ldots + P_n = 0$$

that leads to the following cases.

A. If all the roots are real and different numbers, then the general solution is

$$y = C_1 \exp\left(\mu_1 x\right) + C_2 \exp\left(\mu_2 x\right) + \ldots + C_n \exp\left(\mu_n x\right) \tag{3.12}$$

B. If one of the real roots is a multiple one (repeated k times), then k factors in the general solution (3.12) are substituted by

$$\left(C_1 + C_2 x + \ldots + C_k x^{k-1}\right)\cdot\exp(\mu x)$$

C. If some roots are complex-conjugate pairs $\mu_{1,2} = \alpha \pm j\beta$ repeated k times, then the corresponding k pairs of the factors are substituted with

$$[(C_1 + C_2 x + \ldots + C_k x^{k-1}) \cdot \cos \beta x + (D_1 + D_2 x + \ldots + D_k x^{k-1}) \sin \beta x] \cdot \exp(\alpha x)$$

In case $k = 1$ (the only one pair), the expression above reduces to

$$(C_1 \cos \beta x + D_1 \sin \beta x) \cdot \exp(\alpha x)$$

◆ Example 3.72 Find the solution of the equation [1]:

$$y^{(5)} + y^{(4)} + 2y^{(3)} + 2y'' + y' + y = 0$$

Solution. The characteristic equation

$$\mu^5 + \mu^4 + 2\mu^3 + 2\mu^2 + \mu + 1 = 0$$

has the following roots:

$$\mu_1 = -1; \; \mu_2 = j; \; \mu_3 = -j; \; \mu_4 = j; \; \mu_5 = -j$$

or one real root ($\mu = -1$) and two ($k = 2$) complex-conjugate pairs $\mu = \pm j$. Thus, the general solution is:

$$y = C_1 \cdot e^{-x} + (C_2 + C_3 x) \cos x + (C_4 + C_5 x) \sin x$$

Rule #2. For the nonhomogeneous nth-order constant-coefficient equation

$$y^{(n)} + P_1 \cdot y^{(n-1)} + \ldots + P_n y = R(x)$$

the approach described for the second-order nonhomogeneous equation in Section 3.4.4 (*Rule #3*) can be used.

In the general solution of the corresponding homogeneous equation, all constants are substituted by the unknown function of x, $C_n = C_n(x)$

$$y = C_1 \cdot \varphi_1(x) + C_2 \cdot \varphi_2(x) + \ldots + C_n \cdot \varphi_n(x)$$

The solution is differentiated to find the derivatives y', y'', y''', and so forth, which provides us with the additional equations to determine the derivative of the unknown coefficients $C_n = C_n(x)$.

3.4.6 How to Solve the Set of Equations

The system of the linear equations explicitly expressed with respect to the derivatives

$$
\begin{cases}
\dfrac{dx}{dt} = f(x, y, t) \\[2ex]
\dfrac{dy}{dt} = \varphi(x, y, t)
\end{cases}
$$

can be solved by excluding one of the unknown variables and bringing it to the linear equation of higher-orders with respect to a single unknown variable

$$
\frac{dx^k}{dt^k} = \phi(x, t)
$$

as it is shown in the example below.

♦ <u>Example 3.73</u> Find the solution of the set of equations [1]:

$$
dx/dt = x - y + (3/2)t^2 \tag{3.13}
$$

$$
dy/dt = -4x - 2y + 4t + 1 \tag{3.14}
$$

Solution.
Step 1. To exclude y and dy/dt, we differentiate (3.13) that results in

$$
\frac{d^2x}{dt^2} = \frac{dx}{dt} - \frac{dy}{dt} + 3t \tag{3.15}
$$

Step 2. From (3.13), we find the expression for y as the function of x, dx/dt, t

$$
y = x - \frac{dx}{dt} + \frac{3}{2}t^2 \tag{3.16}
$$

Step 3. Substituting (3.16) into (3.14), we find the expression for dy/dt as the function of x, dx/dt, t

$$\frac{dy}{dt} = -4x - 2\left(x - \frac{dx}{dt} + \frac{3}{2}t^2 \right) + 4t + 1 = -6x - 2\frac{dx}{dt} - 3t^2 + 4t + 1 \qquad (3.17)$$

Step 4. Substituting (3.17) into (3.15), we obtain the second-order linear equation with respect to x :

$$\frac{d^2 x}{dt^2} + \frac{dx}{dt} - 6x = 3t^2 - t - 1 \qquad (3.18)$$

Step 5. Using Rule #3 from Section 3.4.4, we find the general solution of (3.18)

$$x = C_1 \cdot e^{2t} + C_2 \cdot e^{-3t} - \frac{1}{2}t^2 \qquad (3.19)$$

Step 6. Substituting (3.19) into (3.16), we find:

$$y = -C_1 \cdot e^{2t} + 4C_2 \cdot e^{-3t} + t^2 + t$$

3.4.7 The Easiest Way to Solve Differential Equations

As we can see from the previous sections, it might be very difficult to solve even the simplest differential equations in analytical form. The easiest and most mistake-free way to solve them is to use a computer to obtain numerical solutions (see Chapter 9).

References

[1] Vigodskiy, M. Ya., *Mathematical Handbook*, (in Russian), Moscow: Nauka, 1972.

[2] James, G. (ed.), *Modern Engineering Mathematics*, Harlow, England: Pearson Education Ltd., 2001.

Selected Bibliography

Bolton, W., *Ordinary Differential Equations*, New York: Longman, 1996.

Bronshtein I. N., and K. A. Semendyaev, *Mathematical Handbook for Engineers and Students*, (in Russian), Moscow: Nauka, 1980.

Courant, R., *Lectures on Differential and Integral Calculus*, Berlin: Springer-Verlag, 1971.

Creese, T. M., and R. M. Harlock, *Differential Equations for Engineers*, New York: McGraw-Hill, 1978.

Croft, T., et al., *Engineering Mathematics: A Foundation for Electronic, Electrical, Communications and System Engineers*, Englewood Cliffs, NJ: Prentice Hall, 2000.

Kurtz, M., *Handbook of Applied Mathematics for Engineers and Scientists*, New York: McGraw-Hill, 1992.

Polianin, A. D., *Handbook of Linear Partial Differential Equations for Engineers and Scientists*, Boca Raton, FL: CRC Press, 2001.

Rade, L., and B. Westergen, *Mathematics Handbook for Science and Engineering*, Cambridge, MA: Birkhauser, 1995.

Stephenson, G., and P. M., Radmore, *Advanced Mathematical Methods for Engineering and Science Students*, Cambridge, England: Cambridge University Press, 1990.

Chapter 4

COMPLEX NUMBERS AND FUNCTIONS

4.1 Complex Numbers

4.1.1 Definitions

A *complex number* z is a combination of two real numbers x and y written in the form

$$z = x + j \cdot y$$

where $j = \sqrt{-1}$.

The number x is called a *real* part, $x = \mathrm{Re}\{z\}$, and the number y is called an *imaginary* part, $y = \mathrm{Im}\{z\}$.

The *modulus* of a complex number z is a positive number

$$|z| = \rho = \sqrt{x^2 + y^2}$$

The *argument* of a complex number z is:

$$\arg(z) = \varphi = A\tan(y/x) = a\tan(y/x) + 2k\rho, \quad k = 0, \pm 1, \pm 2, \ldots$$

The complex number can be represented through its modulus and argument as:

$$z = \rho \cos\varphi + j \cdot \rho \sin\varphi = \rho(\cos\varphi + j \cdot \sin\varphi)$$

Geometrically, complex numbers can be represented as points on a plane when X is the real axis and Y is the imaginary axis (*Argand diagram*, Figure 4.1).

137

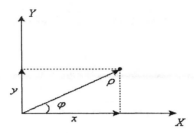

Figure 4.1 Argand diagram.

If $\text{Im}(z)=0$, $z=x$, and the number z is called *purely real*. If $\text{Re}(z)=0$, $z=jy$, and the number z is called *purely imaginary*. If $x=y=0$, then $z=0$.

4.1.2 Equality

Two complex numbers $z_1 = x_1 + jy_1$; $z_2 = x_2 + jy_2$ are equal when they represent the same points on the Argand diagram. Thus, $x_1 = x_2$, $y_1 = y_2$.

4.1.3 Addition and Subtraction

To add or subtract two complex numbers, the operations have to be performed on their corresponding real and imaginary parts. Thus, if $z_1 = x_1 + jy_1$ and $z_2 = x_2 + jy_2$, then

$$z_1 + z_2 = (x_1 + x_2) + j(y_1 + y_2); \quad z_1 - z_2 = (x_1 - x_2) + j(y_1 - y_2)$$

♦ <u>Example 4.1</u> Find the sum and difference of $z_1 = 4 + j3$, $z_2 = 5 + j2$.
Solution.

$$z_1 + z_2 = (4+5) + j(3+2) = 9 + j5; \quad z_1 - z_2 = (4-5) + j(3-2) = -1 + j.$$

4.1.4 Multiplication

To multiply two complex numbers, we have to apply the normal rules for multiplication of variables in brackets. Thus, if $z_1 = x_1 + jy_1$, $z_2 = x_2 + jy_2$,

$$z_1 \cdot z_2 = (x_1 + jy_1) \cdot (x_2 + jy_2) = x_1 x_2 + jy_1 x_2 + jx_1 y_2 + j^2 y_1 y_2$$

Since $j^2 = -1$, $z_1 \cdot z_2 = (x_1 x_2 - y_1 y_2) + j(y_1 x_2 + x_1 y_2)$

♦ Example 4.2 Find the product of $z_1 = 3 + j2$ and $z_2 = 5 + j3$ [1].

Solution. $z_1 \cdot z_2 = (3 + j2)(5 + j3) = (15 - 6) + j(10 + 9) = 9 + j19$.

4.1.5 Complex Conjugates

Two complex numbers $z = x + jy$ and $z^* = x - jy$ are called *complex conjugates* with respect to each other. For complex conjugate numbers:

- $z + z^* = 2x = 2\,\text{Re}(z)$ • $z - z^* = j2y = j \cdot 2\,\text{Im}(z)$ • $z \cdot z^* = x^2 + y^2 = \rho^2$

4.1.6 Division

To divide two complex numbers $z_1 = x_1 + jy_1$, $z_2 = x_2 + jy_2$, the quotient has to be multiplied by z_2^* :

$$\frac{z_1}{z_2} = \frac{x_1 + jy_1}{x_2 + jy_2} = \frac{(x_1 + jy_1)(x_2 - jy_2)}{(x_2 + jy_2)(x_2 - jy_2)} = \frac{(x_1 x_2 + y_1 y_2) + j(x_2 y_1 - x_1 y_2)}{x_2^{\,2} + y_2^{\,2}}$$

$$= \frac{x_1 x_2 + y_1 y_2}{x_2^{\,2} + y_2^{\,2}} + j\,\frac{x_2 y_1 - x_1 y_2}{x_2^{\,2} + y_2^{\,2}}$$

♦ Example 4.3 Find the quotient of $z_1 = 2 + j3$, $z_2 = 1 + j2$.

Solution. $\dfrac{z_1}{z_2} = \dfrac{2 + j3}{1 + j2} = \dfrac{2 + 6}{5} + j\,\dfrac{3 - 4}{5} = \dfrac{8}{5} - j\,\dfrac{1}{5}$

4.1.7 Arithmetic with a Modulus and an Argument

If two complex numbers are represented through its modulus and argument (in a polar form, see Figure 4.1)

$$z_1 = \rho_1(\cos\varphi_1 + j\sin\varphi_1); \ z_2 = \rho_2(\cos\varphi_2 + j\sin\varphi_2);$$

then multiplication and division gives the following results:

- $z_1 \cdot z_2 = \rho_1 \cdot \rho_2[\cos(\varphi_1 + \varphi_2) + j\sin(\varphi_1 + \varphi_2)]$

- $\left| z_1 \cdot z_2 \right| = \rho_1 \rho_2$; $\arg(z_1 \cdot z_2) = \varphi_1 + \varphi_2$

- $\dfrac{z_1}{z_2} = \dfrac{\rho_1}{\rho_2}[\cos(\varphi_1 - \varphi_2) + j\sin(\varphi_1 - \varphi_2)]$

- $\left| \dfrac{z_1}{z_2} \right| = \dfrac{\rho_1}{\rho_2}$ • $\arg\left(\dfrac{z_1}{z_2}\right) = \varphi_1 - \varphi_2$

4.1.8 Power of Complex Numbers

Since, as we saw in the previous section, the modulus of the product of complex numbers is equal to the product of the modulus of the factors, and argument of the product is equal to the sum of the arguments of the factors, then:

$$z^n = \rho^n \left(\cos\varphi + j\sin\varphi\right)^n = \rho^n \left(\cos n\varphi + j\sin n\varphi\right)$$

In general,

$$\cos n\varphi = \cos^n\varphi - C_n^2 \cos^{n-2}\varphi \sin^2\varphi + C_n^4 \cos^{n-4}\varphi \sin^4\varphi - \ldots$$

$$\sin n\varphi = C_n^1 \cos^{n-1}\varphi \sin\varphi - C_n^3 \cos^{n-3}\varphi \sin^3\varphi + \ldots$$

where

$$C_n^p = \frac{n(n-1)\ldots(n-p+1)}{p!}$$

Thus, for $n=2$:

$$\cos 2\varphi = \cos^2\varphi - \sin^2\varphi; \quad \sin 2\varphi = 2\sin\varphi\cos\varphi$$

4.1.9 Euler's Formula

Euler's formula links the exponential and trigonometric functions as:

$$e^{j\varphi} = \cos\varphi + j\sin\varphi$$

Thus, a complex number $z = \rho(\cos\varphi + j\sin\varphi)$ can be written very concisely in the *exponential form*:

$$z = \rho e^{j\varphi}$$

Important results are:

- $e^{\pm 2\pi \cdot j} = 1$ • $e^{\pm \pi \cdot j} = -1$ • $e^{\pm \frac{\pi}{2} j} = \pm j$ • $e^{\pm \frac{\pi}{4} j} = \dfrac{1 \pm j}{\sqrt{2}}$

It is worthwhile to remember the conversion rule: $-j = \dfrac{1}{j}$

(since $\dfrac{1 \cdot -j}{j \cdot -j} = \dfrac{-j}{1} = -j$)

Euler's formula also establishes the link between trigonometric and hyperbolic functions. Since

$$e^{j\varphi} = \cos\varphi + j\sin\varphi \text{ and } e^{-j\varphi} = \cos\varphi - j\sin\varphi,$$

then

- $\cos\varphi = \dfrac{e^{j\varphi} + e^{-j\varphi}}{2}$ • $\sin\varphi = \dfrac{e^{j\varphi} - e^{-j\varphi}}{2j}$

- $\cosh jx = \dfrac{e^{jx} + e^{-jx}}{2} = \cos x$ • $\cos jx = \dfrac{e^{-x} + e^{x}}{2} = \cosh$

- $\sinh jx = \dfrac{e^{jx} - e^{jx}}{2} = j\sin x$ • $\sin jx = \dfrac{e^{-x} - e^{x}}{2} = j\sinh x$

- $\tanh jx = j\tan x$ • $\tan jx = j\tanh x$

◆ Example 4.4   Express the number $2 + j2$ in exponential form [1].

Solution. $\rho = \sqrt{2^2 + 2^2} = \sqrt{8} = 2\sqrt{2}$; $\arg(z) = a\tan(2/2) = a\tan(1) = \dfrac{\pi}{4}$

Thus,

$$z = 2\sqrt{2}e^{j\left(\frac{\pi}{4}\right)}$$

4.1.10 Logarithms of Complex Numbers

Natural logarithm of a complex number can be evaluated as:

$$\ln z = \ln|z| + j \cdot \arg(z) = \ln \rho + j \left(\varphi + 2\kappa\pi \right)$$

4.1.11 Roots of Complex Numbers

The root on the nth power of a complex number z can be evaluated as:

$$z^{\frac{1}{n}} = \rho^{\frac{1}{n}} \left[\cos\left(\frac{\varphi}{n} + \frac{2\pi\kappa}{n} \right) + j \sin\left(\frac{\varphi}{n} + \frac{2\pi\kappa}{n} \right) \right] = \rho^{\frac{1}{n}} \exp\left[j\left(\frac{\varphi}{n} + \frac{2\pi\kappa}{n} \right) \right]$$

Thus, a root of nth power from a complex number has n different roots, corresponding to $\kappa = 0, 1, 2, ..., n-1$.

4.1.12 Powers of Trigonometric Functions

The functions $\sin^n \theta$ and $\cos^n \theta$ can be expressed in terms of multiple angles based on Euler's formula that leads to identities:

- $2 \cos n\theta = z^n + z^{-n}$ • $2j \sin n\theta = z^n - z^{-n}$

4.1.13 Differentiation and Integration with Respect to the Argument

The derivative and the integral of a complex number $z = \rho e^{j\varphi}$ is:

$$\frac{dz}{d\varphi} = j \cdot \rho e^{j\varphi} = \rho \cdot e^{j\frac{\pi}{2}} \cdot e^{j\varphi} = \rho \cdot e^{j\left(\varphi + \frac{\pi}{2} \right)}$$

$$\int z\, d\varphi = \rho \int e^{j\varphi}\, d\varphi = \rho \frac{1}{j} e^{j\varphi} = \rho \cdot (-j) \cdot e^{j\varphi} = \rho \cdot e^{-j\frac{\pi}{2}} \cdot e^{j\varphi} = \rho \cdot e^{j\left(\varphi - \frac{\pi}{2} \right)}$$

4.2 Functions of a Complex Variable

4.2.1 Definitions

If $U(x, y)$ and $V(x, y)$ are the functions of real variables x, y, then the expression

$$U(x, y) + jV(x, y)$$

is considered to be the function of complex variable $z = x + jy$, and is written as $f(z)$.

The function $f(z)$ is *continuous* when $z = a$, if for any positive infinitesimal ε there is a positive number δ so that the inequality $|z - a| < \delta$ leads to the inequality $|f(z) - f(a)| < \varepsilon$.

The function $f(z)$ is *single-valued* if for each value of the argument only one value of the function can be assigned.

♦ <u>Example 4.5</u> The function $u = \sqrt[m]{z}$ is not single-valued, since if $z = \rho \cdot e^{j\varphi}$, $u = r \cdot e^{j\theta}$, then $r = \sqrt[m]{\rho}$, and $\theta = \varphi/m + (2\pi/m)\kappa$. Thus, for one value of z, there are m values of u .

4.2.2 Analytic and Holomorphic Functions

The derivative of a single-valued complex function (or derivative of one of its values if the function is multiple-valued) is defined as:

$$\lim_{\Delta z \to 0} \frac{f(z + \Delta z) - f(z)}{\Delta z}$$

Let us consider the function

$$f(z) = U(x, y) + V(x, y)$$

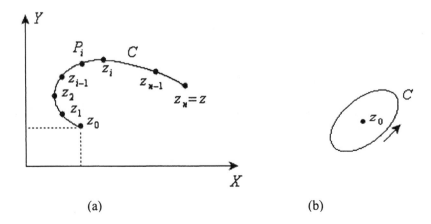

(a)   (b)

Figure 4.2 (a) A concept of a curved integral and (b) a closed-loop contour.

If the partial derivatives

$$\frac{\partial U}{\partial x} = \frac{\partial V}{\partial y} \text{ and } \frac{\partial U}{\partial y} = -\frac{\partial V}{\partial x}$$

exist and they are continuous, then these are necessary and sufficient conditions for the function $f(z)$ to be an *analytic* one.

The single-valued analytic function of a complex variable is called a *holomorphic function*.

4.2.3 A Curved Integral of a Function of a Complex Variable

Let us consider a continuous curve C starting at point z_0 on the Argand diagram [Figure 4.2(a)]. The curve can be divided into an arbitrary number of intervals with intermediate points $z_1, z_2, ..., z_{n-1}$, $z_n = z$. Let us consider a point P_i between z_{i-1} and z_i, $\Delta z_i = z_i - z_{i-1}$. Then, a curved integral of the function $f(z)$ along the curve C is a limit:

$$\int_C f(z)dz = \lim_{n \to \infty} \sum_{i=1}^{n} f(P_i)\Delta z_i$$

If

$$f(z) = U(x, y) + jV(x, y), \ dz = dx + jdy$$

then

$$\int_C f(z)dz = \int_C (Udx - Vdy) + j \int_C (Udy + Vdx)$$

4.2.4 Cauchy's Theorem and Formula

The *Cauchy's theorem* states that for any closed-loop contour C within the region where function $f(z)$ is holomorphic, the following equation is valid:

$$\int_C f(z)\,dz = 0$$

For any interior point in this region z_0 [Figure 4.2(b)], the value of the function $f(z_0)$ in this point is given by *Cauchy's formula:*

$$f(z_0) = \frac{1}{2\pi j} \int \frac{f(z)}{z - z_0}\,dz$$

when we move along the contour C in the positive direction (counterclockwise).

The derivative of nth order is:

$$f^{(n)}(z_0) = \frac{n!}{2\pi j} \int_C \frac{f(z)}{(z - z_0)^{n+1}}\,dz$$

4.2.5 Taylor's Series

Let us consider the function $f(z)$, which is holomorphic within a circle C centered at the point a, and z_0 is a point inside this circle.

The *Taylor series* for such a function can be written as:

$$f(z_0) = \frac{1}{2\pi j} \int_C \frac{f(z)}{z - z_0}\,dz = f(a) + \frac{z_0 - a}{1!} f'(a) + \dots \frac{(z_0 - a)^n}{n!} f^{(n)}(a) + \dots$$

When $a = 0$ (the center of the circle coincides with the origin of the coordinate frame), the Taylor series is reduced to the *Maclaurin series:*

$$f(z_0) = f(0) + \frac{z_0}{1!} f'(0) + \dots \frac{z_0^n}{n!} f^{(n)}(z_0) + \dots$$

Cauchy, Augustin (1789–1857). French mathematician who contributed to the theory of differentiation, integration, finite groups, and analysis of the functions of the complex variables. He wrote 789 papers, a quantity exceeded only by Euler and Cayley. He invented the name for the *determinant,* and gave definitions of *limit, continuity,* and *convergence* that we use today.

146 Mathematical Handbook for Electrical Engineers

4.2.6 Critical Points and Poles

A point at which an analytic function can be represented via the Taylor series is called an *ordinary* point. If the point is not an ordinary one, it is called a *critical* point.

The *poles* are isolated critical points in the vicinity of which $f(z)$ remains a single-valued function, and which are ordinary points for the reciprocal functions $1/f(z)$.

The *essential critical points* are the points in the vicinity of which $f(z)$ remains a single-valued function, but they are also critical points for the reciprocal function $1/f(z)$.

The *branch points* are points in the vicinity of which the function $f(z)$ does not remain a single-valued one.

4.2.7 Loraine's Series

The function $f(z)$, which has a pole or essential critical point $z = a$, can be represented as a *Loraine series* in the vicinity of this point:

$$f(z_0) = ... + \frac{A_{-n}}{(z_0 - a)^n} + ... + \frac{A_{-1}}{(z_0 - a)} + A_0 + A_1(z_0 - a) + ... + A_n(z_0 - a)^n ...$$

where

$$A_n = \frac{1}{2\pi j} \int_C \frac{f(z)}{(z - a)^{n+1}} dz ; \qquad A_{-n} = \frac{1}{2\pi j} \int_C f(z)(z - a)^{n-1} dz$$

4.2.8 Theorem of Residues

If a is a *pole* for the function $f(z)$, which is holomorphic in all points within a contour C but a, then the following formula is valid:

$$\int_C f(z)dz = 2\pi j \cdot A_{-1}$$

where A_{-1} is the coefficient at $\frac{1}{z - a}$ when $f(z)$ is represented by the Loraine's series (see Section 4.2.7).

The number A_{-1} is called a *residue* $\text{Res}(a)$ of the function $f(z)$ with respect to a critical point a.

4.2.9 How to Evaluate a Residue

Typically, the method of derivatives is used to evaluate the residues for the functions of the $f(z)/g(z)$ type with a pole of nth order in a point a. In this case, $A_{-1} = \text{Res}(a)$ is calculated as:

$$\text{Res}(a) = \frac{1}{(n-1)!} \frac{d^{n-1}}{dz^{n-1}} \left[(z-a)^n \frac{f(z)}{g(z)} \right]_{z=a}$$

♦ Example 4.6 Find a residue of the function $\dfrac{z \cdot e^z}{(z-a)^3}$ with respect to $z = a$.

Solution. The function has the pole of the third order in the point a.

$$\text{Res}(a) = \frac{1}{2!} \cdot \frac{d^2}{dz^2} \left(z e^z \right) \Big|_{z=a} \; ; \quad \frac{d}{dz} \left(z e^z \right) = (z+1)e^z \; ; \quad \frac{d^2}{dz^2} = (z+2) \cdot e^z \; ;$$

Thus, for $z = a$:

$$\text{Res}(a) = \frac{1}{2} (a+2) \cdot e^a = \left(1 + \frac{a}{2} \right) \cdot e^a$$

Reference

[1] James, G. (ed.), *Modern Engineering Mathematics*, Harlow, England: Pearson Education Ltd., 2001.

Selected Bibliography

Ablowitz, M. J., and A. S. Fokas, *Complex Variables: Introduction and Applications* , Cambridge, England: Cambridge University Press, 1997.

Krzyz, J. G., *Problems in Complex Variable Theory*, New York: American Elsevier, 1972.

Kwok, Y. K., *Applied Complex Variables for Scientists and Engineers*, Cambridge, England: Cambridge University Press, 2002.

Le Page, W. R., *Complex Variables and Laplace Transforms for Engineers*, New York: Dover Publications, 1980.

Chapter 5

FOURIER SERIES

5.1 The Trigonometric Series

The *trigonometric series* is the one that has the form:

$$\frac{a_0}{2} + a_1 \cos x + b_1 \sin x + a_2 \cos 2x + b_2 \sin 2x + \ldots a_n \cos nx + b_n \sin nx + \ldots$$

$$(5.1)$$

The set of functions $\cos nx$ and $\sin nx$ is *orthogonal* on the interval $[-\pi, \pi]$ since

$$\int_{-\pi}^{\pi} \cos nx \cdot \sin kx = 0 \quad \begin{array}{l} n = 0,1,2,\ldots \\ k = 0,1,2,\ldots \end{array}$$

Thus, any real periodic function $f(x)$ with the period 2π for which the integral

$$\int_{-\pi}^{\pi} |f(x)| dx$$

exists, can be represented via trigonometric series (5.1) with the coefficients

$$a_n = \frac{1}{\pi} \int_{-\pi}^{\pi} f(x) \cos nx \, dx; \quad b_n = \frac{1}{\pi} \int_{-\pi}^{\pi} f(x) \sin nx \, dx$$

$$(5.2)$$

that is, $a_0 = \dfrac{1}{\pi} \displaystyle\int_{-\pi}^{\pi} f(x) dx$

For the function $f(x)$ of any period T, the trigonometric series takes the form:

$$\frac{a_0}{2} + \sum_{n=1}^{\infty} a_n \cos \frac{2\pi x}{T} \cdot n + \sum_{n=1}^{\infty} b_n \sin \frac{2\pi x}{T} \cdot n = \frac{a_0}{2} + \sum_{n=1}^{\infty} a_n \cos n\omega x$$

$$+ \sum_{n=1}^{\infty} b_n \sin n\omega x$$

where $\omega = 2\pi/T$ is the circular frequency. The coefficients a_n, b_n are given by Euler's formulas

$$a_n = \frac{2}{T} \int_{-T/2}^{T/2} f(x)\cos n\omega x\,dx; \quad b_n = \frac{2}{T} \int_{-T/2}^{T/2} f(x)\sin n\omega x\,dx, \quad n = 0,1,2,3,\ldots \qquad (5.3)$$

The coefficients (5.3) converge to coefficients (5.2) when $T = 2\pi$.

5.2 Fourier Series Expansion

The representation of the function $f(x)$ as:

$$f(x) = \frac{a_0}{2} + \sum_{n=1}^{\infty} a_n \cos n\omega x + \sum_{n=1}^{\infty} b_n \sin n\omega x \qquad (5.4)$$

with coefficients a_n, b_n given by (5.3), is called the *Fourier series expansion,* and coefficients a_n, b_n are called *Fourier coefficients.*

The Fourier series expansion has a fundamental role in electrical engineering, typically in the tasks when $f(x) = s(t)$ is a time-domain signal. The coefficients a_n, b_n are typically referred to as *in-phase* and *quadrature* components of the nth harmonic.

Fourier, Joseph (1768–1830).   French mathematician. Mainly known for discovery that any periodic function can be represented as a superposition of sinusoidal and cosinusoidal factors. He was a friend and advisor of Napoleon. Fourier believed that his health would be improved by wrapping himself up in blankets, and in this state he tripped down the stairs in his house and killed himself.

The Fourier series expansion is the result of the Fourier theorem, which states that a periodic function that satisfies certain conditions can be expressed as the sum of a number of sine functions of different amplitude, phases, and periods:

$$f(x) = A_0 + A_1 \sin(\omega x + \varphi_1) + \ldots + A_n \sin(n\omega x + \varphi_n) + \ldots$$

where A_n and φ_n are constant.

The term $A_1 \sin(\omega x + \varphi_1)$ has the same frequency ω as the parent function $f(x)$, and is called the *first harmonic* or the *fundamental mode*.

The term $A_n \sin(n\omega x + \varphi_n)$ is called the *n*th *harmonic* (A_n is the amplitude and φ_n is the phase of the harmonic).

Since

$$A_n \sin(n\omega x + \varphi_n) = A_n \cos\varphi_n \cdot \sin n\omega x + A_n \sin\varphi_n \cdot \cos n\omega x$$

$$= b_n \sin n\omega x + a_n \cos n\omega x; \quad a_n = A_n \sin\varphi_n, \quad b_n = A_n \cos\varphi$$

the Fourier series expansion can be written in the form (5.4).

Using Euler's formula (see Section 4.1.9),

$$e^{jn\omega x} = \cos n\omega x + j \sin n\omega x$$

the Fourier coefficients can be represented as:

$$a_n + jb_n = \frac{2}{T} \int\limits_{-T/2}^{T/2} f(x) e^{jn\omega x} \, dx$$

♦ <u>Example 5.1</u> Find the Fourier series expansion of the periodic function $f(x) = x$ of period $T = 2\pi$ [1].

Solution. The Fourier coefficients can be found as follows: ($\omega = 2\pi/T = 1$)

$$a_0 = \frac{1}{\pi} \int\limits_0^{2\pi} f(x) \, dx = \frac{1}{\pi} \int\limits_0^{2\pi} x \, dx = \frac{1}{\pi} \frac{x^2}{2} \bigg|_0^{2\pi} = 2\pi$$

$$a_n = \frac{1}{\pi} \int\limits_0^{2\pi} x \cos n x \, dx \; ; \; b_n = \frac{1}{\pi} \int\limits_0^{2\pi} x \sin n x \, dx$$

With the aid of a computer (see Section 9.3), we can evaluate the integrals above and find:

$$a_n = \frac{1}{\pi}\left[\frac{\cos(2\pi n) + 2n\sin(2\pi n)\cdot\pi}{n^2} - \frac{1}{n^2}\right]$$

$$b_n = -\frac{1}{\pi}\left[\frac{-\sin(2\pi n) + 2n\cos(2\pi n)\cdot\pi}{n^2}\right]$$

Since $\sin(2\pi n) = 0$, $\cos(2\pi n) = 1$, $a_n = 1/n^2 - 1/n^2 = 0$, $b_n = -\frac{2}{n}$. Thus,

$$f(x) = \pi - \sum_{n=1}^{\infty}\frac{2}{n}\sin nx$$

5.3 Fourier Series for Even and Odd Functions

From the definition of integration for the *even* and *odd* functions $f(x)$:

$$\int_{-a}^{a} f(x)dx = 2\int_{0}^{a} f(x)dx \text{, if } f(x) \text{ is even}$$

$$\int_{-a}^{a} f(x)dx = 0 \text{, if } f(x) \text{ is odd}$$

Thus, for an even function $f(x)$, the Fourier coefficients are:

$$a_n = \frac{4}{T}\int_{0}^{T/2} f(x)\cos n\omega x\, dx; \quad b_n = 0$$

and the Fourier series expansion takes the form:

$$f(x) = \frac{a_0}{2} + \sum_{n=1}^{\infty} a_n \cos n\omega x$$

For an odd function, the Fourier coefficients are:

$$a_n = 0 \; ; \quad b_n = \frac{4}{T} \int_0^{T/2} f(x) \sin n\omega x \, dx, \quad n = 1, 2, 3, \ldots$$

and the Fourier series expansion takes the form

$$f(x) = \sum_{n=1}^{\infty} b_n \sin n\omega x$$

The useful properties of even and odd functions:
- The sum of two or more odd functions is an odd function.
- The product of two even functions is an even function.
- The product of two odd functions is an even function.
- The product of an odd and an even function is an odd function.
- The derivative of an even function is an odd function.
- The derivative of an odd function is an even function.

5.4 Linearity of the Fourier Series

Let us consider the function $f(x) = \alpha\varphi(x) + \beta\psi(x)$, where $\varphi(x), \psi(x)$ are periodic functions of period T, and α, β are arbitrary constants. Then, if the Fourier series expansions for $\varphi(x), \psi(x)$ are:

$$\varphi(x) = \frac{a_0}{2} + \sum_{n=1}^{\infty} a_n \cos n\omega x + \sum_{n=1}^{\infty} b_n \sin n\omega x$$

$$\psi(x) = \frac{c_0}{2} + \sum_{n=1}^{\infty} c_n \cos n\omega x + \sum_{n=1}^{\infty} d_n \sin n\omega x$$

then the Fourier series expansion for $f(x)$ is

$$f(x) = \frac{\alpha a_0 + \beta c_0}{2} + \sum_{n=1}^{\infty} (\alpha a_n + \beta c_n) \cos n\omega x + \sum_{n=1}^{\infty} (\alpha b_n + \beta d_n) \sin n\omega x$$

5.5 Fourier Series for a Discontinuous Function

If $f(x)$ is a bounded periodic function that in any period has a finite number of isolated minima and maxima (including zero), and a finite number of points of finite discontinuity (this is known as *Dirichlet condition*), then

- At points x_i, where $f(x)$ is the discontinuous function, the Fourier series expansion is equal to:

$$\frac{1}{2}\left[f(x_i - 0) + f(x_i + 0)\right]$$

where $f(x_i - 0)$ is the limit to which $f(x)$ tends when $x \to x_i$ on the left, and $f(x_i + 0)$ is the limit to which $f(x)$ tends when $x \to x_i$ on the right.
- At points where $f(x)$ is the continuous function, the Fourier series expansion converges to $f(x)$.

5.6 Differentiation and Integration of Fourier Series

If function $f(x)$ is a periodic function that satisfies the Dirichlet condition (see Section 5.5), then its derivative $f'(x)$, wherever it exists, may be found by term-by-term differentiation of the Fourier series of $f(x)$, if it is continuous everywhere and the function $f'(x)$ has a Fourier series expansion. In this case, for the function

$$f(x) = \frac{a_0}{2} + \sum_{n=1}^{\infty} (a_n \cos n x + b_n \sin n x), \quad T = 2\pi \tag{5.5}$$

its derivative is:

$$f'(x) = \sum_{n=1}^{\infty} (nb_n \cos n x - na_n \sin n x)$$

A Fourier series expansion of a periodic function $f(x)$ that satisfies Dirichlet's condition may be integrated term by term, and the integrated series converges to the integral of the function $f(x)$. In this case, for the function (5.5) for $-\pi \le x_1 < x \le \pi$

Dirichlet, Peter (1805–1859).   German mathematician who is said to have slept with Gauss' *Disquisitiones Arithmeticae* under his pillow. He studied analysis and differential equations, giving his name to the Dirichlet boundary conditions, produced Dirichlet's theorem on primes, and gave the first proof of Bertrand's postulate. He was best known as the one who gave the first set of conditions sufficient to guarantee the convergence of the Fourier series.

$$\int_0^x f(x)\,dx = \int_{x_1}^x \frac{a_0}{2}\,dx + \sum_{n=1}^{\infty} \int_{x_1}^x (a_n \cos nx + b_n \sin nx)\,dx$$

$$= \frac{a_0(x - x_1)}{2} + \sum_{n=1}^{\infty} \left[\frac{b_n}{n}(\cos nx_1 - \cos nx) + \frac{a_n}{n}(\sin nx - \sin nx_1) \right]$$

Reference

[1] James, G., et al. *Modern Engineering Mathematics*, Harlow, England: Pearson Education Ltd., 2001.

Selected Bibliography

Bolton, W., *Fourier Series*, Reading, MA: Longman, 1996.

Cartwright, M., *Fourier Methods for Mathematicians, Scientists and Engineers*, New York: Ellis Horwood, 1990.

Chapter 6

MATRIX ALGEBRA

6.1 Matrix Algebra with Real Numbers

6.1.1 Definitions

Any array of real numbers

$$A = \begin{vmatrix} a_{11} & a_{12} & \cdots & a_{1n} \\ a_{21} & a_{22} & \cdots & a_{2n} \\ \cdots & \cdots & \cdots & \cdots \\ a_{m1} & a_{m2} & \cdots & a_{mn} \end{vmatrix}$$

is called a *matrix* with m rows and n columns, or an $m \times n$ matrix.

The entry a_{ij} denotes the *element* of a matrix in the ith row and jth column.

If $m = n$, the matrix is called a *square* one.

If $n = 1$, the matrix is called a *column vector*. If $m = 1$, the matrix is called a *row vector*.

If all elements of a matrix are equal to zero, it is called the *zero* or *null* matrix.

♦ Example 6.1

$$A = \begin{vmatrix} 1 & 0 & -1 \\ 2 & 2 & 4 \\ 3 & 7 & 3 \end{vmatrix} \text{ is a } 3 \times 3 \text{ square matrix;}$$

$$B = \begin{vmatrix} x \\ y \\ z \end{vmatrix} \text{ is a } 3 \times 1 \text{ column vector;}$$

$$C = \begin{vmatrix} 0.15 & 4 & -11 & 26 \end{vmatrix} \text{ is a } 1 \times 4 \text{ row vector.}$$

157

6.1.2 Equality

Two matrices A and B are equal if their elements are the same:

$$a_{ij} = b_{ij}$$

6.1.3 Addition

The sum of two matrices A and B is the matrix C, such that its elements are the sum of matrices A and B elements:

$$c_{ij} = a_{ij} + b_{ij}$$

Note. Only matrices of the same size can be added: thus, only an $m \times n$ matrix can be added to another $m \times n$ matrix.

♦ Example 6.2

$$\begin{vmatrix} 1 & 2 & 3 \\ 4 & 5 & 6 \end{vmatrix} + \begin{vmatrix} 1 & -2 & 0 \\ 3 & -6 & -10 \end{vmatrix} = \begin{vmatrix} 2 & 0 & 3 \\ 7 & -1 & -4 \end{vmatrix}$$

The following laws are valid for matrix addition:

• Commutative law: $A + B = B + A$.
• Associative law: $(A + B) + C = A + (B + C)$.

6.1.4 Multiplication by a Number

If matrix A is multiplied by a number b, each element is multiplied by this number. Thus, the matrix bA has the element ba_{ij}.

♦ Example 6.3

$$5 \cdot \begin{vmatrix} 1 & 3 \\ 2 & 4 \end{vmatrix} = \begin{vmatrix} 5 & 15 \\ 10 & 20 \end{vmatrix}$$

The distributive law is valid for this operation:

$$b(A + B) = bA + bB$$

6.1.5 Multiplication of the Matrices

If A is an $m \times q$ matrix with elements a_{ij}, and B is a $q \times n$ matrix with elements b_{ij}, then the product is an $m \times n$ matrix with the elements:

$$e_{ij} = \sum_{k=1}^{q} a_{ik} \cdot b_{kj}, \quad i = 1, \ldots, m, \quad j = 1, \ldots, n$$

In other words, the ith row of the first matrix A is multiplied term by term with the jth column of matrix B, and the products are added to form the element c_{ij}.

♦ Example 6.4

$$A = \begin{vmatrix} a_{11} & a_{12} & a_{13} \\ a_{21} & a_{22} & a_{23} \\ a_{31} & a_{32} & a_{33} \end{vmatrix} = \begin{vmatrix} 1 & 1 & 0 \\ -3 & -2 & 1 \\ 2 & 0 & 1 \end{vmatrix}; \ B = \begin{vmatrix} b_{11} & b_{12} & b_{13} \\ b_{21} & b_{22} & b_{23} \\ b_{31} & b_{32} & b_{33} \end{vmatrix} = \begin{vmatrix} 0 & 1 & 4 \\ 1 & 3 & 2 \\ 5 & 0 & 1 \end{vmatrix}$$

The matrix $C = A \cdot B$ is:

$$C = \begin{vmatrix} c_{11} & c_{12} & c_{13} \\ c_{21} & c_{22} & c_{23} \\ c_{31} & c_{32} & c_{33} \end{vmatrix}$$

with the elements:

$$c_{11} = a_{11} \cdot b_{11} + a_{12} \cdot b_{21} + a_{13} \cdot b_{31} = 1$$
$$c_{12} = a_{11} \cdot b_{12} + a_{12} \cdot b_{22} + a_{13} \cdot b_{32} = 4$$
$$c_{13} = a_{11} \cdot b_{13} + a_{12} \cdot b_{23} + a_{13} \cdot b_{33} = 6$$

$$c_{21} = a_{21} \cdot b_{11} + a_{22} \cdot b_{21} + a_{23} \cdot b_{31} = 3$$
$$c_{22} = a_{21} \cdot b_{12} + a_{22} \cdot b_{22} + a_{23} \cdot b_{32} = -9$$
$$c_{23} = a_{21} \cdot b_{13} + a_{22} \cdot b_{23} + a_{23} \cdot b_{33} = -15$$

$$c_{31} = a_{31} \cdot b_{11} + a_{32} \cdot b_{21} + a_{33} \cdot b_{31} = 5$$
$$c_{32} = a_{31} \cdot b_{12} + a_{32} \cdot b_{22} + a_{33} \cdot b_{32} = 2$$
$$c_{33} = a_{31} \cdot b_{13} + a_{32} \cdot b_{23} + a_{33} \cdot b_{33} = 9$$

Thus,

$$C = \begin{vmatrix} 1 & 4 & 6 \\ 3 & -9 & -15 \\ 5 & 2 & 9 \end{vmatrix}$$

♦ Example 6.5

$$A = \begin{vmatrix} a_{11} & a_{12} & a_{13} \end{vmatrix} = \begin{vmatrix} 1 & -1 & 4 \end{vmatrix}; \quad B = \begin{vmatrix} b_{11} \\ b_{21} \\ b_{31} \end{vmatrix} = \begin{vmatrix} 1 \\ 3 \\ -1 \end{vmatrix}$$

The elements of matrix $C = A \cdot B$ are:

$$c_{11} = a_{11} \cdot b_{11} + a_{12} \cdot b_{21} + a_{13} \cdot b_{31} = -6$$

Thus, $C = -6$.

6.1.6 A Symmetrical Matrix

For a square $n \times n$ matrix, the diagonal containing elements $a_{11}, a_{22}, \ldots, a_{nn}$ is called a *main (principal, leading) diagonal.*

If the elements positioned symmetrically to the main diagonal are equal, that is,

$$a_{ij} = a_{ji}$$

the matrix is called a *symmetrical* one.

♦ Example 6.6 The following matrix A is a symmetric matrix:

$$A = \begin{vmatrix} x & 2 & 6 \\ 2 & y & 3 \\ 6 & 3 & z \end{vmatrix}$$

If $a_{ij} = -a_{ji}$, the matrix is called a *skew-symmetric.* The diagonal elements of a skew-symmetric matrix must all be zeros.

6.1.7 Diagonal and Unit Matrices

A *diagonal matrix* is a square matrix that has its nonzero elements only along the main diagonal. The diagonal matrix with $a_{11} = a_{22} = \ldots a_{nn} = 0$ is a zero matrix.

♦ Example 6.7

Matrix $A = \begin{vmatrix} a_{11} & 0 & 0 \\ 0 & a_{22} & 0 \\ 0 & 0 & a_{33} \end{vmatrix}$ is a diagonal matrix.

Elements a_{11}, a_{22}, a_{33} can be any real numbers.

A unit matrix I is the matrix for which $a_{11} = a_{22} = \ldots a_{nn} = 1$

$$I = \begin{vmatrix} 1 & 0 & 0 & \ldots & 0 \\ 0 & 1 & 0 & \ldots & 0 \\ \ldots & \ldots & \ldots & \ldots & \ldots \\ 0 & 0 & 0 & \ldots & 1 \end{vmatrix}$$

Sometimes the unit matrix is called the *identity* matrix.

6.1.8 A Transposed Matrix

The matrix $B = A^T$ is called a *transposed* with respect to matrix A, if its rows are the columns of matrix A; that is,

$$b_{ij} = a_{ji}$$

♦ Example 6.8   For the matrix

$A = \begin{vmatrix} 2 & 4 & 5 \\ 3 & 2 & 8 \\ 7 & 1 & 1 \end{vmatrix}$ the transposed matrix is $A^T = \begin{vmatrix} 2 & 3 & 7 \\ 4 & 2 & 1 \\ 5 & 8 & 1 \end{vmatrix}$

♦ Example 6.9. For the row vector $B = \begin{vmatrix} 2 & 9 & 1 \end{vmatrix}$, the transposed matrix is a column vector B^T :

$$B^T = \begin{vmatrix} 2 \\ 9 \\ 1 \end{vmatrix}$$

Thus, a row vector is transposed to a column vector, and vice versa.

6.1.9 Properties of Matrix Multiplication

- Associate law $A(BC) = (AB)C$;
- Distributive law over multiplication by a number $(bA)B = A(bB) = bAB$;
- Distributive law over addition $(A + B)C = AC + BC$; $A(B + C) = AB + AC$;
- Multiplication by unit matrices $I \cdot A = A \cdot I = A$;
- Transpose of a product $(AB)^T = B^T \cdot A^T$.

6.1.10 Determinant

Let us consider 2×2 and 3×3 square matrices. The determinant $|A|$ of a 2×2 matrix A is:

$$A = \begin{vmatrix} a_{11} & a_{12} \\ a_{21} & a_{22} \end{vmatrix}; \quad |A| = a_{11} \cdot a_{22} - a_{12} \cdot a_{21}$$

The determinant $|A|$ of a 3×3 matrix A is:

$$A = \begin{vmatrix} a_{11} & a_{12} & a_{13} \\ a_{21} & a_{22} & a_{23} \\ a_{31} & a_{32} & a_{33} \end{vmatrix}$$

$$|A| = a_{11} \cdot \begin{vmatrix} a_{22} & a_{23} \\ a_{32} & a_{33} \end{vmatrix} - a_{12} \begin{vmatrix} a_{21} & a_{23} \\ a_{31} & a_{33} \end{vmatrix} + a_{13} \begin{vmatrix} a_{21} & a_{22} \\ a_{31} & a_{32} \end{vmatrix}$$

$$= a_{11}(a_{22} \cdot a_{33} - a_{32} \cdot a_{23}) - a_{12}(a_{21} \cdot a_{33} - a_{31} \cdot a_{23}) + a_{13}(a_{21} \cdot a_{32} - a_{31} \cdot a_{22})$$

The general rule to evaluate a determinant for $n \times n$ matrix A is as follows:

- In the given determinant, we delete row i and column j, and then what is left is called the *minor* M_{ij}.
- The minor multiplied by the appropriate sign is called the *cofactor* of the element a_{ij} :

$$A_{ij} = (-1)^{i+j} M_{ij}$$

- The determinant $|A|$ is evaluated as:

$$|A| = \sum_{j=1}^{n} a_{ij} \cdot A_{ij} = \sum_{j=1}^{n} (-1)^{i+j} a_{ij} M_{ij}$$

♦ Example 6.10 Evaluate the determinant of the matrix A

$$A = \begin{vmatrix} 1 & 2 & 4 \\ -1 & 0 & 3 \\ 3 & 1 & -2 \end{vmatrix}$$

Solution.

$$|A| = 1 \begin{vmatrix} 0 & 3 \\ 1 & -2 \end{vmatrix} - 2 \begin{vmatrix} -1 & 3 \\ 3 & -2 \end{vmatrix} + 4 \begin{vmatrix} -1 & 0 \\ 3 & 1 \end{vmatrix} = 7$$

6.1.11 An Adjoint Matrix

An *adjoint matrix adjA* for the matrix A is the transpose of the matrix of cofactors A_{ij} for the matrix A. Thus, for 3×3 matrix A:

$$adjA = \begin{vmatrix} A_{11} & A_{12} & A_{13} \\ A_{21} & A_{22} & A_{23} \\ A_{31} & A_{32} & A_{33} \end{vmatrix}^{T}$$

♦ Example 6.11 Find the adjoint matrix $adjA$ for $A = \begin{vmatrix} 1 & 3 \\ 2 & 8 \end{vmatrix}$.

Solution. Cofactors for matrix A are:

$$A_{11} = 8;\ A_{12} = -2;\ A_{21} = -3;\ A_{22} = 1$$

Thus,

$$C = \begin{vmatrix} A_{11} & A_{12} \\ A_{21} & A_{22} \end{vmatrix} = \begin{vmatrix} 8 & -2 \\ -3 & 1 \end{vmatrix}$$

and

$$adjA = C^{T} = \begin{vmatrix} 8 & -3 \\ -2 & 1 \end{vmatrix}$$

The useful formulas are

- $A \cdot adjA = |A| \cdot I$ • $adj(A \cdot B) = adj(B) \cdot adj(A)$

Sometimes an adjoint matrix is called an *adjugate* matrix.

6.1.12 The Inverse Matrix

If $|A| \neq 0$, the *inverse* matrix A^{-1} with respect to matrix A is:

$$A^{-1} = \frac{adjA}{|A|}$$

The inverse of a product of two matrices is:

$$(A \cdot B)^{-1} = B^{-1} \cdot A^{-1}$$

♦ <u>Example 6.12</u> Find the inverse matrix for $A = \begin{vmatrix} 1 & 2 \\ 2 & 3 \end{vmatrix}$.

Solution.

$|A| = 3 - 4 = -1;$ $A_{11} = 3;$ $A_{12} = -2;$ $A_{21} = -2;$ $A_{22} = 1;$

$$adjA = \begin{vmatrix} A_{11} & A_{12} \\ A_{21} & A_{22} \end{vmatrix}^{T} = \begin{vmatrix} 3 & -2 \\ -2 & 1 \end{vmatrix}; \quad A^{-1} = \frac{adjA}{|A|}$$

Thus,

$$A^{-1} = \begin{vmatrix} -3 & 2 \\ 2 & -1 \end{vmatrix}$$

6.1.13 How to Use Matrices to Solve Linear Equations

Any set of linear equations

$$a_{11}x_1 + a_{12}x_2 + \ldots a_{1n}x_n = b_1$$
$$a_{21}x_1 + a_{22}x_2 + \ldots a_{2n}x_n = b_2$$
$$\ldots\ldots\ldots\ldots\ldots\ldots\ldots\ldots$$
$$a_{n1}x_1 + a_{n2}x_2 + \ldots a_{nn}x_n = b_n$$

that looks very bulky in a conventional notation can be written very concisely in matrix notation:

$$A \cdot x = B$$

where

$$A = \begin{vmatrix} a_{11} & a_{12} & \cdots & a_{1n} \\ \cdots & \cdots & \cdots & \cdots \\ a_{n1} & a_{n2} & \cdots & a_{nn} \end{vmatrix}; \quad B = \begin{vmatrix} b_1 \\ b_2 \\ \cdots \\ b_n \end{vmatrix}; \quad x = \begin{vmatrix} x_1 \\ x_2 \\ \cdots \\ x_n \end{vmatrix}$$

In a general case, when $A \neq 0$, $B \neq 0$, the solution is given as:

$$x = A^{-1} \cdot B$$

♦ Example 6.13 Find the solution for a set of equations:

$$\begin{cases} 2x + y = 5 \\ x - 2y = -5 \end{cases}$$

Solution.

Step 1. $A = \begin{vmatrix} 2 & 1 \\ 1 & -2 \end{vmatrix}, \quad B = \begin{vmatrix} 5 \\ -5 \end{vmatrix}$

Step 2. $|A| = -4 - 1 = -5$; $A_{11} = -2$, $A_{12} = -1$, $A_{21} = -1$, $A_{22} = 2$

Step 3. $A^{-1} = -\dfrac{1}{5} \begin{vmatrix} -2 & -1 \\ -1 & 2 \end{vmatrix}^T = -\dfrac{1}{5} \begin{vmatrix} -2 & -1 \\ -1 & 2 \end{vmatrix} = \begin{vmatrix} \dfrac{2}{5} & \dfrac{1}{5} \\ \dfrac{1}{5} & -\dfrac{2}{5} \end{vmatrix}$

Step 4. $\begin{vmatrix} x \\ y \end{vmatrix} = \begin{vmatrix} \dfrac{2}{5} & \dfrac{1}{5} \\ \dfrac{1}{5} & -\dfrac{2}{5} \end{vmatrix} \cdot \begin{vmatrix} 5 \\ -5 \end{vmatrix} = \begin{vmatrix} 1 \\ 3 \end{vmatrix}$

Answer: $x = 1$, $y = 3$.

The set of equations in Example 6.13 is a fairly simple one, and can be easily solved with the elimination method (see Section 1.3.8). It was chosen to demonstrate the basic steps to solve a set of linear equations. When the set of equations is more complicated for greater n, basic steps stay the same but inverse matrix A^{-1} and product $A^{-1} \cdot B$ can be calculated with the aid of a computer (see Chapter 9).

6.1.14 How to Use Matrices in Coordinate Conversions

If we know the components of the vector $\vec{u}$ in any n-dimensional system of coordinates $\vec{u} = \left(u_1, u_2, \ldots u_n\right)$ and matrix of coordinate conversion C, then the components of this vector in the new system of coordinates $\vec{v} = \left(v_1, v_2, \ldots, v_n\right)$ after conversion will be given as:

$$\vec{v} = C \cdot \vec{u}$$

♦ Example 6.14 Let us consider the vector $\vec{u} = \left(x, y, z\right)$ with three coordinates x, y, z in a right-handed Cartesian coordinates [Figure 6.1(a)]. After rotation of the coordinates with respect to one of the axes new coordinates of the vector are $\vec{v} = \left(x_1, y_1, z_1\right)$.

The new coordinates can be found as: $\vec{v} = C \cdot \vec{u}$, or

$$\begin{vmatrix} x_1 \\ y_1 \\ z_1 \end{vmatrix} = \begin{vmatrix} c_{11} & c_{12} & c_{13} \\ c_{21} & c_{22} & c_{23} \\ c_{31} & c_{32} & c_{33} \end{vmatrix} \cdot \begin{vmatrix} x \\ y \\ z \end{vmatrix}$$

If the coordinates are rotated with respect to axis OZ by angle φ :

$$C_\varphi = \begin{vmatrix} \cos\varphi & \sin\varphi & 0 \\ -\sin\varphi & \cos\varphi & 0 \\ 0 & 0 & 1 \end{vmatrix}$$

If rotation is with respect to axis OX by angle ψ :

$$C_\psi = \begin{vmatrix} 1 & 0 & 0 \\ 0 & \cos\psi & \sin\psi \\ 0 & -\sin\psi & \cos\psi \end{vmatrix}$$

If rotation is with respect to axis OY by angle θ :

$$C_\theta = \begin{vmatrix} \cos\theta & 0 & \sin\theta \\ 0 & 1 & 0 \\ -\sin\theta & 0 & \cos\theta \end{vmatrix}$$

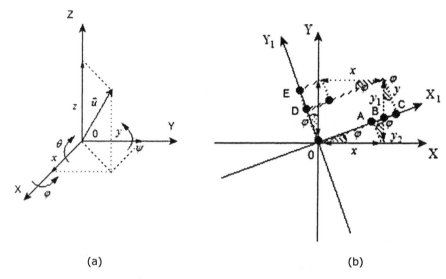

(a)   (b)

Figure 6.1 (a) Rotation of axes in Cartesian coordinates and (b) explanation
of derivation of the conversion matrix.

If the coordinates are rotated with respect to all three axes, by angles
φ, θ, ψ , then (see Appendix 6)

$$C = C_\varphi \cdot C_\theta \cdot C_\psi$$

Note. Cartesian coordinate conversion is very sensitive to the orientation of the
coordinate axes (right-handed or left-handed; see Section 7.1.6) and positive
directions of the rotation angles. That is why, when applying them to solving
practical tasks (e.g., an electronic sensor is mounted on a ship with surveillance
of the upper hemisphere, or on the spacecraft with surveillance of the lower
hemisphere, and so forth), you need to understand how conversion matrices
are derived in order to avoid erroneous results. An example of how to derive
the elements of the matrix C_φ is given below.

♦ Example 6.15   Derive elements of the matrix C_φ when coordinates are
rotated with respect to the axis OZ by the angle with positive direction, as
shown in Figure 6.1(b).
Solution. The new coordinate is:

$$x_1 = OC = OA + AB + BC .$$

Since $OA = x \cdot \cos \varphi$, $AB = y_2 \cdot \sin \varphi$, $BC = y_1 \cdot \sin \varphi$, and $y_1 + y_2 = y$,
we find
$$x_1 = x \cdot \cos \varphi + y \cdot \sin \varphi$$

Another new coordinate is:

$$y_1 = OD = OE - ED$$

Since $OE = y \cdot \cos \varphi, \ ED = x \cdot \sin \varphi$, we find

$$y_1 = -x \cdot \sin \varphi + y \cdot \cos \varphi$$

Thus:

$$\begin{vmatrix} x_1 \\ y_1 \end{vmatrix} = \begin{vmatrix} \cos \varphi & \sin \varphi \\ -\sin \varphi & \cos \varphi \end{vmatrix} \cdot \begin{vmatrix} x \\ y \end{vmatrix}$$

for specified set of coordinates and positive directions of φ.

A quick sanity check can be done for some obvious values. For example, for $\varphi = 0$ $x_1 = x$, $y_1 = y$, and for $\varphi = 90°$, $x_1 = y$, $y_1 = -x$ that looks right for Figure 6.1 (b) setup.

More about coordinate conversions are provided in Section 1.9 and Appendix 6. Examples of how to use coordinate conversion formulas to solve practical problems are given in Chapter 10.

6.2 Matrix Algebra for Complex Numbers

6.2.1 The Hermitian Matrix

The *Hermitian matrix* is one in which elements that are symmetrical with respect to the main diagonal are complex-conjugate numbers

$$a_{ij} = a_{ji}^{*}$$

and the main diagonal consists of real numbers only.

♦ Example 6.16 The matrix H is the Hermitian matrix.

Hermite, Charles (1822–1901). French mathematician. Hermite did pioneering work on Abelian functions, studied algebraic invariants, investigated what is now called the Hermite differential equation, and discovered some of the properties of Hermitian matrices. He was known to be very generous in helping young mathematicians during his career.

$$H = \begin{vmatrix} 1 & 3+4j & -j \\ 3-4j & 3 & 5 \\ j & 5 & -6 \end{vmatrix}$$

If the matrix with complex elements is transposed, and the resulting elements are substituted with conjugate ones, the matrix is called a *Hermitian-conjugate matrix* A^+.

♦ Example 6.17 Find the Hermitian-conjugate matrix A^+ for matrix:

$$A = \begin{vmatrix} 1+j & 2 & j \\ 3 & 0 & 1-j \\ 3+2j & -3 & -j \end{vmatrix}$$

Solution.

$$A^+ = \begin{vmatrix} 1-j & 3 & 3-2j \\ 2 & 0 & -3 \\ -j & 1+j & j \end{vmatrix}$$

For a product of two matrices A and B, the following rule is valid:

$$(A \cdot B)^+ = B^+ \cdot A^+$$

The same rule is valid for a product of several matrices $A, B,...,Q$

$$(A \cdot B \cdot ... \cdot Q)^+ = Q^+ ... \cdot B^+ \cdot A^+$$

The Hermitian matrix H is equal to its Hermitian-conjugate H^+.

♦ Example 6.18 Find a Hermitian-conjugate matrix H^+ for the matrix from Example 6.16.
Solution.

$$H^+ = \begin{vmatrix} 1 & 3+4j & -j \\ 3-4j & 3 & 5 \\ j & 5 & -6 \end{vmatrix}$$

We can see that $H^+ = H$.

6.2.2 Modulus and Scalar Product

The square of modulus of the vector $\vec{u} = (u_1, u_2, \ldots, u_n)$, where $u_i, i = \overline{1,n}$ are complex numbers, can be written as:

$$\left|\vec{u}\right|^2 = u^+ \cdot u$$

that gives

$$\left|\vec{u}\right|^2 = \left| u_1^* \cdot u_2^* \cdot \ldots u_n^* \right| \cdot \begin{vmatrix} u_1 \\ u_2 \\ \ldots \\ u_n \end{vmatrix}$$

or

$$\left|\vec{u}\right|^2 = u_1^* \cdot u_1 + u_2^* \cdot u_2 + \ldots + u_n^* \cdot u_n$$

◆ Example 6.19 Find the modulus of the vector $\vec{u} = (j, 2-j)$.

Solution.

$$\left|\vec{u}\right|^2 = -j \cdot j + (2+j)(2-j) = -j \cdot j + 4 - j \cdot j = 6; \quad \left|\vec{u}\right| = \sqrt{6}$$

The scalar product of two matrices $\vec{u} = (u_1, u_2, \ldots, u_n)$ and $\vec{v} = (v_1, v_2, \ldots, v_n)$, with complex elements is equal to

$$\vec{u} \cdot \vec{v} = u^+ \cdot v = u_1^* \cdot v_1 + \ldots + u_n^* \cdot v_n$$

6.2.3 The Characteristic Equations, Eigenvalues, and Eigenvectors

Consider the set of equations,

$$A \cdot x = \lambda x$$

where A is an $n \times n$ matrix and $\vec{x} = (x_1, \ldots, x_n)^T$ is a column vector.
This set can be written as:

$$(\lambda I - A) \cdot x = 0$$

where I is a unit matrix.

The solution (rather than $x = 0$) exists if

$$\Delta(\lambda) = |\lambda I - A| = 0$$

The above equation is called the *characteristic equation* of A.

The values of the number λ, which are the roots of the characteristic equation, are called *eigenvalues* of the matrix A. Corresponding solutions for $\bar{x}$ are called *eigenvectors*.

♦ Example 6.20   Find the characteristic equation, eigenvalues, and eigenvectors for the matrix $A = \begin{vmatrix} 0 & -1 \\ 1 & 0 \end{vmatrix}$ [1].

Solution.

$$|\lambda I - A| = \begin{vmatrix} \lambda - 0 & 0 - (-1) \\ 0 - 1 & \lambda - 0 \end{vmatrix} = \begin{vmatrix} \lambda & 1 \\ -1 & \lambda \end{vmatrix} = \lambda^2 + 1$$

Thus, the characteristic equation is

$$\lambda^2 + 1 = 0$$

which has two roots and thus two eigenvalues:

$$\lambda_1 = j; \lambda_2 = -j$$

For the eigenvalue $\lambda_1 = 1$ equation to find eigenvectors takes the form:

$$(\lambda I - A) \begin{vmatrix} x_1 \\ x_2 \end{vmatrix} = \begin{vmatrix} j & 1 \\ -1 & j \end{vmatrix} \cdot \begin{vmatrix} x_1 \\ x_2 \end{vmatrix} = 0$$

or

$$jx_1 + x_2 = 0$$

$$-x_1 + jx_2 = 0$$

that gives the solutions for eigenvectors corresponding to $\lambda_1 = j$ as $x_1 = j, x_2 = 1$.

Correspondingly, for $\lambda_2 = -j$, the equation takes the form:

$$\begin{vmatrix} -j & 1 \\ -1 & -j \end{vmatrix} \cdot \begin{vmatrix} x_3 \\ x_4 \end{vmatrix} = 0$$

Thus, for eigenvalue $\lambda_2 = -j$, the eigenvector is given by the components $x_3 = 1, x_4 = j$.

Answer. Characteristic equation is $\lambda^2 + 1 = 0$; the eigenvalues are $\lambda_1 = j$, $\lambda_2 = -j$; the eigenvectors are $\vec{x}_1 = \big| j, 1 \big|^T$ for $\lambda_1 = j$, $\vec{x}_2 = \big| 1, j \big|^T$ for $\lambda_2 = -j$.

For the Hermitian matrix, all eigenvalues are real numbers and all eigenvectors are orthogonal. Since the symmetrical matrix is a partial case of a Hermitian matrix, all eigenvectors for a symmetrical matrix are also orthogonal. Physically, matrix A elongates or compresses the eigenvector $\vec{x}$ with eigenvalue $\vec{\lambda}$ being the coefficients of this elongation or compression.

6.2.4 How to Use Matrices in Analysis of Two-Port Circuits

Many circuits and devices in electrical engineering (active and passive electrical circuits, filters, amplifiers, and so forth) can be considered as two-port circuits. The first port is the input of the circuit or device, and the second is the output (Figure 6.2).

Let us denote the input and output currents and voltages as i_1, u_1 and i_2, u_2, respectively.

If there are no transient effects and currents/voltages are harmonic functions of time t, they can be written as:

$$i_1 = \dot{I}_1 \cdot \exp(j\omega t) \quad i_2 = \dot{I}_2 \cdot \exp(j\omega t)$$
$$u_1 = \dot{U}_1 \cdot \exp(j\omega t) \quad u_2 = \dot{U}_2 \cdot \exp(j\omega t)$$

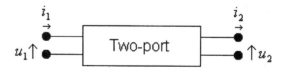

Figure 6.2 Two-port circuit.

where ω is the angular frequency related to the linear frequency f as $\omega = 2\pi f$, and $\dot{I}_1, \dot{I}_2, \dot{U}_1, \dot{U}_2$ are complex amplitudes (envelopes) of the input and output currents and voltages:

$$\dot{I}_1 = I_1 \cdot e^{j\varphi} \qquad \dot{I}_2 = I_2 \cdot e^{j\varphi}$$
$$\dot{U}_1 = U_1 \cdot e^{j\varphi} \qquad \dot{U}_2 = U_2 \cdot e^{j\varphi}$$

Here φ is the phase, and I_1, I_2, U_1, U_2 are real amplitudes of the currents and voltages.

The values $\dot{I}_1, \dot{I}_2, \dot{U}_1, \dot{U}_2$ are not independent, and linked as:

$$\dot{I}_1 = \alpha_{11} \cdot \dot{U}_1 + \alpha_{12} \cdot \dot{U}_2$$
$$\dot{I}_2 = \alpha_{21} \cdot \dot{U}_1 + \alpha_{22} \cdot \dot{U}_2$$

or, in the matrix form

$$\dot{I} = \alpha \cdot \dot{U}$$

If $\alpha \neq 0$, this results in

$$\dot{U} = \alpha^{-1} \cdot \dot{I} = \dot{Z} \cdot \dot{I}$$

The matrix Z is called the *impedance* matrix:

$$\dot{Z} = \frac{1}{|\alpha|} \begin{vmatrix} \alpha_{22} & -\alpha_{12} \\ -\alpha_{21} & \alpha_{11} \end{vmatrix}$$

Thus, the following sets of equations can be written:

$$\dot{U}_1 = Z_{11} \cdot \dot{I}_1 + \dot{Z}_{12} \cdot I_2; \quad \dot{I}_1 = \dot{Y}_{11} \cdot \dot{U}_1 + \dot{Y}_{12} \cdot \dot{U}_2$$
$$\dot{U}_2 = \dot{Z}_{21} \cdot \dot{I}_1 + \dot{Z}_{22} \cdot I_2; \quad \dot{I}_2 = \dot{Y}_{21} \cdot \dot{U}_1 + \dot{Y}_{22} \cdot \dot{U}_2$$
$$\dot{U}_1 = \dot{H}_{11} \cdot I_1 + \dot{H}_{12} \cdot \dot{U}_2; \quad \dot{U}_1 = \dot{A}_{11} \cdot \dot{U}_2 + A_{12} \cdot \dot{I}_2$$
$$\dot{I}_2 = \dot{H}_{21} \cdot I_1 + \dot{H}_{22} \cdot \dot{U}_2; \quad \dot{I}_1 = \dot{A}_{21} \cdot \dot{U}_2 + \dot{A}_{22} \cdot I_2$$

or, in the matrix form:

$$\begin{vmatrix} \dot{U}_1 \\ \dot{U}_2 \end{vmatrix} = \dot{Z} \cdot \begin{vmatrix} \dot{I}_1 \\ \dot{I}_2 \end{vmatrix}, \quad \begin{vmatrix} \dot{I}_1 \\ \dot{I}_2 \end{vmatrix} = \dot{Y} \cdot \begin{vmatrix} \dot{U}_1 \\ \dot{U}_2 \end{vmatrix}, \quad \begin{vmatrix} \dot{U}_1 \\ \dot{I}_2 \end{vmatrix} = \dot{H} \cdot \begin{vmatrix} \dot{I}_1 \\ \dot{U}_2 \end{vmatrix}, \quad \begin{vmatrix} \dot{U}_1 \\ \dot{I}_1 \end{vmatrix} = \dot{A} \cdot \begin{vmatrix} \dot{U}_2 \\ \dot{I}_2 \end{vmatrix}$$

In electrical engineering, the matrices $\dot{Z}, \dot{Y}, \dot{H}, \dot{A}$ are used to evaluate different parameters of two-port devices. The elements of the latter matrix sometimes are denoted as $A_{11} = A$; $A_{12} = B$; $A_{21} = C$; $A_{22} = D$, and called $ABCD$ parameters.

Another important matrix is a *scattering matrix S*, which is a popular tool for representing lossy passive microwave networks. It represents the relationship between the variable b proportional to the outgoing wave and variable a proportional to the incoming wave by equations:

$$b_1 = S_{11} \cdot a_1 + S_{12} \cdot a_2$$
$$b_2 = S_{21} \cdot a_1 + S_{22} \cdot a_2$$

or, in matrix form

$$\begin{vmatrix} b_1 \\ b_2 \end{vmatrix} = S \cdot \begin{vmatrix} a_1 \\ a_2 \end{vmatrix}$$

Examples of how to calculate some parameters of two-port RF devices are given in Chapter 10.

Reference

[1] James, G., et al., *Modern Engineering Mathematics*, Harlow, England: Pearson Education Ltd., 2001.

Selected Bibliography

Deif, A. S., *Advanced Matrix Theory for Scientists and Engineers*, New York: Halsted Press, 1982.

Graham, A., *Matrix Theory and Applications for Engineers and Mathematicians*, New York: Halsted Press, 1979.

Krause, A. D., *Matrices for Engineers*, New York: Oxford University Press, 2002.

Chapter 7

VECTOR ALGEBRA

7.1 Vectors

7.1.1 Definitions

A *true scalar* is completely defined by the number only, has no directions associated with it, and does not depend on the orientation of the reference coordinates.

♦ <u>Example 7.1</u> Temperature and mass are true scalars (e.g., 10°C or 35 kg).

A *pseudo-scalar* is also defined by the number only. The modulus of this number does not depend on the orientation of the reference coordinates. However, the sign of this number does depend on the choice of the positive directions of the axes.

♦ <u>Example 7.2</u> Volume, surface, area, and angle are pseudo-scalars.

A *vector* depends on both a magnitude (vector modulus) and direction. In order to qualify as vectors, the quantities must also satisfy some particular rule of combination cited later. An example of the vector in three dimensions is given in Figure 7.1.

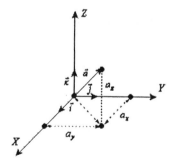

Figure 7.1 Vector $\vec{a}$ in three-dimensional Cartesian coordinates.

175

In Cartesian coordinates, the vector $\vec{a}$ can be represented by its projections a_x, a_y, a_z, which are called the *Cartesian components*.

The *modulus*, or length of the vector $\vec{a}$, is written as $|\vec{a}|$. A vector with a modulus equal to 1 is called a *unit vector*.

If we denote mutually perpendicular unit vectors in Cartesian coordinates as $\vec{i}, \vec{j}, \vec{k}$ (see Figure 7.1), the vector $\vec{a}$ can be represented as:

$$\vec{a} = a_x \vec{i} + a_y \vec{j} + a_z \vec{k}$$

The modulus of the vector is:

$$|\vec{a}| = \sqrt{a_x^2 + a_y^2 + a_z^2}$$

Two vectors $\vec{a}$ and $\vec{b}$ are equal if they have the same modulus, direction, and sense. Vectors $\vec{a}$ and $\vec{b}$ in Figure 7.2(a, b) are not equal, while vectors in Figure 7.2(c) are equal.

If vectors belong to the parallel lines, they are called *parallel,* otherwise they are called *antiparallel.* The *zero* or *null* ($\vec{0}$) vector has a zero modulus.

7.1.2 The Product of a Vector and a Scalar

The product of a vector $\vec{a}$ and a scalar C is a vector with a modulus $C \cdot |\vec{a}|$. It is parallel to the vector $\vec{a}$ and has the same direction if $C > 0$, or has the opposite direction if $C < 0$.

7.1.3 Addition and Subtraction of the Vectors

Addition of the vectors follows the *parallelogram rule*, which states that the sum of two vectors $\vec{a}$ and $\vec{b}$ is the vector $\vec{c}$ represented by the diagonal of the

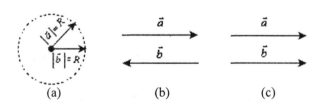

(a)   (b)   (c)

Figure 7.2 Vectors which (a, b) are not equal, and (c) equal.

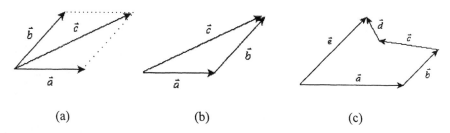

(a)   (b)   (c)

Figure 7.3 (a) Parallelogram, (b) triangle, and (c) polygon rules.

parallelogram with $\vec{a}$ and $\vec{b}$ as adjacent sides [Figure 7.3(a)], or equivalent *triangle rule* [Figure 7.3(b)]. For the sum of several vectors $\vec{e} = \vec{a} + \vec{b} + \vec{c} + \vec{d}$, the triangle rule becomes the *rule of polygon* [Figure 7.3(c)].

Addition of the vectors follows the laws:

- Commutative law $\vec{a} + \vec{b} = \vec{b} + \vec{a}$;
- Associative law $(\vec{a} + \vec{b}) + \vec{c} = \vec{a} + (\vec{b} + \vec{c})$;
- Distributive law $C(\vec{a} + \vec{b}) = C \cdot \vec{a} + C \cdot \vec{b}$.

In order to subtract vector $\vec{b}$ from vector $\vec{a}$, the vector $-\vec{b}$ has to be added to vector $\vec{a}$:

$$\vec{a} - \vec{b} = \vec{a} + (-\vec{b})$$

7.1.4 The Scalar Product

The scalar product c of two vectors $\vec{a}$ and $\vec{b}$ is a scalar, and defined as:

$$c = \vec{a} \cdot \vec{b} = |\vec{a}| \cdot |\vec{b}| \cdot \cos(\vec{a}, \vec{b})$$

where $\cos(\vec{a}, \vec{b}) = \cos\theta$ is the angle between two vectors ($0 \le \theta \le \pi$).

♦ Example 7.3   The length of vectors $\vec{a}$ and $\vec{b}$ are 3 cm and 2 cm correspondingly. The angle $\theta = 120^0$. Find the scalar product.
Solution. $\cos(120^0) = -0.5$. Thus, $\vec{a} \cdot \vec{b} = 3 \cdot 2 \cdot -0.5 = -3\,\mathrm{cm}^2$

If the vector is represented by its components:

$$\vec{a} = a_x \vec{i} + a_y \vec{j} + a_z \vec{k} = (a_x, a_y, a_z)$$

$$\vec{b} = b_x \vec{i} + b_y \vec{j} + b_z \vec{k} = (b_x, b_y, b_z)$$

then the scalar product is:

$$\vec{a} \cdot \vec{b} = a_x b_x + a_y b_y + a_z b_z$$

Thus, the angle θ between the two vectors $\vec{a}$ and $\vec{b}$ can be found as:

$$\cos \theta = \frac{\vec{a} \cdot \vec{b}}{|\vec{a}| \cdot |\vec{b}|} = \frac{a_x b_x + a_y b_y + a_z b_z}{\sqrt{a_x^2 + a_y^2 + a_z^2} \cdot \sqrt{b_x^2 + b_y^2 + b_z^2}}$$

♦ Example 7.4   Find the angle between vectors $\vec{a} = (-2, 1, 2)$ and $\vec{b} = (-2, -2, 1)$ [1].

Solution. The lengths of the vectors are:

$$|\vec{a}| = \sqrt{(-2)^2 + 1^2 + 2^2} = 3, \quad |\vec{b}| = \sqrt{(-2)^2 + (-2)^2 + 1^2} = 3$$

The scalar product $\vec{a} \cdot \vec{b} = -2 \cdot -2 + 1 \cdot -2 + 2 \cdot 1 = 4$.

Thus, $\cos \theta = \dfrac{4}{3 \cdot 3} = \dfrac{4}{9} \cong 0.44$; $\theta = a\cos(0.44) \cong 63.6^0$.

The definition of the scalar product results in two important statements:

- The scalar product of the same vector is equal to its modulus squared: $\vec{a} \cdot \vec{a} = |\vec{a}|^2$. Thus, $|\vec{a}| = \sqrt{\vec{a}^2}$.
- If the scalar product of two nonzero vectors is equal to zero, these vectors are orthogonal ones ($\theta = \pi/2$). $\vec{a} \cdot \vec{b} = |\vec{a}| \cdot |\vec{b}| \cdot \cos \dfrac{\pi}{2} = 0$.

The scalar product sometimes is also called the *dot* or *inner* product.

7.1.5 Properties of the Scalar Product

- The scalar product is equal to zero if the length of one of the vectors is zero or they are perpendicular (i.e., orthogonal, $\theta = \pi/2$)
- Commutative law:

$\vec{a} \cdot \vec{b} = \vec{b} \cdot \vec{a}$

• Distributive law:

$\vec{a}(\vec{b} + \vec{c}) = \vec{a} \cdot \vec{b} + \vec{a} \cdot \vec{c}$

$\vec{a}(C\vec{b}) = (C\vec{a}) \cdot \vec{b} = C(\vec{a} \cdot \vec{b})$ where C is a scalar.

• The scalar products of the basic unit vectors are:

$\vec{i} \cdot \vec{i} = \vec{i}^{2} = 1 \quad \vec{j} \cdot \vec{j} = \vec{j}^{2} = 1 \quad \vec{k} \cdot \vec{k} = \vec{k}^{2} = 1$

$\vec{i} \cdot \vec{j} = \vec{j} \cdot \vec{i} = 0 \quad \vec{j} \cdot \vec{k} = \vec{k} \cdot \vec{j} = 0 \quad \vec{k} \cdot \vec{i} = \vec{i} \cdot \vec{k} = 0$

7.1.6 The Right-Handed and Left-Handed Sets of Vectors

For three nonzero and nonparallel vectors $\vec{a}, \vec{b}, \vec{c}$ with common origin point O taken in the order as specified ($\vec{a}$ first, $\vec{b}$ second, and $\vec{c}$ third), the set is called *right-handed*, if in order to move vector $\vec{a}$ to the position of vector $\vec{b}$ along the shortest path, it has to be rotated counterclockwise as observed from the end of vector $\vec{c}$ [Figure 7.4(a)]. Otherwise (if the rotation is clockwise), the set is called *left-handed* [Figure 7.4(b)].

7.1.7 The Vector Product

The vector product of two vectors $\vec{a}$ and $\vec{b}$ is the third vector $\vec{c} = \vec{a} \times \vec{b}$ evaluated according to the following rules:

• The modulus of vector $\vec{c}$ is:

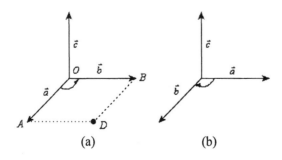

(a)   (b)

Figure 7.4 The (a) right-handed and (b) left-handed sets of vectors.

$$| \vec{c} | = | \vec{a} | \cdot | \vec{b} | \cdot \sin(\vec{a}, \vec{b}) = | \vec{a} | \cdot | \vec{b} | \cdot \sin \theta$$

- The direction of vector $\vec{c}$ is such that it is perpendicular to both $\vec{a}$ and $\vec{b}$.

- The set of vectors $\vec{a}, \vec{b}, \vec{c}$ is right-handed.

The modulus of the vector product $\vec{a} \times \vec{b}$ is equal to the area of the parallelogram $OADB$ [Figure 7.4 (a)].

♦ Example 7.5. Find the vector products $\vec{i} \times \vec{j}$ and $\vec{j} \times \vec{i}$ for three-dimensional Cartesian coordinates (Figure 7.5) [1].

Solution. $\vec{i} \times \vec{j} = \vec{k}$; $\vec{j} \times \vec{i} = -\vec{k}$.

If vectors are specified by its components:

$\vec{a} = (a_x, a_y, a_z)$, $\vec{b} = (b_x, b_y, b_z)$, then

$$\vec{a} \times \vec{b} = \left(\begin{vmatrix} a_y & a_z \\ b_y & b_z \end{vmatrix}, \begin{vmatrix} a_z & a_x \\ b_z & b_x \end{vmatrix}, \begin{vmatrix} a_x & a_y \\ b_x & b_y \end{vmatrix} \right)$$

where $\begin{vmatrix} a_y & a_z \\ b_y & b_z \end{vmatrix} = a_y b_z - b_y a_z$ $\quad \begin{vmatrix} a_z & a_x \\ b_z & b_x \end{vmatrix} = a_z b_x - b_z a_x$

$$\begin{vmatrix} a_x & a_y \\ b_x & b_y \end{vmatrix} = a_x b_y - b_x a_y$$

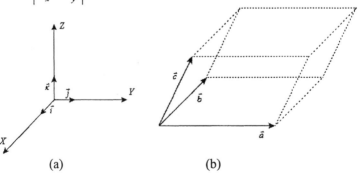

(a)   (b)

Figure 7.5 (a) Unit vectors in Cartesian coordinates, and (b) triple scalar product.

are determinants of the second order (see Section 6.1.10). Thus, the matrix form for the vector product is:

$$\vec{a} \times \vec{b} = \begin{vmatrix} \vec{i} & \vec{j} & \vec{k} \\ a_x & a_y & a_z \\ b_x & b_y & b_z \end{vmatrix}$$

For the basic unit vectors, the vector products are as follows:

$$\vec{i} \times \vec{i} = 0 \quad \vec{i} \times \vec{j} = \vec{k} \quad \vec{i} \times \vec{k} = -\vec{j}$$
$$\vec{j} \times \vec{i} = -\vec{k} \quad \vec{j} \times \vec{j} = 0 \quad \vec{j} \times \vec{k} = \vec{i}$$
$$\vec{k} \times \vec{i} = \vec{j} \quad \vec{k} \times \vec{j} = -\vec{i} \quad \vec{k} \times \vec{k} = 0$$

7.1.8 Properties of the Vector Product

- $\vec{a} \times \vec{a} = 0$

- Anticommutative law

$$\vec{a} \times \vec{b} = -\vec{b} \times \vec{a}$$

- Nonassociative law for multiplication

$$\vec{a} \times (\vec{b} \times \vec{c}) \neq (\vec{a} \times \vec{b}) \times \vec{c}$$

- Distributive law

$$\vec{a} \times (\vec{b} + \vec{c}) = (\vec{a} \times \vec{b}) + (\vec{a} \times \vec{c})$$
$$\vec{a} \times (C\vec{b}) = C(\vec{a} \times \vec{b}) = (C\vec{a}) \times \vec{b} \text{ , where } C \text{ is a scalar.}$$

- If $\vec{a}$ and $\vec{b}$ are parallel or antiparallel ($\theta = 0$ or $\theta = \pi$), then $\vec{a} \times \vec{b} = 0$.

7.1.9 The Triple Scalar Product

The *triple scalar product* $\vec{a} \cdot (\vec{b} \times \vec{c})$ is the scalar V, equal to the volume of the parallelepiped mounted on $\vec{a}, \vec{b}, \vec{c}$ [Figure 7.5(b)].

$$V = \vec{a} \cdot (\vec{b} \times \vec{c}) = a_x(b_y c_z - b_z c_y) + a_y(b_z c_x - b_x c_z) + a_z(b_x c_y - b_y c_x)$$

or, in the matrix form

$$V = \vec{a} \cdot (\vec{b} \times \vec{c}) = \begin{vmatrix} a_x & a_y & a_z \\ b_x & b_y & b_z \\ c_x & c_y & c_z \end{vmatrix}$$

7.1.10 Properties of the Triple Scalar Product

• The signs of scalar multiplication (dot) and vector multiplication (cross) can be interchanged

• $(\vec{a} \times \vec{b}) \cdot \vec{c} = \vec{a} \cdot (\vec{b} \times \vec{c})$.

• If $(\vec{a} \times \vec{b}) \cdot \vec{c} = 0$, then either $\vec{a} = 0$, $\vec{b} = 0$, or $\vec{c} = 0$, or any two vectors are parallel, or all three vectors are coplanar (i.e., belong to the same plane).

7.1.11 The Triple Vector Product

The triple vector product is given as: $\vec{a} \times (\vec{b} \times \vec{c}) = (\vec{a} \cdot \vec{c}) \cdot \vec{b} - (\vec{a} \cdot \vec{b}) \cdot \vec{c}$. It is a vector coplanar with vectors $\vec{b}$ and $\vec{c}$.

7.2 Differentiation and Integration of Vectors

7.2.1 A Derivative of a Vector

Let us consider a vector $\vec{a}(t)$ as a function of a variable t. The *derivative* of a vector $\vec{a}(t)$ with respect to t is the limit:

$$\frac{d\vec{a}(t)}{dt} = \lim_{\Delta t \to 0} \frac{\Delta \vec{a}}{\Delta t} = \lim_{\Delta t \to 0} \frac{\vec{a}(t + \Delta t) - \vec{a}(t)}{\Delta t}$$

The function $\vec{a}(t)$ is a continuous function with respect to t, if the modulus of the vector $\Delta \vec{a}$ tends to zero when Δt tends to zero.

The derivative of the function $\vec{a}(t)$ with components $a_x(t)$, $a_y(t)$, $a_z(t)$ can be written as:

$$\frac{d\vec{a}}{dt} = \vec{i} \frac{da_x}{dt} + \vec{j} \frac{da_y}{dt} + \vec{k} \frac{da_z}{dt}$$

7.2.2 A Derivative of a Point

Let us consider a point P in three-dimensional space which is a function of variable t (Figure 7.6). The *derivative* of a point P is the limit:

$$\frac{dP(t)}{dt} = \lim_{\Delta t \to 0} \frac{\Delta \bar{P}(t)}{\Delta t} = \lim_{\Delta t \to 0} \frac{P(t + \Delta t) - P(t)}{\Delta t}$$

Actually, the derivative of the point P is the same as the derivative of the vector $\overrightarrow{OP}$.

7.2.3 The Derivative of a Vector with Respect to Another Vector

For two vectors $\bar{a} = (a_x, a_y, a_z)$ and $\bar{b} = (b_x, b_y, b_z)$, the derivative of the vector $\bar{a}$ with respect to the vector $\bar{b}$ is the vector $\bar{c}$:

$$\bar{c} = \frac{d\bar{a}}{d\bar{b}} = \frac{\partial \bar{a}}{\partial x} b_x + \frac{\partial \bar{a}}{\partial y} b_y + \frac{\partial \bar{a}}{\partial z} b_z$$

The components of the vector $\bar{c}$ (projections to the axes OX, OY, OZ) are:

$$c_x = \frac{\partial a_x}{\partial x} b_x + \frac{\partial a_x}{\partial y} b_y + \frac{\partial a_x}{\partial z} b_z$$

$$c_y = \frac{\partial a_y}{\partial x} b_x + \frac{\partial a_y}{\partial y} b_y + \frac{\partial a_y}{\partial z} b_z$$

$$c_z = \frac{\partial a_z}{\partial x} b_x + \frac{\partial a_z}{\partial y} b_y + \frac{\partial a_z}{\partial z} b_z$$

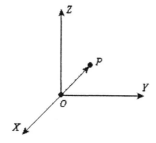

Figure 7.6 A point in three-dimensional space.

7.2.4 Properties of the Vector Derivatives

- If $\vec{d} = \vec{a} + \vec{b} + ... + \vec{u}$, $\quad \vec{d}' = \vec{a}' + \vec{b}' + ... + \vec{u}'$.

- If $\vec{b} = C\vec{a}$, then $\vec{b}' = C\vec{a}'$, where C is a scalar.

- If $\vec{u}(t) = \lambda(t) \cdot \vec{a}(t)$, then $\vec{u}'(t) = \lambda(t) \cdot \vec{a}'(t) + \lambda'(t) \cdot \vec{a}(t)$.

- If $\vec{c} = \vec{a} \cdot \vec{b}$ (a scalar product), then $\vec{c}' = \vec{a} \cdot \vec{b}' + \vec{a}' \cdot \vec{b}$.

- If $\vec{c} = \vec{a} \times \vec{b}$ (a vector product), then $\vec{c}' = \vec{a} \times \vec{b}' + \vec{a}' \times \vec{b}$.

Note. In the latter formula watch the succession of the factors since $\vec{b} \times \vec{a}' = -\vec{a}' \times \vec{b}$.

7.2.5 Integral of a Vector

For a vector $\vec{a}(t) = \vec{i} \cdot a_x(t) + \vec{j} \cdot a_y(t) + \vec{k} \cdot a_z(t)$, the definite integral with limits t_0 and t_1 is:

$$\int_{t_0}^{t_1} \vec{a}(t)\, dt = \vec{i} \int_{t_0}^{t_1} a_x(t)\, dt + \vec{j} \int_{t_0}^{t_1} a_y(t)\, dt + \vec{k} \int_{t_0}^{t_1} a_z(t)\, dt$$

7.2.6 Gradient

The *gradient* of a scalar function $f(x, y, z)$ is the vector with coordinates $\frac{\partial f}{\partial x}, \frac{\partial f}{\partial y}, \frac{\partial f}{\partial z}$:

$$\mathrm{grad} f = (\frac{\partial f}{\partial x}, \frac{\partial f}{\partial y}, \frac{\partial f}{\partial z})$$

Thus,

$$\mathrm{grad} f = \vec{i} \frac{\partial f}{\partial x} + \vec{j} \frac{\partial f}{\partial y} + \vec{k} \frac{\partial f}{\partial z}$$

If the scalar function $f(m, n, ...)$ is a composite function of several scalars $m(x, y, z)$, $n(x, y, z), ...$ each of which is a function of other scalars x, y, z (e.g., coordinates), then the gradient can be written as:

$$grad f = \frac{\partial f}{\partial m} grad m + \frac{\partial f}{\partial n} grad n + ...$$

7.2.7 Divergence and Curl

Let us consider a point $M(x, y, z)$ with coordinates x, y, z, and the vector function $\vec{a}$ of this point with coordinates a_x, a_y, a_z.

Divergence of the vector $\vec{a}$ is a scalar $div\,\vec{a}$ defined as:

$$div\,\vec{a} = \frac{\partial a_x}{\partial x} + \frac{\partial a_y}{\partial y} + \frac{\partial a_z}{\partial z}$$

The divergence can also be expressed as a sum of the scalar products:

$$div\,\vec{a} = \vec{i} \cdot \frac{\partial \vec{a}}{\partial x} + \vec{j} \cdot \frac{\partial \vec{a}}{\partial y} + \vec{k} \cdot \frac{\partial \vec{a}}{\partial z}$$

Curl of the vector $\vec{a}$ is a vector $rot\,\vec{a}$ defined as:

$$rot\,\vec{a} = \vec{i}\left(\frac{\partial a_z}{\partial y} - \frac{\partial a_y}{\partial z}\right) + \vec{j}\left(\frac{\partial a_x}{\partial z} - \frac{\partial a_z}{\partial x}\right) + \vec{k}\left(\frac{\partial a_y}{\partial x} - \frac{\partial a_x}{\partial y}\right)$$

$$= \vec{i} \cdot (rot\,\vec{a})_x + \vec{j} \cdot (rot\,\vec{a})_y + \vec{k} \cdot (rot\,\vec{a})_z$$

where $(rot\,\vec{a})_x$, $(rot\,\vec{a})_y$, $(rot\,\vec{a})_z$ are projections of the vector $rot\,\vec{a}$ to axes OX, OY, OZ, correspondingly.

Curl can also be expressed as a sum of the vector products:

$$rot\,\vec{a} = \vec{i} \times \frac{\partial \vec{a}}{\partial x} + \vec{j} \times \frac{\partial \vec{a}}{\partial y} + \vec{k} \times \frac{\partial \vec{a}}{\partial z}$$

Sometimes, the notation $curl\,\vec{a}$ is used instead of $rot\,\vec{a}$.

7.2.8 Laplace Operator

The *Laplace operator* is defined as:

$$\Delta = \frac{\partial^2}{\partial x^2} + \frac{\partial^2}{\partial y^2} + \frac{\partial^2}{\partial z^2}$$

If we apply the Laplace operator to a scalar function f, then:

$$\Delta f = \frac{\partial^2 f}{\partial x^2} + \frac{\partial^2 f}{\partial y^2} + \frac{\partial^2 f}{\partial z^2}$$

Let us apply the Laplace operator to a vector $\bar{a}$. Thus,

$$\Delta\bar{a} = \frac{\partial^2 \bar{a}}{\partial x^2} + \frac{\partial^2 \bar{a}}{\partial y^2} + \frac{\partial^2 \bar{a}}{\partial z^2}$$

Since $\bar{a} = \bar{i}a_x + \bar{j}a_y + \bar{k}a_z$, then $\Delta\bar{a} = \bar{i}\Delta a_x + \bar{j}\Delta a_y + \bar{k}\Delta a_z$.

7.2.9 Hamilton Operator (del)

The *Hamilton operator* (sometimes called *del*) is defined as:

$$\vec{\nabla} = \bar{i}\frac{\partial}{\partial x} + \bar{j}\frac{\partial}{\partial y} + \bar{k}\frac{\partial}{\partial z}$$

Thus, gradient of a scalar f can be written as:

Laplace, Pierre (1749–1827).   French physicist and mathematician. He contributed to mathematical astronomy, systematized and elaborated probability theory, and developed *Laplace transform*. In this work, he frequently omitted derivations, leaving only results with the remark "it is easy to see." It is said that later he himself could not remember the derivations without days of work. Once Napoleon is said to have questioned Laplace on his neglect to mention God in his works. Laplace replied that to come up with the solution he had no need for that hypothesis.

Hamilton, William Rowan (1805–1865). Irish mathematician. At age 17 he discovered an error in Laplace's work *Celestial Mechanics*. Contributed to anticommutative algebra and theory of refraction in biaxial crystals.

$$grad\ f = \vec{\nabla}f = \vec{i}\ \frac{\partial f}{\partial x} + \vec{j}\ \frac{\partial f}{\partial y} + \vec{k}\ \frac{\partial f}{\partial z}$$

The scalar product of the vectors $\vec{\nabla}$ and $\vec{a}$ is:

$$\vec{\nabla} \cdot \vec{a} = \frac{\partial a_x}{\partial x} + \frac{\partial a_y}{\partial y} + \frac{\partial a_z}{\partial z} = div\ \vec{a}$$

The vector product of the vectors $\vec{\nabla}$ and $\vec{a}$ is:

$$\nabla \times \vec{a} = \vec{i}\left(\frac{\partial a_z}{\partial y} - \frac{\partial a_y}{\partial z}\right) + \vec{j}\left(\frac{\partial a_x}{\partial z} - \frac{\partial a_z}{\partial x}\right) + \vec{k}\left(\frac{\partial a_y}{\partial x} - \frac{\partial a_x}{\partial y}\right) = rot\ \vec{a}$$

The following notations are often used interchangeably:

$$\vec{\nabla}f = grad\ f\ ;\ \vec{\nabla} \cdot \vec{a} = div\,\vec{a}\ ;\ \vec{\nabla} \times \vec{a} = rot\,\vec{a}\ ;\ \Delta f = \vec{\nabla}^2 f;\ \Delta\vec{a} = \vec{\nabla}^2\vec{a}$$

Common formulas for gradient, divergence, and curl are given in <u>Appendix 11</u>.

7.2.10 Circulation and Flow: Green's and Stokes' Formulas

If $\vec{a}$ is the vector function of a point P, the *circulation* of a vector $\vec{a}$ along the curve AB (Figure 7.7) is defined as the curved integral:

$$\int_A^B \vec{a} \cdot d\vec{P}$$

If S is a double-sided surface with a point P on it, $d\sigma$ is the element of this surface around P, $d\vec{\sigma}$ is a vector with the length equal to $d\sigma$ and directed along the positive direction of the normal to this element, then the *flow* of the vector $\vec{a}$ is defined as:

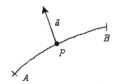

Figure 7.7 Circulation of a vector along the curve.

$$\int_S \bar{a} \cdot d\bar{\sigma}$$

For any continuous scalar f and vector $\bar{a}$ functions of the point P within the volume V and its surface S the following formulas are valid:

- $$\int_V grad\, f\, dV = \int_S f\, d\bar{\sigma}$$

- $$\int_V div\, \bar{a}\, dV = \int_S \bar{a} \cdot d\bar{\sigma}$$

- $$\int_V rot\, \bar{a}\, dV = \int_S d\bar{\sigma} \times \bar{a}$$

These formulas allow the substitution of the triple integrals with the double integrals.

If S is a closed surface with volume V within, and p, q are two scalar points within this volume, then *Green's formula* is valid:

$$\int_V (p\Delta q - q\Delta p)\, dV = \int_S (p\, grad\, q - q\, grad\, p) \cdot d\bar{\sigma}$$

If $\bar{n}$ is the unit vector of the normal to the surface S, then *Stokes' formula* is valid:

$$\int_S \bar{n} \cdot rot\, \bar{a}\, dV = \int_C \bar{a} \cdot d\bar{P}$$

where C is the closed-loop curve limiting surface S.

Green, George (1793–1841). English mathematician and physicist. He developed the Green's theorem (which was discovered independently in Russia by Ostrogradskiy), introduced Green's functions as a means of solving boundary value problems, and developed integral transform theorems.

Stokes, George (1819–1903). Irish mathematician and physicist. He systematically studied fluid mechanics, elastic solids, waves in elastic media, and diffraction. He also formulated the three-dimensional analog of Green's theorem, known as Stokes' theorem, and was a pioneer in the use of divergent series.

Reference

[1] Vigodskiy, M. Ya., *Mathematical Handbook*, (in Russian), Moscow: Nauka, 1972.

Selected Bibliography

Hinchey, F. A., *Vectors and Tensors for Engineers and Students*, New York: John Wiley and Sons, 1976.

James, G., (ed.), *Modern Engineering Mathematics*, Harlow, England: Pearson Education Ltd., 2001.

Lewis, P. E., and J. P. Ward, *Vector Analysis for Scientists and Engineers*, Reading, MA: Addison-Wesley, 1989.

Chapter 8

PROBABILITY THEORY, RANDOM FUNCTIONS, AND APPLIED STATISTICS

8.1 Random Events and Variables

8.1.1 A Random Event

A *random event* is any occurrence that can happen or not happen as the result of a test. The *probability* of the event is the quantitive measure of the likelihood that this event will occur. It ranges from 0 to 1. The more likely for the event to occur, the higher the probability is.

The event can be totally deterministic (*D*), impossible (*I*), or random (*A, B, C,...*). Probabilities of the events *D, I,* and *A* are:

$$P(D) = 1; \quad P(I) = 0; \quad 0 \le P(A) \le 1$$

♦ Example 8.1 Dice have six sides, with numbers ranging from 1 to 6. When the die is cast what is the probability of the event *A* that a player scores 7 points, and event *B* that he scores less than 7?
Solution. Event *A* is impossible, thus *P(A)* = *P(I)* = 0. Event *B* is deterministic, thus *P(B)* = *P(D)* = 1.

8.1.2 Equivalent Events

If event *B* occurs any time when event *A* occurs, it is denoted $A \subset B$. If $A \subset B$ and at the same time $B \subset A$, the events *A* and *B* are called the *equivalent events:*

$$A = B \quad P(A) = P(B)$$

8.1.3 Sum and Product of Random Events

The *sum* of events $A_1, A_2, A_3, ..., A_n$ is the event *A* that occurs when at least one event from the series happens:

191

$$A = \sum_n A_n = A_1 \bigcup A_2 \bigcup A_3 \ldots \bigcup A_n = \bigcup_n A_n$$

The *product* of events $A_1, A_2, A_3, \ldots, A_n$ is the event A that occurs when all events from the series happen simultaneously:

$$A = \prod_n A_n = A_1 \bigcap A_2 \bigcap A_3 \ldots \bigcap A_n = \bigcap_n A_n$$

The following rules are applicable to the operations of addition and multiplication of the random events:

$$A + A = A \quad A + D = D \quad A + I = A \quad A + B = B + A$$

$$(A + B) + C = A + (B + C)$$

$$AA = A \quad AD = D \quad AI = A \quad AB = BA \quad (AB)C = A(BC)$$

$$(A + B)C = AC + BC \quad AB + C = (A + C)(B + C)$$

8.1.4 Disjoint Events

If the mutual occurrence of events A and B is impossible, the events are called *disjoint* or *mutually exclusive* ones. In this case:

$$A \cdot B = I$$

8.1.5 Complete Group of Events

Events A, B, C,... form the *complete group* of events if, as the result of the test, at least one of them definitely occurs:

$$A + B + C + \ldots = D$$

8.1.6 Complement Events

Two events are called *complement* events if they are mutually exclusive and form the complete group: ($\overline{A}$ is a complement of A):

$$P(\overline{A}) = 1 - P(A)$$

♦ Example 8.2 One shot is made in an attempt to hit a target. The probability of hitting the target is 0.8. What is the probability of missing the target?
Solution. Events A to hit the target and B to miss the target form the complete group and are mutually exclusive. Event A has the probability $P(A)$ = 0.8. Thus, the probability of missing the target:
$P(B) = P(\overline{A}) = 1 - P(A) = 0.2$

The following rules are applicable to complement events:

$$\overline{\overline{A}} = A \quad \overline{D} = I \quad \overline{I} = D \quad A + \overline{A} = D \quad A\overline{A} = I$$

$$\overline{A + B} = \overline{A} \cdot \overline{B} \quad \overline{AB} = \overline{A} + \overline{B} \quad A + B = A + \overline{A} \cdot B$$

$$A + B = \overline{\overline{A} \cdot \overline{B}} \quad AB = \overline{\overline{A} + \overline{B}}$$

8.1.7 Conditional Probability

The probability that event A occurs under the condition that event B has already occurred is denoted $P(A|B)$ and is called a *conditional probability* of the event A :

$$P(A|B) = P(A \cdot B)/P(B)$$

8.1.8 Independent Events

If event A statistically does not depend on event B, events are called *independent*:

$$P(A \cdot B) = P(A) \cdot P(B); \quad P(A|B) = P(A); \quad P(B|A) = P(B)$$

♦ Example 8.3 The probability of a radio transmitter failing (event A) is 0.1, and of a radio receiver failing (event B) is 0.2. What is the probability that both the transmitter and the receiver fail simultaneously (event C), if causes of their failure are independent?
Solution. Since events A and B are independent, the probability of simultaneous failure is: $P(C) = P(A \cdot B) = P(A) \cdot P(B) = 0.1 \cdot 0.2 = 0.02$.

8.1.9 Probability of the Sum of the Events

The probability of the *sum* of two events A and B :

$$P(A+B) = P(A) + P(B) - P(AB)$$

If the events are disjoint:

$$P(A+B) = P(A) + P(B)$$

In the general case:

$$P\left(\sum_{n=1}^{N} A_n\right) = \sum_{n=1}^{N} P(A_n) - \sum_{n=1}^{N-1} \sum_{j=n+1}^{N} P(A_n A_j) +$$

$$+ \sum_{n=1}^{N-2} \sum_{j=n+1}^{N-1} \sum_{i=j+1}^{N} P(A_n A_j A_i) - \ldots + (-1)^{N-1} P\left(\prod_{n=1}^{n} A_n\right)$$

The sum of disjoint events that forms a complete group is equal to unity:

$$\sum_{n=1}^{N} P(A_n) = 1$$

The sum of two complement events is equal to unity:

$$P(A) + P(\overline{A}) = 1$$

8.1.10 Probability of the Product of Events

The probability of the *product* of two events A and B is

$$P(AB) = P(A) \cdot P(B \mid A) = P(B) \cdot P(A \mid B)$$

For independent events:

$$P(AB) = P(A) \cdot P(B)$$

In the general case:

$$P(A_1 A_2, \ldots, A_n) = P(A_1) P(A_2 \mid A_1) P(A_3 \mid A_1 A_2), \ldots, P(A_n \mid A_1 A_2, \ldots, A_{n-1})$$

For N independent events:

$$P(\prod_{n=1}^{N} A_n) = \prod_{n=1}^{N} P(A_n)$$

8.1.11 An Event Conditional on the Set of Other Events

If event B occurs as the result of an event A occurrence, and event B is a set of N events $B_1, B_2, ..., B_N$, then

$$P(A) = \sum_{i=1}^{N} P(B_i) \cdot P(A \mid B_i)$$

8.1.12 Bayes' Rule

Let us assume we have the series of hypotheses $B_1, B_2, ..., B_N$ about the possible occurrence of some event, and we know the probabilities of these hypotheses $P(B_i)$, $i = 1...N$, before a test. After the test was done, the event A had occurred.

Bayes' rule (or theorem of hypotheses) states that for the full group of disjoint hypotheses (events) $B_1, B_2, ..., B_N$, the conditional probability of each hypothesis after the test will be:

$$P(B_i / A) = \frac{P(B_i) \cdot P(A \mid B_i)}{\sum_{i=1}^{N} P(B_i) \cdot P(A \mid B_i)} = \frac{P(B_i) P(A \mid B_i)}{P(A)}$$

♦ Example 8.4 Two missiles were launched to hit a target. The probability of the first missile hitting the target is 0.8, and the probability of the second missile hitting the target is 0.4. The target was hit with the amount of damage showing that only one missile hit the target. Find the probabilities that the target was hit by the first missile and the second missile.

Solution. Before launch, the following hypotheses are possible: both missiles miss B_1; both missiles hit B_2; the first missile hits, the second misses B_3; and the first missile misses, the second hits B_4. The probabilities of these hypotheses:

$P(B_1) = (1 - 0.8)(1 - 0.4) = 0.12;$

$P(B_2) = 0.8 \cdot 0.4 = 0.32;$

$P(B_3) = 0.8 \cdot (1 - 0.4) = 0.48$

$P(B_4) = (1 - 0.8) \cdot 0.4 = 0.08$

The conditional probabilities of the occurred event A that the target was hit by only one missile are:

$$P(A/B_1) = 0 \; ; \; P(A/B_2) = 0 \; ; \; P(A/B_3) = 1 \; ; \; P(A/B_4) = 1$$

After the test, hypotheses B_1 and B_2 become impossible. Probabilities of the hypotheses $P(B_3/A)$ and $P(B_4/A)$ are:

$$P(B_3/A) = \frac{0.48 \cdot 1}{0.48 \cdot 1 + 0.08 \cdot 1} = 0.857$$

$$P(B_4/A) = \frac{0.08 \cdot 1}{0.48 \cdot 1 + 0.08 \cdot 1} = 0.143$$

Thus, the probability that the target was hit by the first missile is 0.857, and that it was hit by the second one is 0.143.

8.1.13 A Random Variable

A *random variable* is a variable that can take a random value in each trial. The basic types of random variables are *discrete random variables* and *continuous random variables*. The discrete random variable can take only a particular finite or countable infinite set of values. For continuous random variables, the set of values cannot be defined, and it fills in some interval continuously.

8.1.14 Probability Distribution Function

The complete statistical description of a random variable can be given by a probability distribution law. For a discrete random variable X, this law defines the linkage between possible values x_i the variable X can take, and its probabilities $p_i = p(x_i)$. The general description both for discrete and continuous variables is based on the concept of a *probability distribution function* (PDF). For a one-dimensional random variable, this function $F_1(x)$ defines the probability P of the fact that a random variable X will be less than some value x :

$$F_1(x) = P(X < x)$$

The probability distribution function has the following properties:

- $F_1(-\infty) = 0$ and $F_1(+\infty) = 1$

- $F_1(x_2) \geq F_1(x_1)$ if $x_2 > x_1$

- $P(x_1 \leq X < x_2) = F(x_2) - F(x_1)$

The PDF of a discrete random variable is a stepped function with the steps in points $x_1, x_2, \ldots,$ and the PDF of a continuous random variable is a continuous function (Figure 8.1).

8.1.15 Probability Density Function

If a probability distribution function has a derivative over the entire range of the possible values of random variables, the concept of a *probability density function* (pdf) is introduced:

$$f_1(x) = d\,F_1(x)/d\,x$$

A probability density function has the following basic properties:

- $f_1(x) \geq 0$

- $P(x_1 \leq X < x_2) = \int\limits_{x_1}^{x_2} f_1(x)\,d\,x = F_1(x_2) - F_1(x_1)$

- $\int\limits_{-\infty}^{\infty} f_1(x)\,d\,x = 1$

For a discrete random variable:

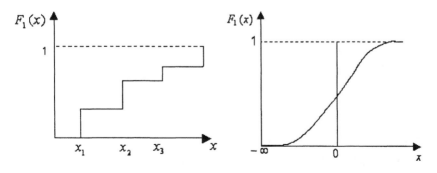

Figure 8.1 Probability distribution functions of discrete and continuous random variables.

$$f_1(x) = \sum_{i=1}^{N} p_i \cdot \delta(x - x_i)$$

where x_i are the possible values of random variables (N values), p_i is the probability of the possible values, and $\delta(u - u_0)$ is the delta-function (see Appendix 3):

$$\delta(u - u_0) = \begin{cases} \infty, & \text{if } u = u_0 \\ 0, & \text{if } u \ne u_0 \end{cases}$$

$$\int_{U_0 - \alpha}^{U_0 + \alpha} \delta(u - u_0) \, d u = 1 \quad \text{for any } \alpha > 0$$

$$\int_{U_0 - \alpha}^{U_0 + \alpha} f(u) \cdot \delta(u - u_0) \, d u = f(u_0)$$

The common probability distribution functions (PDF) and probability density functions (pdf) for discrete and continuous random variables are given in Appendix 14.

8.1.16 Gaussian Probability Distribution and Density Functions

One of the most important distribution laws used in electrical engineering is the *Gaussian* (normal) distribution. In this case, the probability that a random variable with mathematical expectation (mean) m and variance σ^2 will fall within the interval (a, b) is:

$$P(a \le X < b) = \Phi\left(\frac{b - m}{\sigma}\right) - \Phi\left(\frac{a - m}{\sigma}\right)$$

where $\Phi(x)$ is the error function integral.

Typically, different definitions of $\Phi(x)$ and integral forms for Gaussian pdfs are used in different texts. The common definitions of $\Phi(x)$ and formulas to transform the integral forms from one to another are given in Appendix 15.

The PDF $F_1(x)$ and pdf $f_1(x)$ for a random variable X with Gaussian distribution ($-\infty < x < \infty$) are:

$$F_1(x) = \frac{1}{\sigma\sqrt{2\pi}} \int\limits_{-\infty}^{x} e^{-\frac{(t-m)^2}{2\sigma^2}} dt = \Phi\left(\frac{x-m}{\sigma}\right); \quad f_1(x) = \frac{1}{\sigma\sqrt{2\pi}} \exp\left[-\frac{(x-m)^2}{2\sigma^2}\right]$$

The fact that a variable X has the Gaussian (normal) distribution with parameters m and σ is often denoted as $X \subset N(m,\sigma)$.

When $m = 0$ and $\sigma = 1$, the distribution is called the *standard Gaussian* distribution $X \subset N(0,1)$. In this case:

$$F_1(x) = \frac{1}{\sqrt{2\pi}} \int\limits_{-\infty}^{x} e^{-t^2/2} dt = \Phi(x); \quad f_1(x) = \frac{1}{\sqrt{2\pi}} \cdot e^{-x^2/2}$$

8.1.17 Statistical Parameters

In practice, sometimes it is difficult or often impossible to find analytical distribution functions for many classes of random variables of interest. In this case, the description of a random variable can be done at the level of some numerical values (nonrandom numbers), called *statistical parameters*. The most important statistical parameters are:

- *Mathematical expectation*:

$$m_x = M(x) = \sum_{n=1}^{N} x_n \cdot p_n \qquad \text{(discrete variable)}$$

$$m_x = M(x) = \int\limits_{-\infty}^{\infty} x f_1(x) dx \qquad \text{(continuous variable)}$$

Gauss, Johann Carl Friedrich (1777–1855). German mathematician who is sometimes called the "prince of mathematics." He contributed to the theory of mathematics (algebra, differential geometry, numbers theory, mathematical statistics, astronomy, geodesy, and terrestrial magnetism. Since early childhood (at the age of three, Gauss informed his father of an arithmetical error in a complicated payroll calculation and stated the correct answer), until his death at age 78, he worked every day on numerous scientific problems. There is a story that in 1807 when he was interrupted in the middle of a problem and told that his wife was dying, he replied: "Tell her to wait a moment 'til I'm through."

• *Variance:*

$$D_x = \sigma_x^2 = M\left[(x - m_x)^2\right] = \sum_{n=1}^{N} (x_n - m_x)^2 \cdot p_n \qquad \text{(discrete variable)}$$

$$D_x = \sigma_x^2 = \int_{-\infty}^{\infty} (x - m_x)^2 f_1(x) dx \qquad \text{(continuous variable)}$$

• The *median* is the value of a random variable when

$$P(X < M_I) = P(X > M_I) = 0.5$$

Thus, $F_1(M_I) = 0.5$ and

$$\int_{-\infty}^{M_I} f_1(x) dx = \int_{M_I}^{\infty} f_1(x) dx$$

• The *mode* is the value of a random variable, when, for a discrete variable $P(X = M)$ and for a continuous variable $f_1(M)$, reaches a maximum. The distribution is called *unimodal* if a single mode exists, and it is called *multimodal* if there are several modes in the distribution function.

• The *fractile* is the value $x = x_P$ when $F_1(x_P) = P$.

♦ Example 8.5 Find the mathematical expectation of the random variable with Gaussian distribution with parameters *m* and σ.
Solution. The pdf of the random variable with the Gaussian distribution is:

$$f_1(x) = \frac{1}{\sigma\sqrt{2\pi}} \exp\left[-\frac{(x - m)^2}{2\sigma^2}\right]$$

Mathematical expectation can be found as:

$$m_x = \int_{-\infty}^{\infty} x f_1(x) dx = \frac{1}{\sigma\sqrt{2\pi}} \int_{-\infty}^{\infty} x \cdot \exp[-\frac{(x - m)^2}{2\sigma^2}] dx$$

Let us denote $x - m = t$. Thus, $x = m + t$, $dx = dt$, and the expression can be rewritten as:

$$m_x = \frac{1}{\sigma\sqrt{2\pi}} \int_{-\infty}^{\infty} (m+t) \cdot \exp[-\frac{t^2}{2\sigma^2}] dt =$$

$$\frac{1}{\sigma\sqrt{2\pi}} \{ \int_{-\infty}^{\infty} m \cdot \exp[-\frac{t^2}{2\sigma^2}] dt + \int_{-\infty}^{\infty} t \cdot \exp[-\frac{t^2}{2\sigma^2}] dt \}$$

Let us denote $y = \frac{t}{\sqrt{2}\sigma}$. Thus, $dt = \sqrt{2}\sigma \cdot dy$. Using a computer (see Section 9.3) we find that:

$$\int_{-\infty}^{\infty} \exp[-y^2] \cdot dy = \sqrt{\pi} \cdot$$

Thus, the first factor $\frac{\sqrt{2}\sigma \cdot m}{\sigma\sqrt{2\pi}} \int_{-\infty}^{\infty} \exp[-y^2] \cdot dy = m.$

The second factor is:

$$\frac{1}{\sigma\sqrt{2\pi}} \int_{-\infty}^{\infty} t \cdot \exp[-\frac{t^2}{2\sigma^2}] dt = \frac{2\sigma}{\sigma\sqrt{2\pi}} \int_{-\infty}^{\infty} y \cdot \exp[-y^2] dy$$

Using a computer we find that:

$$\int_{-\infty}^{\infty} y \exp[-y^2] \cdot dy = 0 \cdot$$

Thus, the mathematical expectation $m_x = m$.

8.1.18 Moments of Distribution

Moments are the generalized numerical values used for characterizing a random variable. The moment of order k of a random variable X with respect to an arbitrary point a is a mathematical expectation:

$$m_k(a) = M\{(X - a)^k\}$$

If $a = 0$, the moment m_k is called the *initial* one, and if $a = m_x$, the moment M_k is called a *central* one.

Typically, the following moments are used to describe a random variable:

- $m_1 = $ *mathematical expectation* (m)

- $M_2 = $ *variance* (σ^2)

- M_3 leading to the *asymmetry coefficient* γ_1:

$$\gamma_1 = M_3 \Big/ \sqrt{M_2^3} = M_3 \Big/ \sigma^3$$

- M_4 leading to the *coefficient of excess* γ_2:

$$\gamma_2 = \frac{M_4}{M_2^2} - 3 = \frac{M_4}{\sigma^4} - 3$$

The latter describes the deviation of a distribution function from a Gaussian one for which $\gamma_2 = 0$.

The formulas to calculate the moments of a random variable and the moments for common distribution laws are given in Appendix 16.

8.1.19 Functional Transform of Random Variables

A common task is to determine the pdf $f_1(y)$ of the continuous random variable Y when the pdf $f_1(x)$ of the random variable X and deterministic dependence $Y = \varphi(X)$ are known.

If the reciprocal function $X = h(Y)$ is single-valued:

$$f_1(y) = f_1(x) \cdot \left| \frac{dx}{dy} \right| = f_1[h(y)] \cdot \left| \frac{dh(y)}{dy} \right|$$

If the reciprocal function is ambiguous:

$$f_1(y) = \sum_{n=1}^{N} f_1[h_n(y)] \cdot \left| \frac{dh_n(y)}{dy} \right|$$

where n is the number of ambiguities. The functional transform of the discrete random variable does not change the distribution function while the numerical values of the distribution are changed.

The moments of the random variable $Y = \varphi(X)$ can be determined in two ways:

- By applying the pdf $f_1(y)$;
- By averaging dependence $\varphi(X)$.

Applying these approaches to mean value and variance results in:

$$m_y = \int_{-\infty}^{\infty} y f_1(y) dy = \int_{-\infty}^{\infty} \varphi(x) f_1(x) dx$$

$$\sigma_y^2 = \int_{-\infty}^{\infty} (y - m_y)^2 f_1(y) dy = \int_{-\infty}^{\infty} [\varphi(x) - m_y]^2 f_1(x) dx$$

For a discrete random variable:

$$m_y = M[\varphi(X)] = \sum_n \varphi(x_n) \cdot f_n$$

$$\sigma_y^2 = D[\varphi(X)] = \sum_n [\varphi(x_n) - m_y]^2 \cdot f_n$$

The probability distribution functions for some common functional transforms of random variables are given in Appendix 17.

8.1.20 The Characteristic Function

There is a special type of functional transform that is widely used:

$$Y = e^{j \vartheta x}$$

The characteristic function $\theta_1(j\vartheta)$ is defined as a mathematical expectation of a random variable $e^{j\vartheta x}$

$$\theta_1(j\vartheta) = M\left(e^{j\vartheta x}\right) = \int_{-\infty}^{\infty} e^{j\vartheta x} f_1(x) dx$$

where ϑ is a real variable, $j = \sqrt{-1}$.

$$\theta_1\left(j\vartheta\right)=\sum_{n=1}^{N}p_i\cdot e^{j\vartheta x_n}$$

Then:

$$f_1\left(x\right)=\frac{1}{2\pi}\int_{-\infty}^{\infty}\theta_1\left(j\vartheta\right)e^{-j\vartheta x}d\vartheta$$

The moments can be defined as:

$$m_k=\frac{1}{j^k}\frac{d^k\,\theta_1\left(j\vartheta\right)}{d\vartheta^k}\bigg|\,\vartheta=0$$

The characteristic functions for common distributions are given in Appendix 14.

8.1.21 A Multidimensional Random Variable

A *multidimensional random* variable is a system (set) of two or more one-dimensional random variables: $X_1,X_2,...,X_n$. It is described by the probability distribution function of *n*th order :

$$P\{X_1<x_1,X_2<x_2,...,X_n<x_n\}=F_n\left(x_1,x_2,...,x_n\right)$$

and corresponding probability density function:

$$f_n\left(x_1,x_2,...,x_n\right)=\frac{d^n\,F_n\left(x_1,x_2,...,x_n\right)}{d\left(x_1,x_2,...,x_n\right)}$$

8.1.22 Two-Dimensional Distribution Functions

Two-dimensional probability distribution functions $F_2\left(x_1,x_2\right)$ and corresponding density functions $f_2\left(x_1,x_2\right)$ have the following properties, hereinafter $F_2\left(x,y\right)=F_2\left(x_1,x_2\right),\ f_2\left(x,y\right)=f_2\left(x_1,x_2\right)$:

1. $F_2\left(x_2, y\right) \geq F_2\left(x_1, y\right)$ if $x_2 > x_1$

    $F_2\left(x, y_2\right) \geq F_2\left(x, y_1\right)$ if $y_2 > y_1$

2. $F_2\left(x, -\infty\right) = F_2\left(-\infty, y\right) = F_2\left(-\infty, -\infty\right) = 0$

3. $F_2\left(\infty, \infty\right) = 1$

4. $F_2\left(x, \infty\right) = F_1\left(x\right)$ $F_2\left(\infty, y\right) = F_1\left(y\right)$

5. $F_2\left(x, y\right) = \int\limits_{-\infty}^{x}\int\limits_{-\infty}^{y} f_2\left(u, \vartheta\right) \, du \, d\vartheta$

6. $P\left(x_1 \leq X \leq x_2, \ y_1 \leq Y \leq y_2\right) =$

    $F_2\left(x_2, y_2\right) - F_2\left(x_1, y_2\right) - F_2\left(x_2, y_1\right) + F_2\left(x_1, y_1\right)$

7. $f_2\left(x, y\right) \geq 0$

8. $\int\limits_{-\infty}^{\infty}\int\limits_{-\infty}^{\infty} f_2\left(x, y\right) dx dy = 1$

9. $\theta_2\left(j\vartheta_1, j\vartheta_2\right) = M\left[e^{j\left(\vartheta_1 x + \vartheta_2 y\right)}\right] = \int\limits_{-\infty}^{\infty}\int\limits_{-\infty}^{\infty} e^{j\left(\vartheta_1 x + \vartheta_2 y\right)} \cdot f_2\left(x, y\right) dx dy$

The one-dimensional distribution and density functions are expressed through two-dimensional ones as follows:

$$F_1\left(x\right) = F_2\left(x, \infty\right) = \int\limits_{-\infty}^{x}\int\limits_{-\infty}^{\infty} f_2\left(x, y\right) dx dy$$

$$f_1\left(x\right) = \frac{dF_1\left(x\right)}{dx} = \int\limits_{-\infty}^{\infty} f_2\left(x, y\right) dy$$

$$F_1(y) = F_2(\infty, y) = \int_{-\infty}^{y} \int_{-\infty}^{\infty} f_2(x, y)\, dx\, dy$$

$$f_1(y) = \frac{dF_1(y)}{dy} = \int_{-\infty}^{\infty} f_2(x, y)\, dx$$

8.1.23 The Conditional Probability Density Function

The *conditional pdfs* are defined as:

$$f_1(x|y) = \frac{f_2(x, y)}{f_1(y)} = \frac{f_2(x, y)}{\displaystyle\int_{-\infty}^{\infty} f_2(x, y)\, dx}$$

$$f_1(y|x) = \frac{f_2(x, y)}{f_1(x)} = \frac{f_2(x, y)}{\displaystyle\int_{-\infty}^{\infty} f_2(x, y)\, dy}$$

If X and Y are independent random variables:

$$f_2(x, y) = f_1(x) \cdot f_2(y)$$

8.1.24 The Moments: Mathematical Expectation, Variance, and Correlation

The *moments* are defined as:

- The initial moment of order $k_1 + k_2$

$$m_{k_1, k_2} = M\left[X^{k_1} Y^{k_2}\right] = \sum_i \sum_j x_i^{k_1} y_j^{k_2} p_{ij} \qquad \text{(discrete variable)}$$

$$m_{k_1, k_2} = \int_{-\infty}^{\infty} \int_{-\infty}^{\infty} x^{k_1} \cdot y^{k_2} \cdot f_2(x, y)\, dx\, dy \qquad \text{(continuous variable)}$$

- The central moment of order $k_1 + k_2$

$$M_{k_1,k_2} = M\left[X_0^{k_1}Y_0^{k_2}\right] = \sum_i\sum_j (x_i - m_x)^{k_1}(y_j - m_y)^{k_2}p_{ij}$$

$$M_{k_1,k_2} = \int\limits_{-\infty}^{\infty}\int\limits_{-\infty}^{\infty}(x - m_x)^{k_1}(y - m_y)^{k_2}f(x,y)dxdy$$

In practice, the widely used moments are:

$$m_{10} = M\left(X^1Y^0\right) = m_x \qquad m_{01} = M\left(X^0Y^1\right) = m_y$$

that is, *mathematical expectations* of random variables X and Y;

$$M_{20} = M\left[(X - m_x)^2(Y - m_y)^0\right] = D_x = \sigma_x^2$$

$$M_{02} = M\left[(X - m_x)^0(Y - m_y)^2\right] = D_y = \sigma_y^2$$

that is, the *variances* of random variables X and Y, and

$$M_{11} = M\left[(X - m_x)\cdot(Y - m_y)\right] = K_{xy}$$

that is, *correlation moment* of random variables X and Y, that can be written as:

$$K_{xy} = \sum_i\sum_j (x_i - m_x)(y_j - m_y)p_{ij} \qquad\qquad \text{(discrete variable)}$$

$$K_{xy} = \int\limits_{-\infty}^{\infty}\int\limits_{-\infty}^{\infty}(x - m_x)(y - m_y)f_2(x,y)dxdy \qquad\qquad \text{(continuous variable)}$$

The dimensionless value $R_{xy} = \dfrac{K_{xy}}{\sigma_x\sigma_y}$ is called a *correlation coefficient* for variables X and Y.

8.1.25 The Conditional Moments

The conditional moments of the random variable X with respect to the random variable Y are defined as follows:

• *conditional mathematical expectation*:

$$m(X|y) = \int_{-\infty}^{\infty} x f_1(x|y)dx = \frac{\int_{-\infty}^{\infty} x f_2(x,y)dx}{\int_{-\infty}^{\infty} f_2(x,y)dx}$$

• *conditional variance*:

$$D(X|y) = \int_{-\infty}^{\infty} [x - M(X|y)]^2 f_1(x|y)dx$$

8.1.26 The Numerical Parameters

The basic numerical parameters of a set of n random variables $X_1, X_2, ..., X_n$ are *mathematical expectations* (means) $m(X_k) = m_k$, *variances* $D(X_k) = D_k$, $k = 1, 2, ..., n$, and *correlation moments* $K(X_i X_j) = K_{ij}$, or corresponding *correlation coefficients* $R_{ij} = K_{ij}/(\sigma_i \sigma_j)$. For the sake of convenience, the correlation moments and coefficients are typically written in a matrix form:

$$\vec{K}_{XY} = \begin{vmatrix} K_{11} & K_{12}...K_{1n} \\ K_{21} & K_{22}...K_{2n} \\ .... \\ K_{n1} & K_{n2}...K_{nn} \end{vmatrix}; \quad \vec{R}_{XY} = \begin{vmatrix} 1 & R_{12}...R_{1n} \\ R_{21} & 1...R_{2n} \\ .... \\ R_{n1} & R_{n2}...1 \end{vmatrix}$$

8.1.27 Functional Transform of Two Random Variables

If random variables X_1, X_2 have to be transformed into random variables Y_1, Y_2, and the functions

$$Y_1 = g_1(X_1, X_2) \quad Y_2 = g_2(X_1, X_2)$$

are known (g_1, g_2 are deterministic functions), probability density functions are linked as:

$$f_2\left(y_1, y_2\right) = \frac{d^2 F_2\left(y_1, y_2\right)}{dy_1\, dy_2}$$

$$f_2\left(y_1, y_2\right) = f_2\left[h_1\left(y_1, y_2\right), h_2\left(y_1, y_2\right)\right] \cdot \left|\frac{d\left(x_1, x_2\right)}{d\left(y_1, y_2\right)}\right|$$

if the reciprocal functions $X_1 = h_1\left(Y_1, Y_2\right)$, $X_2 = h_2\left(Y_1, Y_2\right)$ are single-valued. If ambiguity originates several branches h_i, the sum for each branch has to be taken in the previous formula for $f_2\left(y_1, y_2\right)$.

The common task is to find the pdf for two random variables X_1 and X_2 when the following is known:

1. Mutual pdf $f_2\left(x_1, x_2\right)$
2. Function $Y = Y_1 = g_1\left(X_1, X_2\right)$
3. Function $Y_2 = g_2\left(X_1, X_2\right) = X_2$ (or X_1)

If the reciprocal function $X_1 = h_1\left(Y_1, Y_2\right) = h\left(Y_1, Y_2\right)$ is single-valued, then:

$$f_1\left(y\right) = \int_{-\infty}^{\infty} f_2\left(y_1, y_2\right) dy_2 = \int_{-\infty}^{\infty} f_2\left[h\left(y, y_2\right), y_2\right] \cdot \left|\frac{dh\left(y, y_2\right)}{dy}\right| dy_2$$

The probability distribution functions of sum, difference, product, and quotient of two random variables are given in <u>Appendix 17</u>.

8.1.28 The Moments of Transformed Variables

The moments of the variables Z_1, Z_2 that are the functions of two random variables X, Y

$$Z_1 = g_1\left(X, Y\right) \quad Z_2 = g_2\left(X, Y\right)$$

can be found directly from mutual pdf $f_2\left(x, y\right)$:

- $m_{Z_k} = M\left(Z_k\right) = \sum_i \sum_j g_k\left(x_i, y_j\right) \cdot p_{ij}$ (discrete variable)

- $m_{Z_k} = M(Z_k) = \int\limits_{-\infty}^{\infty} \int\limits_{-\infty}^{\infty} g_k(x,y) f_2(x,y) \, dxdy$ (continuous variable)

- $D(Z_k) = \sum_i \sum_j \left[g_k(x_i,y_j) - m_{Z_k} \right]^2 \cdot p_{ij}$ (discrete variable)

- $D(Z_k) = \int\limits_{-\infty}^{\infty} \int\limits_{-\infty}^{\infty} \left[g_k(x,y) - m_{Z_k} \right]^2 \cdot f_2(x,y) \, dxdy$ (continuous variable)

- $K_{Z_1 Z_2} = \sum_i \sum_j \left[g_1(x_i,y_j) - m_{Z_1} \right]\left[g_2(x_i,y_j) - m_{Z_2} \right] \cdot p_{ij}$ (discrete variable)

- $K_{Z_1 Z_2} = \int\limits_{-\infty}^{\infty} \int\limits_{-\infty}^{\infty} \left[g_1(x,y) - m_{Z_1} \right]\left[g_2(x,y) - m_{Z_2} \right] \cdot f_2(x,y) \, dxdy$

                                                          (continuous variable)

The following formulas that are often used in practical applications can be derived based on these expressions:

1. $M(C) = C \quad M(C \cdot X) = C \cdot M(X)$

where C is a nonrandom parameter.

2. $M(X \pm Y) = M(X) \pm M(Y)$

3. $M\left(\sum\limits_{i=1}^{n} a_i X_i + b \right) = \sum\limits_{i=1}^{n} a_i \cdot M(x_i) + b$

where a_i, b are nonrandom coefficients.

4. $M(XY) = M(X) \cdot M(Y) + K_{xy}$

If X and Y are uncorrelated, $(K_{xy} = 0)$, then $M(XY) = M(X) \cdot M(Y)$.

5. $D(C) = 0$, where C is a nonrandom parameter.

6. $D(C \cdot X) = C^2 \cdot D(X) \quad \sigma(CX) = |C| \cdot \sigma(X)$

7. $D(X \pm Y) = D(X) + D(Y) \pm 2 \cdot K_{xy}$

If X and Y are uncorrelated, $(K_{xy} = 0)$, then $D(X \pm Y) = D(X) + D(Y)$.

8. $D\left(\sum\limits_{i=1}^{n} a_i \cdot X_i + b\right) = \sum\limits_{i=1}^{n} a_i^2 \cdot D(X_i) + 2\sum\limits_{i<j} a_i a_j K_{ij}$

If $X_1, X_2, ..., X_n$ are uncorrelated, $D\left(\sum\limits_{i=1}^{n} a_i X_i + b\right) = \sum\limits_{i=1}^{n} a_i^2 \cdot D(X_i)$.

9. If X and Y are independent, $D(XY) = D(X)D(Y) + m_x^2 \, D(Y) + m_y^2 \, D(X)$.

8.1.29 The Multidimensional Gaussian Distribution

The multidimensional Gaussian pdf of order n is:

$$f_1(x_1, ... x_n) = \frac{1}{\sigma_1 \sigma_2 ... \sigma_n \sqrt{(2\pi)^n \, D}} \times \exp\left\{-\frac{1}{2D} \sum_{i=1}^{n} \sum_{k=1}^{n} D_{ik} \, \frac{x_i - a_i}{\sigma_i} \cdot \frac{x_k - a_k}{\sigma_k}\right\}$$

$$D = \begin{vmatrix} 1 & R_{12}...R_{1n} \\ R_{21} & 1...R_{2n} \\ .... \\ R_{n1} & R_{n2}...1 \end{vmatrix}, \quad R_{ik} = R_{ki}$$

For the two-dimensional case ($n = 2$):

$$f_2(x_1, x_2) = \frac{1}{2\pi \, \sigma_1 \sigma_2 \sqrt{1 - R^2}}$$

$$\times \exp\left\{-\frac{1}{2(1-R^2)}\left[\frac{(x_1 - a_1)^2}{\sigma_1^2} - 2R\frac{(x_1 - a_1)(x_2 - a_2)}{\sigma_1 \sigma_2} + \frac{(x_2 - a_2)^2}{\sigma_2^2}\right]\right\}$$

8.1.30 The Modulus and Phase of a Random Vector

There is a special type of functional transform that is of major interest in electrical engineering applications:

$$Z = \sqrt{X^2 + Y^2} \quad \Phi = \arctan\left(\frac{Y}{X}\right)$$

The reciprocal transformation is:

$$X = Z \cos \Phi \quad Y = Z \sin \Phi$$

This transform is used to describe the pdf of the *modulus* Z and the *phase* Φ of a random vector with Cartesian coordinates given by the random variables X, Y. The one-dimensional probability density functions are:

$$f_1(z) = z \int_0^{2\pi} f_2(z \cos \varphi, z \sin \varphi) d\varphi, \quad z > 0$$

$$f_1(\varphi) = \int_0^{\infty} z \cdot f_2(z \cos \varphi, z \sin \varphi) dz, \quad 0 \le \varphi \le 2\pi$$

If the coordinates X and Y are independent Gaussian random variables with parameters (a, σ) and (b, σ), the modulus of the random vector follows a *Ricean* pdf:

$$f_1(z) = \frac{z}{\sigma^2} e^{-\frac{z^2 + \alpha^2}{2\sigma^2}} \cdot I_0\left(\frac{\alpha z}{\sigma^2}\right) \quad z > 0, \quad \alpha = \sqrt{a^2 + b^2}$$

where $I_0(u)$ is a Bessel function of zero order:

$$I_0(u) = \frac{1}{2\pi} \int_{-\varphi_0}^{2\pi - \varphi_0} e^{-ju \cos t} dt$$

The phase is uniformly distributed within the interval $(-\pi, \pi)$:

$$f_1(\varphi) = \frac{1}{2\pi} \quad |\varphi| \le \pi$$

For $a = b = 0$ a Ricean pdf $f_1(z)$ reduces to a *Rayleigh* pdf:

$$f_1(z) = \frac{z}{\sigma^2} \cdot e^{-z^2/2\sigma^2}$$

8.2 Random Functions

8.2.1 Random Process

A *random process* is a random function of a single argument: time.

A random process is considered to be determined within the interval $(0, T)$ if, for any n and for any moments of time $t_1, t_2, ..., t_n$, the n-dimensional probability density function $f_n(x_1, x_2, ..., x_n; t_1, t_2, ..., t_n)$ or n-dimensional characteristic function $\theta_n(j\vartheta_1, ..., j\vartheta_n; t_1, ..., t_n)$ is defined.

8.2.2 Probability Density Function

The *probability density function* of a random process must meet the following requirements:

- $f_n(x_1, ..., x_n; t_1, ..., t_n) \geq 0$

- $\int\limits_{-\infty}^{\infty} ... \int\limits_{-\infty}^{\infty} f_n(x_1, ..., x_n; t_1, ..., t_n) = 1$

- $f_n(x_1, ..., x_n; t_1, ..., t_n)$ must be symmetrical with respect to its arguments $x_1, ..., x_n$, (i.e., it should not change when the position of the arguments change);
  - For any $m < n$

$$f_m(x_1, ..., x_m; t_1, ..., t_m) = \int\limits_{-\infty}^{\infty} ... \int\limits_{-\infty}^{\infty} f_n(\xi_1, ..., \xi_m, \xi_{m+1}, ...\xi_n) d\xi_{m+1} ... d\xi_n$$

8.2.3 The Moments

The most important statistical parameters of a random process are the following moments:

- *Initial moment of the first order:*

$$m_1(t) = M[\xi(t)] = \int\limits_{-\infty}^{\infty} x f_1(x, t) dx = m_\xi(t), \qquad \text{[i.e., mathematical expectation}$$

(mean)]

- *Two-dimensional initial moment of the second order:*

$$m_{1,1}(t_1,t_2) = M\left[\xi(t_1)\cdot\xi(t_2)\right] = \int\limits_{-\infty-\infty}^{\infty\,\infty} x_1\, x_2\, f_2\left(x_1,x_2;t_1,t_2\right)dx_1\, dx_2 = B_\xi(t_1,t_2)$$

(i.e., covariation function);

- *Two-dimensional central moment of the second order:*

$$M_{1,1}(t_1,t_2) = M\left\{\left[\xi(t_1)-m_\xi(t_1)\right]\left[\xi(t_2)-m_\xi(t_2)\right]\right\}$$

$$= \int\limits_{-\infty-\infty}^{\infty\,\infty} \left[x_1 - m_\xi(t_1)\right]\times\left[x_2 - m_\xi(t_2)\right]f_2\left(x_1,x_2;t_1,t_2\right)dx_1\, dx_2 = K_\xi(t_1,t_2)$$

(i.e., correlation function);

If $t_1 = t_2 = t$, $K_\xi(t,t) = D_\xi(t)$ is the *variance*, and $\sqrt{D_\xi(t)} = \sigma_\xi(t)$ is the *root-mean-square (rms) deviation.*

8.2.4 A Complex Random Process

For a *complex random process*

$$\xi(t) = \eta(t) + j\cdot\xi(t)$$

where $\eta(t)$ and $\xi(t)$ are real random processes, the major moments are:

$$m_\xi(t) = M\left[\xi(t)\right] = m_\eta(t) + jm_\xi(t)$$

$$B_\xi(t_1,t_2) = M\left[\xi(t_1)\cdot\xi^*(t_2)\right] = K_\xi(t_1,t_2) + m_\xi(t_1)\cdot m_\xi^*(t_2)$$

$$K_\xi(t_1,t_2) = M\left\{\left[\xi(t_1)-m_\xi(t_1)\right]\left[\xi^*(t_2)-m_\xi(t_2)\right]\right\} = B_\xi(t_1,t_2) - m_\xi(t_1)\cdot m_\xi^*(t_2)$$

where the process $\eta^*(t)$ is the complex-conjugate to $\eta(t)$.

8.2.5 A Stationary Random Process

A *stationary random process* is a random process having statistical parameters that are invariant with respect to time shift.

For a stationary random process:

$$m_\xi(t) = M\left[\xi(t)\right] = \int_{-\infty}^{\infty} x\, f_1(x)\, dx = m_\xi$$

$$B(t_1, t_2) = M\left[\xi(t_1)\xi(t_2)\right]$$

$$= \int_{-\infty}^{\infty}\int_{-\infty}^{\infty} x_1 x_2 f_2(x_1, x_2)\, dx_1\, dx_2 = K(t_2 - t_1) + m_\xi^2 = B_\xi(\tau) \quad \tau = t_2 - t_1$$

$$K(t_1, t_2) = M\left\{\left[\xi(t_1) - m_\xi\right]\left[\xi(t_2) - m_\xi\right]\right\}$$

$$= \int_{-\infty}^{\infty}\int_{-\infty}^{\infty} (x_1 - m_\xi)(x_2 - m_\xi) f_2(x_1, x_2)\, dx_1\, dx_2 = B_\xi(\tau) - m_\xi^2 = K_\xi(\tau)$$

The correlation function (sometimes also termed the *autocorrelation function*) of the stationary random processes $\xi(t)$ has the following properties:

- $K_\xi(\tau) = K_\xi(-\tau)$

- $\left|K_\xi(\tau)\right| \le K_\xi(0) = D_\xi$

- $\lim\limits_{\tau \to \infty} K_\xi(\tau) = 0$

The *normalized correlation function* is

$$R_\xi(\tau) = \frac{K_\xi(\tau)}{K_\xi(0)} = \frac{K_\xi(\tau)}{D_\xi} \le 1$$

or, in other terms

$$K_\xi(\tau) = D_\xi \cdot R_\xi(\tau).$$

8.2.6 The Mutual Correlation Function

The *mutual correlation function* describes the statistical linkage between two random processes $\xi(t)$ and $\eta(t)$:

$$K_{\xi\eta}(t_1,t_2)= M\left\{[\xi(t_1)-m_\xi][\eta(t_2)-m_\eta]\right\}$$

$$K_{\eta\xi}(t_1,t_2)= M\left\{[\eta(t_1)-m_\eta][\xi(t_2)-m_\xi]\right\}$$

If mutual correlation function only depends on the difference $\tau = t_2 - t_1$, the processes $\xi(t)$ and $\eta(t)$ are termed *stationary-linked* ones, and in this case:

$$K_{\xi\eta}(t_1,t_2)= K_{\xi\eta}(\tau)= K_{\eta\xi}(-\tau)$$

$$R_{\xi\eta}(\tau)= \frac{K_{\xi\eta}(\tau)}{\sqrt{D_\xi D_\eta}} = \frac{K_{\xi\eta}(\tau)}{\sigma_\xi \sigma_\eta}$$

8.2.7 Correlation Interval

Typically, in electrical engineering applications, the normalized correlation functions are monotonous damping or oscillating-damping functions of τ. The degree of correlation in this case is described by the *correlation interval* τ_c:

$$\tau_c = \frac{1}{2}\int_{-\infty}^{\infty}|R(\tau)|\,d\tau = \int_0^{\infty}|R(\tau)|\,d\tau$$

Sometimes τ_c is defined simply as the time shift for which the normalized correlation function drops below some specified value:

$$|R(\tau)|\le R_0, \quad \text{e.g.,} \quad \frac{|R(\tau_c)|}{R(0)}\le 0.1$$

8.2.8 The Spectral Density

The *spectral density* (power spectrum) $S_\xi(\omega)$ of the stationary random process $\xi(t)$ and its covariation function $B_\xi(\tau)$ are linked by a Fourier transform pair:

$$S_\xi(\omega) = \int\limits_{-\infty}^{\infty} B_\xi(\tau) e^{-j\omega\tau} \, d\tau$$

$$B_\xi(\tau) = \frac{1}{2\pi} \int\limits_{-\infty}^{\infty} S_\xi(\omega) e^{j\omega\tau} \, d\omega$$

Correspondingly, for the centered random process $\xi^0(t) = \xi(t) - m_\xi$

$$S_{\xi^0}(\omega) = \int\limits_{-\infty}^{\infty} K_\xi(\tau) e^{-j\omega\tau} \, d\tau = S_\xi(\omega) - 2\pi m_\xi^2 \delta(\omega)$$

$$K_\xi(\tau) = \frac{1}{2\pi} \int\limits_{-\infty}^{\infty} S_{\xi^0}(\omega) e^{j\omega\tau} \, d\omega$$

Because $B(\tau)$ and $K(\tau)$ are the even functions of argument τ :

$$S_\xi(\omega) = 2 \int\limits_{0}^{\infty} B_\xi(\tau)\cos\, \omega\tau d\tau \; ; \quad S_{\xi^0}(\omega) = 2 \int\limits_{0}^{\infty} K_\xi(\tau)\cos\, \omega\tau d\tau$$

$$B_\xi(\tau) = \frac{1}{\pi} \int\limits_{0}^{\infty} S_\xi(\omega)\cos\, \omega\tau d\omega \; ; \quad K_\xi(\tau) = \frac{1}{\pi} \int\limits_{0}^{\infty} S_{\xi^0}(\omega)\cos\, \omega\tau d\omega$$

Because $S_{\xi^0}(\omega)$ and $S_\xi(\omega)$ differ only in the constant factor $2\pi m_\xi^2 \cdot \delta(\omega)$, hereinafter we will use only the $S(\omega) = S_\xi(\omega)$ representation $[S_{\xi^0}(\omega) = S_\xi(\omega)$ if $m_\xi = 0]$. The main features of the spectral density of the stationary random process are as follows:

- $S(\omega) \geq 0$ for any ω ;

- for real random processes $S(\omega) = S(-\omega)$

Sometimes, instead of the double-sided spectrum, the one-sided "physical" spectral density is used:

$$S(f) = S(\omega) + S(-\omega) = \begin{cases} 2S(\omega) & f \geq 0 \\ 0 & f < 0 \end{cases}$$

which gives the expressions:

$$S(f) = 4 \int_0^\infty B(\tau)\cos 2\pi f\tau\, d\tau \; ; \quad B(\tau) = \int_0^\infty S(f)\cos 2\pi f\tau\, df$$

Note. Using the concept of the spectral density, it is very important to make sure that the proper normalization is used.

The most common numerical parameters of the spectrum are:

- *Effective spectrum width:* $\Delta\omega_e = \dfrac{1}{S(0)} \displaystyle\int_{-\infty}^{\infty} S(\omega)\, d\omega$

- *Mean frequency:* $m_{1\omega} = M\,[\omega] = \dfrac{2}{\sigma^2} \displaystyle\int_0^\infty \omega S(\omega)\, d\omega$

- *Mean-square frequency:* $m_{2\omega} = M\,[\omega^2] = \dfrac{2}{\sigma^2} \displaystyle\int_0^\infty \omega^2 S(\omega)\, d\omega$

- *Root-mean-square width:*

$$\sigma_\omega = \sqrt{M\,[\omega^2] - M^2[\omega]} = \sqrt{\dfrac{2}{\sigma^2} \int_0^\infty (\omega - m_\omega)^2\, S(\omega)\, d\omega}$$

The correspondence between some typical correlation functions and its spectral densities is given in Appendix 19.

8.2.9 A Derivative of a Random Process

The random process $\xi(t)$ has a *derivative* $\dot\xi(t) = d\xi(t)/dt$ if

$$\lim_{T\to 0} M\left\{ \left[\frac{\xi(t+T) - \xi(t)}{T} - \dot\xi(t) \right]^2 \right\} = 0$$

The *mathematical expectation* $m_{\dot{\xi}}(t)$, *covariation* $B_{\dot{\xi}}(t_1,t_2)$, and *correlation function* $K_{\dot{\xi}}(t_1,t_2)$ are defined as:

$$m_{\dot{\xi}}(t) = M\left[\dot{\xi}(t)\right] = \frac{dm_{\xi}(t)}{dt}$$

$$B_{\dot{\xi}}(t_1,t_2) = M\left[\dot{\xi}(t_1)\cdot\dot{\xi}(t_2)\right] = \frac{\partial^2 B_{\xi}(t_1,t_2)}{\partial t_1 \partial t_2}$$

$$K_{\dot{\xi}}(t_1,t_2) = M\left[\dot{\xi}^0(t_1)\cdot\dot{\xi}^0(t_2)\right] = \frac{\partial^2 K_{\xi}(t_1,t_2)}{\partial t_1 \partial t_2}$$

For the *stationary random process*:

$$m_{\dot{\xi}}(t) = 0$$

$$B_{\dot{\xi}}(\tau) = K_{\dot{\xi}}(\tau) = -\frac{d^2 B_{\xi}(\tau)}{d\tau^2} = -\frac{d^2 K_{\xi}(\tau)}{d\tau^2}$$

$$S_{\dot{\xi}}(\omega) = \omega^2 \cdot S_{\xi^0}(\omega)$$

The *mutual covariation* $B_{\xi\dot{\xi}}(t_1,t_2)$ and *correlation* $K_{\xi\dot{\xi}}(t_1,t_2)$ functions are:

$$B_{\xi\dot{\xi}}(t_1,t_2) = M\left[\xi(t_1)\cdot\dot{\xi}(t_2)\right] = \frac{dB_{\xi}(t_1,t_2)}{dt_2}$$

$$K_{\xi\dot{\xi}}(t_1,t_2) = M\left[\xi^0(t_1)\cdot\dot{\xi}^0(t_2)\right] = \frac{dK_{\xi}(t_1,t_2)}{dt_2}$$

For the *stationary random process*:

$$B_{\xi\dot{\xi}}(\tau) = K_{\xi\dot{\xi}}(\tau) = -B_{\dot{\xi}\xi}(\tau) = -K_{\dot{\xi}\xi}(\tau) = \frac{dB_{\xi}(\tau)}{d\tau} = \frac{dK_{\xi}(\tau)}{d\tau}$$

$$S_{\xi\dot{\xi}}(\omega) = j\omega S_{\xi^0}(\omega)$$

Because $dK_\xi(0)/d\tau = 0$, the stationary process $\xi(t)$, and its derivative $\dot\xi(t)$, are always uncorrelated in any coincident moments of time $(\tau = 0)$.

The general expressions for derivatives of the nth order $\xi^{(n)}(t)$ are given in Appendix 20.

8.2.10 An Integral of a Random Process

A random function $\xi(t)$ can be integrated, that is, the random function

$$\eta(t) = \int_0^t \xi(u)\,du$$

exists, if its correlation function can be integrated:

$$K_\eta(t_1, t_2) = \int_0^{t_1}\int_0^{t_2} K_\xi(u, 9)\,du\,d9$$

For the stationary random process $\xi(t)$:

$$K_\eta(t_1, t_2) = \int_0^{t_1}\int_0^{t_2} K_\xi(9 - u)\,du\,d9$$

$$D_\eta(t) = \int_0^t (t - \tau)\big[K_\xi(\tau) + K_\xi(-\tau)\big]\,d\tau$$

These expressions show that in the general case, $\eta(t)$ is not necessarily a stationary process even when $\xi(t)$ is a stationary one. For the real stationary random process $\xi(t)$:

$$D_\eta(t) = 2\int_0^t (t - \tau)K_\xi(\tau)\,d\tau$$

For any random function $\xi(t)$:

$$m_\eta(t) = \int\limits_0^t m_\xi(u)\,du$$

8.2.11 A Gaussian Random Process

A *Gaussian* random process is the most common process used in electrical engineering. The probability density function and characteristic function for this process are given in <u>Appendix 18</u>. There, Δ is the determinant of the *n*th order:

$$\Delta = \left\| K_\xi\big(t_\mu, t_\gamma\big) \right\|$$

where $\Delta_{\mu\gamma}$ is a cofactor of an element $K_\xi\big(t_\mu, t_\gamma\big)$ in this determinant, and $m_\xi\big(t_\mu\big)$ is a mathematical expectation of the random variable $\xi_\mu = \xi\big(t_\mu\big)$. For a stationary process, *m* is the mathematical expectation, $\sigma^2 = D = R(0)$ is a variance of the process $\xi(t)$, and $D = \left\| R\big(\tau_{\mu\gamma}\big) \right\|$ is a determinant of the order *n* composed from the correlation coefficients

$$R\big(\tau_{\mu\gamma}\big) = R\big(\left|t_\mu - t_\gamma\right|\big) = \frac{K_\xi\big(\left|t_\mu - t_\gamma\right|\big)}{\sigma^2}.$$

8.2.12 A Markovian Random Process

The fundamental role in the simulation of the random processes belongs to a *Markovian* random process. There are four basic types of Markovian processes, depending on whether the random variable $\xi(t)$ and its argument *t* are discrete or continuous: the discrete process and discrete time (*Markovian chain*), the continuous process and discrete time (*Markovian sequence*), discrete process and continuous time (*discrete Markovian process*), and continuous process and continuous time (*continuous Markovian process*).

The random process is called a Markovian one if, for any *n* moments of time $t_1 < t_2 <, ..., t_n$ at the interval (0, T), the conditional distribution function of $\xi(t_n)$ depends only on $\xi(t_{n-1})$ under the conditions of fixed $\xi(t_1), \xi(t_2), ... \xi(t_{n-1})$, that is,

$$P\left\{ \xi(t_n) \le \xi_n \,\middle|\, \xi(t_1) = \xi_1, ... \xi(t_{n-1}) = \xi_{n-1} \right\} = P\left\{ \xi(t_n) \le \xi_n \,\middle|\, \xi(t_{n-1}) = \xi_{n-1} \right\}$$

In other words, the status of the Markovian random process in the future (t_{n+1}) depends only on its status in the present (t_n) and does not depend on the status in the past (t_{n-1}). This assumption makes it possible to reduce considerably the number of computations in simulation algorithms.

In practical applications, the process is considered to be the Markovian one if the three following conditions are met:

- The process is a stationary one.
- The process is a Gaussian one.
- Spectral density of the process is a fractional-rational function of frequency:

$$S(\omega) = \frac{|P_m(j\omega)|^2}{|Q_n(j\omega)|^2}, \quad m < n$$

where P_m, Q_n are some polynomials:

$$P_m(u) = \beta_0 u^m + \beta_1 u^{m-1} + \ldots + \beta_m$$

$$Q_n(u) = \alpha_0 u^n + \alpha_1 u^{n-1} + \ldots + \alpha_n$$

and α_i, β_i are some real coefficients.

The random process $\xi(t)$ that is the stationary Gaussian random process with the correlation function

$$K_\xi(t) = \sigma^2 \cdot e^{-\alpha|\tau|}$$

is always a Markovian one.

8.2.13 A Random Flow

Another important class of random processes is a point random process (a *random flow*).

A *point random process* is a sequence of random events (requests, failures, and so forth) with the random moments $t_1, t_2, t_3 \ldots$ of its occurrence.

There are two basic ways to describe a point random process. The first considers a random sequence of points in time. The second considers the integer random process $N(t)$ that is the number of events (points) at the half-interval (0, t]. The sequence of points can be described by n-dimensional pdf $f_n(\tau_1, \ldots, \tau_n)$ of intervals between the points:

$$\tau_i = t_i - t_{i-1} > 0, \quad i = 1, 2, 3, \ldots$$

or distribution function $F_n(t_1,...,t_n)$ of the coordinates of the points:

$$F_n(t_1,...t_n) = f_n(t_1, t_2 - t_1,..., t_n - t_{n-1})$$

$$f_n(\tau_1, \tau_2,..., \tau_n) = F_n(\tau_1, \tau_1 + \tau_2,..., \tau_1 + \tau_2 + ...\tau_n)$$

For the process $N(t)$, if t_i is a coordinate of the ith point that occurred $t_1 < t_2 <...t_i <...$, then $N(t) = 0$ only if $\tau_i > t$, and $N(t) < n$ if $\tau_1 + \tau_2 + ...\tau_n > t$:

$$P\{N(t) = 0\} = P\{\tau_1 > t\} \quad P\{N(t) < n\} = P\{\tau_1 + \tau_2 + ...\tau_n > t\}, \quad n = 1,2,3,...$$

8.2.14 Poisson Flow

Poisson random flow $N(t)$, $0 < t < \infty$, has three basic features:
• The flow is an ordinary one, that is, the probability that more than one event occurs at any small interval Δt is of a higher order being infinitesimal than Δt :

$$P\{N(t + \Delta t) - N(t) = 1\} = P\{N(\Delta t) = 1\} = \gamma \cdot \Delta t + o(\Delta t)$$

$$P\{N(t + \Delta t) - N(t) > 1\} = P\{N(\Delta t) > 1\} = o(\Delta t)$$

where γ is a positive value with a dimension reciprocal to that of time, and $o(t)$ are factors of the series at least an order less than Δt . Thus,

$$P\{N(t + \Delta t) - N(t) = 0\} = P\{N(\Delta t) = 0\} = 1 - \gamma \Delta t + o(\Delta t)$$

• The flow is a stationary one; that is, its probabilistic characteristics do not change when all points are shifted along the time axis by an arbitrary value Δ .
• The flow increments are independent at nonoverlapping time intervals (the absence of aftereffects).
For these conditions, the probability $P_k(t)$ that k points will occur at the half-interval (0, t] is covered by Poisson law:

$$P_k(t) = (\gamma t)^k \cdot \frac{e^{-\gamma t}}{k!}, \quad \begin{matrix} k = 0,1,2,... \\ t \geq 0 \end{matrix}$$

The probability that there are no points ($k = 0$) at some half-interval τ is

$$P_0(\tau) = e^{-\gamma \tau}$$

and that there is only one point ($k = 1$) is

$$P_1(\tau) = \gamma \tau e^{-\gamma \tau}$$

The mathematical expectation m and variance D are equal to each other:

$$m = D = \gamma t$$

Because $\gamma = m/t$, it can be considered to be the average number of points at the unity interval, and is termed a *flow intensity*. The probability distribution function is:

$$f_{tk}(t) = \gamma e^{-\gamma t} \cdot (\gamma t)^{k-1} \cdot \frac{1}{(k-1)!}; \quad m_{tk} = \frac{k}{\gamma} \quad D_{tk} = \frac{k}{\gamma^2}$$

where t_k is the time of kth point occurrence.

The sequence $\tau_i > i = 1,2,3,...$ of the time intervals between adjacent points of Poisson flow is a sequence of independent random variables with exponential pdf:

$$f(\tau) = \gamma \cdot e^{-\gamma \tau}, \quad \tau > 0; \quad m_\tau = 1/\gamma, \quad D_\tau = 1/\gamma^2$$

8.2.15 A Random Field

M-dimensional *random field* is the random function $\xi(\vec{x})$ of m variables: $x_1, x_2,...x_m$.

A one-dimensional random field $\xi(\vec{x})$ is called a *scalar random field*, and an n-dimensional random field $\vec{\xi}(\vec{x}) = [\xi_1(\vec{x}),...\xi_n(\vec{x})]$ is called a *vector random field*. The parameter $\vec{x}$ is called an argument of the random field. In electrical engineering applications the typical case is the random field $\xi(x,y,z,t)$, which is represented as a function of four arguments: x, y, z are three-dimensional Cartesian coordinates of some point in space, and t is time (e.g., in case of phased arrays simulation). When a scalar random field is the function of a single argument that is time t, it is reduced to a *random process*.

8.2.16 Main Characteristics of a Random Field

The main characteristics to describe a vector random field $\vec{\xi}(\vec{x}) = [\xi_1(\vec{x}),...\xi_n(\vec{x})]$ are:

• The multidimensional probability density function is the distribution of the $n \times k$-dimensional vector $\eta = [\xi(\vec{x}_1),...\xi(\vec{x}_k)]$. If the pdf does not depend on the variation of an argument $\Delta \vec{x}$:

$$f_k(\vec{y}_1,...,\vec{y}_k;\vec{x}_1 + \Delta\vec{x},...,\vec{x}_k + \Delta\vec{x}) = f_k(\vec{y}_1,...,\vec{y}_k;\vec{x}_1,...,\vec{x}_k)$$

the random field is called an *isotropic* one.
• The moments are:

$$m(\vec{x}) = M[\vec{\xi}(\vec{x})] = (m_1(\vec{x}),...m_n(\vec{x})) \text{ is a mathematical expectation,}$$

where $m_i(\vec{x}) = M[\xi_i(\vec{x})]$;

$$K_0(\vec{x}_1,\vec{x}_2) = M[\xi(\vec{x}_1) \cdot \xi^T(\vec{x}_2)] \text{ is a covariation function;}$$

$$K(\vec{x},\vec{x}_2) = M\{[\xi(\vec{x}_1) - m(\vec{x}_1)][\xi(\vec{x}_2) - m(\vec{x}_2)]^T\} \text{ is a correlation function.}$$

The vector random field is called an *isotropic in a broad sense* if

$$m(\vec{x}) = m = const, \quad K(\vec{x} + \vec{y}, \vec{y}) = K(\vec{x})$$

that is, the mean is constant and the correlation function depends only upon the difference of arguments. For an isotropic field:

$$K(\vec{x}) = K^T(-\vec{x}) \quad K_{ij}(\vec{x}) = K_{ji}(-\vec{x})$$

$$K_{ij}(\vec{x}) \le \sqrt{K_{ii}(0)K_{jj}(0)}$$

• The spectral density and the correlation function are linked as:

$$S(\vec{u}) = \frac{1}{(2\pi)^m} \int_{R^m} K(\vec{x})e^{-j\vec{u}^T\vec{x}}d\vec{x} \ ; \quad K(\vec{x}) = \int_{R^m} S(\vec{u})e^{j\vec{u}^T\vec{x}}d\vec{u}$$

For each fixed $\vec{u}$:

$$S(\vec{u}) = S^*(\vec{u}) = \overline{S}^T(\vec{u}); \quad S_{kl}(\vec{u}) = \overline{S}_{lk}(\vec{u})$$

where $\bar{u}$ denotes the conjugation operation with respect to u. This leads to the conditions:

$$S(\bar{u}) = \overline{S}(-\bar{u})$$

$$\vec{z}^T S(\bar{u}) \cdot \overline{\vec{z}} = \sum_{k,l=1}^{n} z_k S_{kl}(\bar{u}) z_l \geq 0$$

for any complex vectors $\vec{z} = (z_1, ..., z_n)$.

8.2.17 Main Characteristics of a Scalar Random Field

The main statistical characteristics to describe a *scalar random field* are:
• The N-dimensional probability density function is composed from n sections of the field in the points $\vec{x}_1, \vec{x}_2 ... \vec{x}_n$; that is, the pdf of the random vector $[\xi(x_1), ..., \xi(x_n)]$ is:

$$f_n(y_1, ..., y_n; \vec{x}_1, ..., \vec{x}_n)$$

The scalar random field is called an *isotropic* one if it is invariant to argument shift; that is, for any shift of argument $\Delta \vec{x}$ and any n:

$$f_n(y_1, ... y_n; \vec{x}_1 + \Delta \vec{x}_1, ... \vec{x}_n + \Delta \vec{x}_n) = f_n(y_1, ... y_n; x_1, ... x_n)$$

• The one-dimensional probability density function is the probability distribution of $\xi(\vec{x})$ when $\vec{x}$ is fixed. The one-dimensional pdf of an isotropic field $f_1(y; \vec{x})$ does not depend on argument x.
• The moments are:

$m(\vec{x}) = M[\xi(\vec{x})]$ is a mathematical expectation;

$\sigma^2[\xi(\vec{x})] = M[\xi(\vec{x}) - m(\vec{x})]^2$ is a variance;

$K_0(\vec{x}_1 \vec{x}_2) = M[\xi(\vec{x}_1) \cdot \xi(\vec{x}_2)]$ is a covariation function;

$K(\vec{x}_1, \vec{x}_2) = M\{[\xi(\vec{x}_1) - m(\vec{x}_1)] \cdot [\xi(\vec{x}_2) - m(\vec{x}_2)]\}$ is a correlation function.

In practical tasks, the random field typically is defined by its correlation function. For a random field that is isotropic in a broad sense:

$$m(\vec{x}) = m = \text{const}$$

$$K(\vec{x}_1, \vec{x}_2) = K(\vec{x}), \quad \vec{x} = \vec{x}_1 - \vec{x}_2$$

The random field is called *Gaussian* (normal) if all multidimensional probability distribution functions are described by Gaussian distribution law.

• The spectral density of the scalar random isotropic field is:

$$S(\vec{u}) = \frac{1}{(2\pi)^m} \int_{R^m} K(\vec{x}) e^{-j\vec{u}^T \vec{x}} d\vec{x}$$

where $\vec{u}^T \vec{x} = \sum_{i=1}^{m} u_i x_i$, R^m is the range of existence for argument $\vec{x}$.

For a two-dimensional isotropic field $\xi(\vec{x})$, $\vec{x} = (x, y)$, the spectral density and correlation function are a Hankel transform pair:

$$S(u) = \frac{1}{2\pi} \int_0^\infty J_0(|u| \cdot \rho) \cdot K(\rho) \cdot \rho \cdot d\rho$$

$$K(\rho) = 2\pi \int_0^\infty S(u) \cdot J_0(\rho u) du$$

where $\rho = \sqrt{x^2 + y^2}$; and

$$J_0(z) = \frac{1}{2\pi} \int_0^\infty e^{-iz \cos t} dt \quad \text{is a Bessel function of zero order.}$$

8.3 Stochastic Simulation

8.3.1 Basic Definitions

In electrical engineering, the majority of stochastic simulation tasks deals with reconstructing samples $z_1, ... z_N$ of a random variable $\xi = \{z_n\}$, or with samples $z(t_1), ... z(t_N)$ of a random process $\xi(t) = \{z(t_n) = z_n\}$ with specified probability distribution laws. Sometimes this procedure is called a *statistical synthesis*.

The *uniform* distribution is the basic distribution used to generate random variables with different probability distribution functions. Most modern programming packages have system-supplied library routines to generate the random variable $\xi_u = \{z_u = u\}$ uniformly distributed within the range (0,1). The algorithms to simulate random variables with other distribution laws are typically based on the functional transformation of $z_u = u$.

8.3.2 How to Simulate Random Variables

In order to simulate a random variable with the specified distribution law, the following steps are to be followed.

Step 1. Simulate the sample u of the random variable ξ_u uniformly distributed within the range (0,1).

Step 2. Transform the sample u into the sample z of the random variable ξ with required distribution law based on the mathematical function:

$$z = f(u)$$

Sometimes two or more independent samples $z1_u$, $z2_u$, ... of the variable ξ_u have to be generated to obtain a single sample z of the variable ξ.

Simulation algorithms and transform functions $z = f(u)$ for the common distribution laws are given in Appendix 21.

Figure 8.2 shows block diagrams illustrating the simulation procedure for the random variables with the following distribution laws:

Gaussian pdf $f(z) = \dfrac{1}{\sqrt{2\pi}\,\sigma_z} \cdot \exp\left[-\dfrac{(z - m_z)^2}{2\sigma_z^2} \right]$

Rayleigh pdf $f(z) = \dfrac{z}{\sigma_z} \cdot \exp\left(-\dfrac{z^2}{2\sigma_z^2} \right)$ $z \geq 0$

Ricean pdf $f(z) = \dfrac{z}{\sigma_z} \cdot \exp\left(-\dfrac{z^2 + a^2}{2\sigma_z^2} \right) \cdot I_0(\dfrac{az}{\sigma_z^2})$ $z \geq 0$

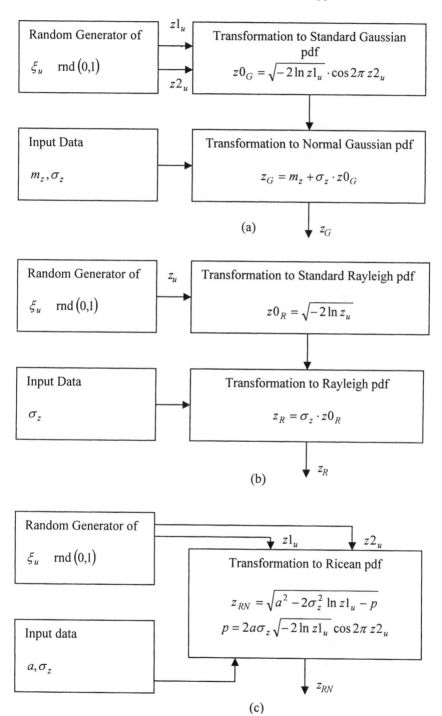

Figure 8.2 (a) Simulation procedures for the random variables with Gaussian, (b) Rayleigh, and (c) Ricean distribution laws.

Examples of how to simulate the random variables with specified distribution functions are given in Chapter 13.

8.3.3 How to Simulate an Uncorrelated Random Process

The simulation of the uncorrelated random process (sometimes called delta-correlated, implying that its correlation function is the delta function) is virtually a simulation of the set of the independent random variables samples with the specified distribution law.

Let us assume we want to simulate the Gaussian random process $\xi_G(t) = \{zG\}$ in 10 uniformly spaced moments of time $t_1, t_2, ..., t_{10}$, and its correlation interval is less than the interval between moments to obtain the samples $\Delta t = t_{n+1} - t_n$; thus, the process is uncorrelated in these moments of time. In this case, based on the procedure cited in Section 8.3.2, we just obtain 10 successive samples zG as shown in Figure 8.2(a) which represent the samples of the Gaussian process in 10 moments of time.

Note. This approach is valid both for stationary and nonstationary processes. For a *stationary* process, basic parameters such as mean $m_\xi(t) = m_\xi$ and variance $D_\xi(t) = \sigma^2$ are time-independent, and thus can be kept constant during simulation of all N samples required. For a *nonstationary* process, these parameters are functions of time and thus have to be varied for each sample obtained [i.e., $m_\xi(t_1), D_\xi(t_1)$, then $m_\xi(t_2)$, $D_\xi(t_2)$, and so forth].

Examples of how to simulate the uncorrelated random processes are given in Chapter 9.

8.3.4 How to Simulate a Correlated Random Process

To simulate the correlated random process there is another variable that has to be taken into account in addition to the mean $m_\xi(t)$ and the variance $D_\xi(t)$. This variable is the correlation function $K_\xi(t_1, t_2)$. The following typical situations can be considered.

- *A Stationary Gaussian Random Process*

This is the most common assumption in electrical engineering tasks, and two parameters — mathematical expectation m_ξ and correlation function $K_\xi(\tau)$— are sufficient to describe such a process $\xi(t)$ completely. Typical techniques to simulate the stationary Gaussian process are as follows.

Step 1. For the process of interest $\xi(t)$, the analytical formula for the correlation function $K_\xi(\tau)$ is derived. Typically, in electrical engineering the correlation functions are approximated as

$$K_\xi(\tau) = \sum_{j=1}^{n} \left[A_j(\tau) \cdot \cos \beta_j \tau + B_j \cdot \sin \left| \beta_j \right| \tau \right] \cdot e^{-\alpha_j |\tau|}$$

Depending on the set of the parameters chosen, this approximation results in typical correlation functions (see <u>Appendix 19</u>).

Step 2. The samples of the process $\xi(t) = \{z(t_0), z(t_1), ..., z(t_{N-1})\}$ in the uniformly spaced moments t_n, $n = 0...N-1$ are simulated based on the recurrent algorithm:

$$z_n = \sum_{j=0}^{l} a_j \cdot z0G_{n-j} - \sum_{j=1}^{m} b_j \cdot z_{n-j}, \quad n = m, m+1, ...$$

Hereinafter, $z_n = z(t_n)$ is the sample of the simulated process in the moment t_n, $z0G \subset N(0,1)$ is the random variable with standard Gaussian pdf ($m = 0$, $\sigma = 1$); l, m, a_j, b_j are parameters defined by the correlation function $K_\xi(\tau)$.

Concrete simulation algorithms for the stationary Gaussian random processes with typical correlation functions are given in <u>Appendix 21</u>.

The block diagram illustrating how to obtain the first three samples of the stationary Gaussian random process with the most common correlation function $K(\tau) = \sigma_\xi^2 \cdot e^{-\alpha |\tau|}$ is given in Figure 8.3, $\Delta t = t_{n+1} - t_n$.

Examples of how to simulate a stationary Gaussian random process are given in Chapters 9 and 13.

- *A Stationary Markovian Gaussian Random Process*

If the stationary Gaussian process is Markovian (see Section 8.2.12), then its status at the future time, or in other words the status in *future* t_{n+1}, depends only on the status in *present* t_n, and does not depend on the status in the *past* t_{n-1}, t_{n-2}, and so forth. This is a very convenient feature for simulation that allows reduced amount of computations.

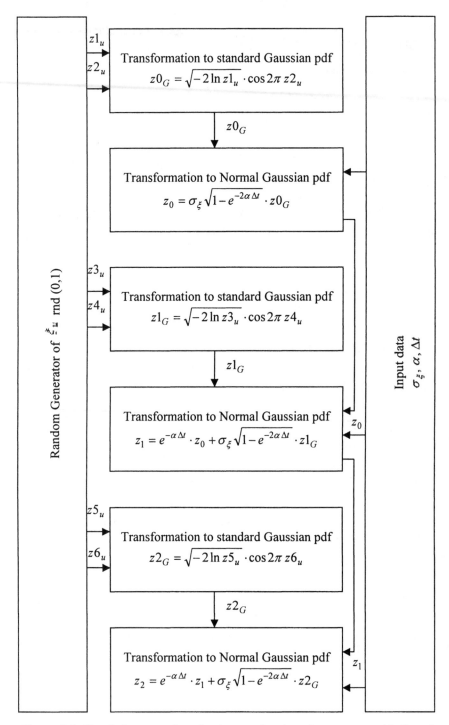

Figure 8.3 Simulation procedure for the correlated random process with Gaussian distribution law.

The algorithm to simulate a three-dimensional Markovian process $\vec{\xi}(t) = \left|\xi_1(t), \xi_2(t), \xi_3(t)\right|$ with a correlation function

$$\vec{K}_\xi(\tau) = \begin{vmatrix} D_1 & \sqrt{D_1 D_2} \cdot \rho_{12} & \sqrt{D_1 D_3}\, \rho_{13} \\ \sqrt{D_2 D_1}\, \rho_{12} & D_2 & \sqrt{D_2 D_3}\, \rho_{23} \\ \sqrt{D_3 D_1}\, \rho_{13} & \sqrt{D_3 D_2}\, \rho_{23} & D_3 \end{vmatrix} \cdot f(\tau)$$

is given in <u>Appendix 21</u>. Here $f(\tau)$ is an arbitrary function of time shift.

The two-dimensional process can be simulated using the first two components $\xi_1(t), \xi_2(t)$, and a scalar stationary Gaussian Markovian process $\xi_1(t)$ can be simulated using the first component only.

Correlation function of the process $\xi_1(t)$ is often approximated as:

$$K_1(\tau) = D \cdot f(\tau) = D \cdot e^{-\alpha|\tau|} = \sigma^2 \cdot e^{-\alpha|\tau|}$$

which gives the following simple algorithm:

$$z_{n+1} = z_n \cdot e^{-\alpha|\tau|} + \sigma \sqrt{1 - e^{-2\alpha|\tau|}} \cdot z0G_{n+1}$$

The initial sample in the moment $n = 0$ is simulated as:

$$z_0 = \sigma \cdot z0G_0$$

$$z_1 = z_0 \cdot e^{-\alpha|\tau|} + \sigma \sqrt{1 - e^{-2\alpha|\tau|}} \cdot z0G_1 \text{ , and so forth.}$$

The stationary Gaussian process with correlation function $K_1(\tau)$ is always a Markovian one.

An example of how to simulate the Markovian random process is given in Chapter 9.

• *A Stationary Non-Gaussian Random Process*

Typically, simulation of a non-Gaussian random process is a much more complicated and nontrivial task. The best developed approach is the method of nonlinear transformation that allows simulation of a non-Gaussian process $\xi(t)$ with the specified one-dimensional pdf $f_1(x)$ and autocorrelation function $K_\xi(\tau)$. Generally speaking, in opposition to the model of a Gaussian process, the model of a non-Gaussian process is not completely defined by $f_1(x)$ and $K_\xi(\tau)$ only. There might be several random processes with different multidimensional pdfs but identical parameters $f_1(x)$ and $K_\xi(\tau)$. Typically, only one process that is more convenient for simulation will be chosen from the set of processes providing the specified one-dimensional probability density function and correlation function.

The basic principles of the method of nonlinear transformation are as follows. A stationary Gaussian process $\varepsilon(t) = \{zG(t)\}$ is chosen to be the initial one. There is always the nonlinear transformation

$$x = \varphi(zG)$$

that transforms the normal pdf $N(m, D)$ associated with $\varepsilon(t)$ into specified pdf $f_1(x)$. If the Gaussian process has correlation function $K_\varepsilon(\tau)$, then some dependence φ_K exists:

$$K_\xi(\tau) = \varphi_K\left[K_\varepsilon(\tau)\right]$$

defined by dependence $x = \varphi(zG)$.

To obtain a required correlation function, one should choose an initial correlation function as:

$$K_\varepsilon(\tau) = \varphi_K^{-1}\left[K_\xi(\tau)\right]$$

where φ_K^{-1} is a function reciprocal to φ_K. The basic steps to implement this method are:

Step 1. Determine the appropriate transformation dependence $x = \varphi(zG)$ based on the specified pdf $f_1(x)$.

Step 2. Determine the dependence

$$K_\xi(\tau) = \varphi_K\left[K_\varepsilon(\tau)\right]$$

based on the function $\varphi(zG)$ that satisfies the equation:

$$K_\xi(\tau) = \frac{1}{2\pi D \sqrt{1 - \left[\dfrac{K_\varepsilon(\tau)}{D}\right]^2}} \int\limits_{-\infty}^{\infty}\int\limits_{-\infty}^{\infty} \varphi(z_1)\varphi(z_2)\cdot\exp(Q)\,dz_1\,dz_2 - m^2$$

where

$$Q = \cfrac{1}{2\left\{1 - \left[\cfrac{K_\varepsilon(\tau)}{D}\right]^2\right\}}$$

$$\times \cfrac{(z_1 - m)^2 + 2\cfrac{K_\varepsilon(\tau)}{D}(z_1 - m)(z_2 - m) + (z_2 - m)^2}{D}$$

Step 3. A computer is used to solve this equation with respect to parameter $K_\varepsilon(\tau)$ in order to determine the correlation function of the initial normal process $\varepsilon(t)$.

Step 4. Simulate the desired non-Gaussian process by simulating the Gaussian process $\varepsilon(t) = \{zG(t)\}$ with the estimated correlation function $K_\varepsilon(\tau)$ and obtain the samples of the non-Gaussian process through the transformation:

$$x(t_n) = \varphi[zG(t_n)]$$

- *A Nonstationary Random Process*

A common approach to simulate a nonstationary process $\xi(t) = \{z(t)\}$ is based on the linear transformation of the stationary process $\eta(t) = \{y(t)\}$:

$$\xi(t) = g(t) + f(t) \cdot \eta(t)$$

where $g(t), f(t)$ are nonrandom functions of time, $\eta(t)$ is the stationary random process with zero mean, correlation function $K_\eta(t_2 - t_1)$, and variance $K_\eta(0) = \sigma_\eta^2 = 1$. The mathematical expectation and correlation function of the process $\xi(t)$ are:

$$m_\xi(t) = g(t); \quad K_\xi(t_1, t_2) = f(t_1)f(t_2) \cdot K_\eta(t_2 - t_1)$$

Thus, the nonstationary process $\xi(t)$ with specified mean $m_\xi(t)$ and correlation function $K_\xi(t_1, t_2)$ is simulated as:

$$z_n = m_\xi(t_n) + \sigma_\xi(t_n) \cdot y_n$$

where $\sigma_\xi(t_n) = \sqrt{K_\xi(t_n, t_n)}$.

An example of how to simulate the nonstationary random process is given in Chapter 9. For more algorithms how to simulate random variables and processes, see the Selected Bibliography in the end of this chapter.

8.4 Applied Statistics

8.4.1 Basic Definitions

In electrical engineering, the methods of applied statistics are typically used to estimate the parameters of different processes which are described by random functions (e.g., signals, noise, interference). These processes can be recorded live from electronic equipment or simulated. In both cases, the task is the same: based on the available sample or samples of the random process, one has to estimate major parameters of this process: mean, variance, correlation function, spectrum, distribution law, and so forth. Sometimes this procedure is called *statistical analysis*, as opposed to the procedure of *statistical synthesis* (simulation) of the random functions considered in Section 8.3. In the former case, the sample is available but statistical parameters are unknown and have to be found. In the latter case, the statistical parameters are known and the samples of the random function with these parameters have to be reconstructed.

In electrical engineering the random function $\xi(t)$ of a single argument time t (a *random process*) is typically considered.

There are two major classes of the random processes: the stationary random process and the nonstationary random process.

For the *stationary* process, statistical parameters are invariant with respect to time shift, and thus the mathematical expectation m_ξ and variance $\sigma^2{}_\xi$ are not functions of time (constant for the entire sample), while the correlation function $K_\xi(\tau)$ is a function of the difference $\tau = t_2 - t_1$ between the two moments of time t_1, t_2 for which it is considered.

For the *nonstationary* process, statistical parameters of the process depend on the time shift; thus, mathematical expectation $m_\xi(t)$ and variance $\sigma^2{}_\xi(t)$ are functions of time, and the correlation function $K_\xi(t_1, t_2)$ depends on the exact position of the moments of time t_1, t_2 on the time scale.

An important class of the stationary processes is an *ergodic stationary process*, for which accuracy of the estimates of statistical parameters obtained by analysis of a single sample converge to the accuracy of the estimates of statistical parameters obtained by analysis of the set of samples. Typical assumption in electrical engineering is that the process considered (e.g., noise) is the *Gaussian ergodic random process*.

8.4.2 How to Estimate Parameters of a Stationary Random Process

For an *ergodic stationary random process* $\xi(t)$, all parameters can be estimated by analyzing a single available sample $x(t)$. The four main parameters of an ergodic stationary process that typically are of practical interest are:

- Mathematical expectation (mean) m_ξ ;

- Variance $D_\xi = \sigma_\xi^2$;

- Autocorrelation function (ACF) $K_\xi(\tau) = \sigma_\xi^2 R_\xi(\tau)$;

- Spectral density (power spectrum) $S_\xi(\omega)$.

When the samples are taken in equal time increments Δt, convenient expressions to estimate the parameters of a stationary ergodic process $\xi(t)$ are given in Appendix 22. There $\hat{u}$ denotes an estimate of u, and N is a number of discrete points in a sample $x(t)$ with the length T, where $T = N \cdot \Delta t$.

8.4.3 How to Estimate Parameters of a Nonstationary Random Process

In the general case, a single sample is not sufficient to estimate statistical parameters of a nonstationary random process. M samples have to be obtained at the interval T; that is, $\xi(t) = \{x_m(t)\}$, $m = \overline{0, M-1}$. These give M independent samples $x_m(n \cdot \Delta t)$, $m = \overline{0, M-1}$; $n = \overline{0, N-1}$ of the discrete random function $\xi(n\Delta t)$. Each sample is taken in N points with an increment Δt at the interval T.

Thus, for each moment of time $n = n \cdot \Delta t$ $(n = 0, 1, 2, ...)$, M samples of the random process $x_m(n)$, $m = \overline{0, M-1}$, exist. Algorithms to estimate the first four moments of a nonstationary random process $\xi(t)$ with independent samples $x(n)$ are given in Appendix 22. These algorithms are valid for any probability distribution function that provides finite moments. For some common distributions, the expressions can be simplified:

- Gaussian pdf:

$$f_1(x) = \frac{1}{\sqrt{2\pi}\sigma} e^{-\frac{(x-a)^2}{2\sigma^2}}$$

$$m_\xi = a, \quad D_\xi = \sigma^2$$

$$\sigma_{\hat{D}_\xi}\big/D_\xi = \sqrt{2/(M-1)}; \quad \sigma_{\hat{\sigma}_\xi}\big/\sigma_\xi \approx 1\big/\sqrt{2M}$$

- Exponential pdf:

$$f_1(x) = \frac{1}{\lambda} e^{-x/\lambda}$$

$$m_\xi = \lambda \quad \sigma_{\hat{m}_\xi} = \lambda\big/\sqrt{M-1} \quad \sigma_{\hat{m}_\xi}\big/m_\xi = 1\big/\sqrt{M-1}$$

- Rayleigh pdf:

$$f_1(x) = \frac{\pi x}{(2m)^2} \cdot e^{-\frac{\pi}{4}\cdot\left(\frac{x}{m}\right)^2}$$

$$m_\xi = m \quad \sigma_{\hat{m}_\xi} = m \cdot \frac{\sqrt{4-\pi}}{\sqrt{\pi(M-1)}} \quad \sigma_{\hat{m}_\xi}\big/m_\xi = \frac{\sqrt{4-\pi}}{\sqrt{\pi(M-1)}}$$

8.4.4 Accuracy of Estimation

Accuracy and reliability of the estimates depend greatly on the choice of the parameters N and Δt. If the samples of the random process are obtained from simulation, these parameters should be defined before simulation starts.

For a *stationary* random process $\xi(t) = \xi(n \cdot \Delta t)$, when its mathematical expectation m_ξ is known, variances of the estimates are:

$$D_{\hat{m}_\xi} = \frac{D_\xi}{N}\left[1 + \frac{2}{N}\sum_{n=1}^{N-1}\left(1-\frac{n}{N}\right)R_\xi(n\Delta t)\right]$$

$$D_{\hat{D}_\xi} = \frac{D_\xi^2}{N}\left\{1 + 4\left[\sum_{n=1}^{N-1}\left(1-\frac{n}{N}\right)(R_\xi(n\Delta t))^2 + D_\xi^2\right]\right\}$$

$$D_{\hat{K}_\xi} = \frac{D_\xi^2}{N-r}\left\{1 + R_\xi^2(r\Delta t) + 2\cdot\sum_{n=1}^{N-1-r}\left(1-\frac{n}{N-r}\right)\times \right.$$
$$\left. \times\left[R_\xi^2(n\Delta t) + R_\xi[(n+r)\Delta t]\cdot R_\xi[(n-r)\Delta t]\right]\right\}$$

where $R_\xi(n\Delta t)$ is the correlation coefficient.

When m_ξ is unknown, the expressions to estimate variances of the estimates become very bulky. That is why, without considerable loss in accuracy, expressions cited previously are usually also used to get the estimates for the process with unknown m_ξ if a sample $x(t)$ is long enough. For the majority of the random processes $\xi(t)$ considered in electrical engineering applications, the following asymptotic equations are valid:

$$D_{\hat{m}_\xi} = \frac{C_m}{T} + o\left(\frac{1}{T}\right) \quad D_{\hat{D}_\xi} = \frac{C_D}{T} + o\left(\frac{1}{T}\right) \quad D_{\hat{k}_\xi} = \frac{C_K}{T} + o\left(\frac{1}{T}\right)$$

where C_m, C_D, C_K are constants and $o(1/T)$ is the infinitesimal value of the order higher than T. These equations show behavior of the estimate accuracy when the length of sample T grows. When $T > 20\,\tau_c$, where τ_c is a correlation interval, the factor $o(T)$ can be omitted (the error introduced does not exceed several percent).

For a *nonstationary* random process $\xi(t)$ with independent samples $x(k)$, accuracy of the estimates for the first four moments are given in Appendix 22.

8.4.5 The Effective Number of Samples

For an ergodic stationary process, the concept of the effective number of samples is often used. The *effective number of samples* is defined as a number N_{ef} of independent values of the process $\xi(t)$ that gives the same variance for the estimate as the entire sample $x(t)$ would give. For $\hat{m}_\xi$ and $\hat{K}_\xi$:

$$N_{ef}(\hat{m}_\xi) = \frac{D_\xi}{D_{\hat{m}_\xi}} ; \quad N_{ef}(\hat{K}_\xi) = \frac{(D_\xi)^2 \cdot \left[1 + (R_\xi(\tau))^2\right]}{D_{\hat{K}_\xi}}$$

8.4.6 How to Choose Parameters of the Sample

Typically, statistical parameters of the process under investigation are unknown a priori (i.e., before analysis starts). That is why it is expedient to divide the procedure into two phases. At the first stage, the tentative values of the parameters $N, \Delta t$ are determined. At the second stage, these values are estimated more accurately by using the test samples. At the first stage, the following approach can be recommended:

Step 1. First, length T of the sample $x(t)$ has to be chosen. Because the correlation function of the process $\xi(t)$ is unknown before the analysis is performed, the autocorrelation function initially can be approximated by the simplest function:

$$R_\varsigma(\tau) = e^{-\alpha|\tau|}$$

For $\left| R_\varsigma(\tau) \right| \leq 0.05$ (less than 5%), the correlation interval is $\tau_c = 3/\alpha$, and the expressions to estimate T are:

$$D_{\hat{m}_\varsigma} = \frac{2D_\varsigma}{\alpha T}; \quad D_{\hat{D}_\varsigma} = \frac{2D_\varsigma^2}{\alpha T}$$

which gives:

$$\hat{T} = 2 \left/ \left(\alpha \cdot \frac{D_{\hat{m}_\varsigma}}{D_\varsigma} \right) \right. \qquad \text{for the mean value;}$$

$$\hat{T} = 2 \left/ \left(\frac{D_{\hat{D}_\varsigma}}{D_\varsigma^2} \right) \right. \qquad \text{for the variance.}$$

These expressions can be used to find the required length of the sample $\hat{T}$ based on accuracy of mean or variance estimate $E = \left(\sigma_{\hat{m}_\varsigma} / \sigma_\varsigma \right)$ or

$$E = \frac{\sigma_{\hat{D}_\varsigma}}{D_\varsigma}; \quad \frac{\hat{T}}{\tau_c} = \frac{2}{3} \cdot \frac{1}{E^2}$$

The dependence of the sample length in fractions of correlation time T/τ_c upon required accuracy E in percent is given in Figure 8.4. Typically, $E = 5\%$ and the length equal to 267 correlation intervals is a reasonable choice.

Step 2. When the estimate of length $\hat{T}$ is defined, the estimate of an increment $\Delta \hat{t}$ has to be selected. After this is done, the required number of samples can be defined as:

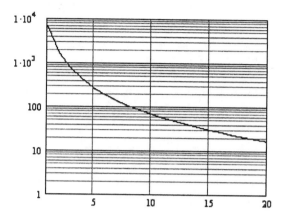

Figure 8.4 Sample length relative to correlation time $\hat{T}/\tau_c$ versus required accuracy of the estimate (percent).

$$\hat{N} = \frac{\hat{T}}{\widehat{\Delta t}}$$

The increment Δt also depends on the correlation interval τ_c. The estimates $\widehat{\Delta t_1}$ for mean, and the estimates $\widehat{\Delta t_2}$ for variance and autocorrelation function, can be found by solving the equations:

$$\frac{4}{9}R_\xi\left(\frac{\widehat{\Delta t_1}}{2}\right) - \frac{5}{18}R_\xi\left(\widehat{\Delta t_1}\right) - \frac{1}{6} = 0 \;; \quad \frac{4}{9}\left[R_\xi\left(\frac{\widehat{\Delta t_2}}{2}\right)\right]^2 - \frac{5}{18}\left[R_\xi\left(\widehat{\Delta t_2}\right)\right]^2 - \frac{1}{6} = 0$$

For the process with autocorrelation function $R_\xi(\tau) = e^{-\alpha|\tau|}$, the solution is:

$$\widehat{\Delta t_1} = 1.02/\alpha \;; \quad \widehat{\Delta t_2} = 0.51/\alpha .$$

The estimate of the spectral density $\hat{S}_\xi(\omega)$ is based on the estimate of the correlation function $K_\xi(\tau)$. The main source of error in this case is the estimation error for an autocorrelation function transformed to the spectrum estimate via model $\hat{S}_\xi(\omega) = $ function $\left(\hat{K}_\xi(\tau)\right)$ in Appendix 22. Analysis of these errors shows that the reliable estimate of the spectrum can be obtained even when the number of samples N_S used to estimate $S_\xi(\omega)$ is much less than the number of samples N_K used to estimate the autocorrelation function. Typically, $N_S \leq 0.1 N_K$. On the other hand, when the number of samples is limited, it shortens the interval $T_S = N_S \cdot \Delta t$ and "smears" the estimate of the function $\hat{S}_\xi(\omega)$. That is why too few samples N_S should not be used. The increment Δt_S is selected to avoid cutoff effects at the higher frequencies:

$$\hat{\Delta t}_S < \frac{\pi}{\beta \cdot \omega_b}$$

where ω_b is a boundary frequency starting from which $S_\xi(\omega_b) = 0$, and $\beta = (2 \div 5)$ is an empirical coefficient.

8.4.7 How to Estimate the Distribution Functions

The basic approach used to estimate the distribution function of the random variable ξ with independent samples $x_1, ..., x_N$ is to form the empirical distribution function:

$$\hat{F}_N(x) = \frac{N_x}{N} = \frac{1}{N} \sum_{n=1}^{N} \theta(x - x_n)$$

where N_x is the number of occurrences ξ_n less than x. Function $\theta(u)$:

$$\theta(u) = \begin{cases} 1, & u > 0 \\ 0, & u \leq 0 \end{cases}$$

This estimate is unbiased:

$$M\{\hat{F}_N(x)\} = F(x)$$

and its variance is:

$$\sigma^2[\hat{F}(x)] = \frac{F(x)[1 - F(x)]}{N}$$

A variety of other methods are used to estimate distribution laws that differ in complexity and accuracy: the method of histograms, the method based on expansion into series, and parametric and nonparametric methods. For more details, see the Selected Bibliography in the end of this chapter.

8.4.8 The Least Mean Square Method

The common task in applied statistics is to find the function

$$y = f(x)$$

that describes some dependence of the parameter y on the parameter x based on the samples $y_n = f(x_n)$, $n = \overline{1, N}$ obtained as the result of simulation or experiment.

The most common approach is the least mean square (LMS) method. In this case, the function $y = f(x)$ should be such that sum of the squares of the deviations of the samples y_n from $f(x_n)$ was minimal:

$$\sum_{n=1}^{N} [y_n - f(x_n)]^2 = \min$$

In the general case, parameters a, b, $c...$ of the function $f(x, a, b, c...)$ can be found through the solution of the set of the equations:

$$
\begin{cases}
\sum_{n=1}^{N} [y_n - f(x_n, a, b, c...)] \left(\dfrac{\partial f}{\partial a} \right)\Bigg|_{x_n} = 0 \\[2ex]
\sum_{n=1}^{N} [y_n - f(x_n, a, b, c...)] \left(\dfrac{\partial f}{\partial b} \right)\Bigg|_{x_n} = 0 \\[2ex]
\sum_{n=1}^{N} [y_n - f(x_n, a, b, c...)] \left(\dfrac{\partial f}{\partial c} \right)\Bigg|_{x_n} = 0 \\[2ex]
\dotfill
\end{cases}
$$

Common approximation functions are as follows.
- *Linear function* $y = ax + b$

  In this case:

$$\frac{\partial f}{\partial a} = x, \quad \frac{\partial f}{\partial a}\bigg|_{x_n} = x_n$$

$$\frac{\partial f}{\partial b} = 1, \quad \frac{\partial f}{\partial b}\bigg|_{x_n} = 1$$

and equations to find a, b are:

$$\begin{cases} \displaystyle\sum_{n=1}^{N}\left[y_n - \left(ax_n + b\right)\right]x_n = 0 \\ \displaystyle\sum_{n=1}^{N}\left[y_n - \left(ax_n + b\right)\right] = 0 \end{cases}$$

The solution gives parameters as follows:

$$a = \frac{K_{xy}}{D_x}, \quad b = m_y - \frac{K_{xy}}{D_x} \cdot m_x$$

where $K_{xy} = \dfrac{1}{N}\displaystyle\sum_{n=1}^{N}\left(x_n - m_x\right)\left(y_n - m_y\right)$

$$D_x = \frac{1}{N}\sum_{n=1}^{N}\left(x_n - m_x\right)^2$$

$$m_x = \frac{1}{N}\sum_{n=1}^{N}x_n, \quad m_y = \frac{1}{N}\sum_{n=1}^{N}y_n$$

- *Parabolic function* $y = ax^2 + bx + c$
  For this function:

$$\left.\frac{\partial f}{\partial a}\right|_{x_n} = x_n^2, \quad \left.\frac{\partial f}{\partial b}\right|_{x_n} = x_n, \quad \left.\frac{\partial f}{\partial c}\right|_{x_n} = 1$$

and the equation is:

$$\sum_{n=1}^{N}\left[y_n - \left(ax_n^2 + bx_n + c\right)\right]x_n^2 = 0$$

$$\sum_{n=1}^{N}\left[y_n - \left(ax_n^2 + bx_n + c\right)\right]x_n = 0$$

$$\sum_{n=1}^{N}\left[y_n - \left(ax_n^2 + bx_n + c\right)\right] = 0$$

This set can be represented in the form convenient for finding the solution with respect to a, b, c with the aid of a computer (matrix form):

$$A \cdot \begin{vmatrix} a \\ b \\ c \end{vmatrix} = \begin{vmatrix} P_1 \\ P_2 \\ P_3 \end{vmatrix}$$

where

$$A = \begin{vmatrix} a_{11} & a_{12} & a_{13} \\ a_{21} & a_{22} & a_{23} \\ a_{31} & a_{32} & a_{33} \end{vmatrix}$$

$$a_{11} = \frac{1}{N}\sum_{n=1}^{N} x_n^4; \quad a_{12} = \frac{1}{N}\sum_{n=1}^{N} x_n^3; \quad a_{13} = \frac{1}{N}\sum_{n=1}^{N} x_n^2$$

$$a_{21} = a_{12}; \quad a_{22} = a_{13}; \quad a_{23} = \frac{1}{N}\sum_{n=1}^{N} x_n$$

$$a_{31} = a_{13}; \quad a_{32} = a_{23}; \quad a_{33} = 1$$

$$P_1 = \frac{1}{N}\sum_{n=1}^{N} x_n^2 y_n; \quad P_2 = \frac{1}{N}\sum_{n=1}^{N} x_n y_n; \quad P_3 = \frac{1}{N}\sum_{n=1}^{N} y_n$$

- The sum of arbitrary functions $f_1(x), f_2(x), ..., f_K(x)$ with coefficients $a_1, a_2, ..., a_K$, for example,

$K = 2$, $f_1(x) = \cos \omega x$, $f_2(x) = \cos 2\omega x$, that results in the function:

$$f(x) = a_1 \cdot \cos \omega x + a_2 \cdot \cos 2\omega x$$

$K = 3$, $f_1(x) = e^{\alpha x}$, $f_2(x) = e^{\beta x}$, $f_3(x) = e^{\gamma x}$, that results in the function:

$$f(x) = a_1 \cdot e^{\alpha x} + a_2 \cdot e^{\beta x} + a_3 \cdot e^{\gamma x}$$

In this case, the set of equations to obtain solutions for $a_k, k = 1, ... K$ is as follows:

$$\sum_{k=1}^{K} a_k \sum_{n=1}^{N} f_1(x_n) \cdot f_k(x_n) = \sum_{n=1}^{N} y_n \, f_1(x_n)$$

$$\sum_{k=1}^{K} a_k \sum_{n=1}^{N} f_2(x_n) \cdot f_k(x_n) = \sum_{n=1}^{N} y_n \, f_2(x_n)$$

$$\cdots\cdots\cdots\cdots\cdots\cdots\cdots\cdots\cdots$$

$$\sum_{k=1}^{K} a_k \sum_{n=1}^{N} f_K(x_n) \cdot f_k(x_n) = \sum_{n=1}^{N} y_n \, f_K(x_n)$$

An example of how to use the LMS method for the function approximation is given in Chapter 13.

Selected Bibliography

Leonov S. A., and A. I. Leonov, *Handbook of Computer Simulation in Radio Engineering, Communications and Radar*, Norwood, MA: Artech House, 2001.

Petruchelli, J. D., et al., *Applied Statistics for Engineers and Scientists*, New York: Prentice Hall, 1999.

Ross, S. M., *Introduction to Probability and Statistics for Engineers and Scientists*, New York: Harcourt/Academic Press, 2000.

Wadsworth, H. M., *Handbook of Statistical Methods for Engineers and Scientists*, New York: McGraw-Hill, 1997.

Walpole, R. E. (ed.), *Probability and Statistics for Engineers and Scientists*, New York: Prentice Hall, 2002.

Yates, R.D., and D. J. Goodman, *Probability and Statistics: A Friendly Introduction for Electrical and Computer Engineers*, New York: John Wiley and Sons, 1998.

Chapter 9

COMPUTER-AIDED COMPUTATIONS

These days for an electrical engineer or a student, life is much easier than it was several decades ago. A number of mathematical software packages are available for computer-aided computations. A convenient one is MATHCAD, because in order to find a derivative, an integral, a solution of the equation, and so forth, a user does not have to do any programming or apply numerical algorithms. He or she writes in the worksheet a mathematical expression and lets the software provide the answer. *Mathcad 2000 Professional* is referred to in this chapter.

9.1 How to Use Computer to Solve Algebraic Equations

The basic steps to solve the equation with a **single variable** and a **single root** are as follows:

1. Guess an initial value of the unknown variable x.

2. Set the whole expression equal to zero, e.g. $f(x) - \phi(x) = 0$.

3. Type $x0 := \text{root}(f(x) - \phi(x), x)$.

4. Type x0 = to see what is the root to the equation.

Example 9.1 Find the root of the equation $x^3 - e^x = 0$.

Solution. Let us choose the initial value $x := 2$

Thus, $x0 := \text{root}(x^3 - e^x, x)$ The root is: $x0 = 1.857$

The basic steps to solve the equation with a **single variable** and **multiple roots** are as follows:

247

1. Type the right side of the equation (there is no need to set the expression to zero)

2. Select the occurrence of the variable for which you are solving

3. Choose *Variable* and *Solve* from *Symbolic* menu

Example 9.2 Find the root of the equation $x^2 + 10x + 9 = 0$.

Solution. Type: $x^2 + 10 \cdot x + 9$

Highlight x and choose *Variable, Solve* from *Symbolic* menu

The answer is: $\begin{pmatrix} -9 \\ -1 \end{pmatrix}$ (compare to Example 1.19, Chapter 1)

The basic steps to solve a **system of equations** are given in Section 9.6.

9.2 How to Use the Computer for Differentiation

The basic steps to evaluate the **derivative** are as follows:

1. Choose the *Calculus* menu and click on the *Derivative* sign $\dfrac{d}{d\blacksquare}\blacksquare$ or $\dfrac{d^n}{dx^n}\blacksquare$

2. In the placeholder write the function you want to differentiate.

3. Select (highlight) the whole expression for the derivative.

4. Choose *Symbolics, Evaluate* and then *Symbolically*.

5. If the derivative has to be evaluated for a specified value of the argument put the sign = after the expression for a derivative.

Example 9.3 Find the derivatives for some common functions

(a) $y := \sin(x)$ (b) $y := x^n$ (c) $y := \ln(x)$

Solution. (a) Write the derivative $\dfrac{d}{dx}\sin(x)$

Evaluate symbolically. The result is: $\cos(x)$

(b) Write the derivative $\dfrac{d}{dx}x^n$

Evaluate symbolically. The result is: $x^n \cdot \dfrac{n}{x}$

(which is the same as more common form $n \cdot x^{n-1}$).

(c) Write the derivative $\dfrac{d}{dx}\ln(x)$

Evaluate symbolically. The result is: $\dfrac{1}{x}$

__Example 9.4__ Find the derivative of the function $y := \sin\left(\dfrac{1}{x}\right) \cdot x$

Solution. Write the derivative $\dfrac{d}{dx}\sin\left(\dfrac{1}{x}\right) \cdot x$

Evaluate symbolically. The result is: $\dfrac{-\cos\left(\dfrac{1}{x}\right)}{x} + \sin\left(\dfrac{1}{x}\right)$

(compare to Example 3.11, Chapter 3).

__Example 9.5__ Find the derivative of the function $y := \sin\left(\dfrac{1}{x}\right) \cdot x + \dfrac{\cos(x)}{\sqrt{1-x^2}}$

for $x := \dfrac{\pi}{8}$

Solution. Write the derivative $\dfrac{d}{dx}\left(\sin\left(\dfrac{1}{x}\right) \cdot x + \dfrac{\cos(x)}{\sqrt{1-x^2}}\right)$

Evaluate symbolically. The result is:

$$\dfrac{-\cos\left(\dfrac{1}{x}\right)}{x} + \sin\left(\dfrac{1}{x}\right) - \dfrac{\sin(x)}{\left(1-x^2\right)^{\left(\frac{1}{2}\right)}} + \dfrac{\cos(x)}{\left(1-x^2\right)^{\left(\frac{3}{2}\right)}} \cdot x$$

Evaluate the derivative for $x := \dfrac{\pi}{8}$

$$\frac{d}{dx}\left(\sin\left(\frac{1}{x}\right)\cdot x + \frac{\cos(x)}{\sqrt{1 - x^2}}\right) = 2.72$$

9.3 How to Use the Computer for Integration

The basic steps to evaluate the **indefinite integral** are as follows:

1. Choose the *Calculus* menu and click on the *Integral* sign $\displaystyle\int$ ∎ d∎

2. In the placeholders write the function you want to integrate and the argument of integration.

3. Select (highlight) the whole expression for the integral.

4. Choose *Symbolics, Evaluate* and then *Symbolically*.

Example 9.6 Find the integrals for some common functions:

(a) $y := \sin(x)$ (b) $y := x^n$ (c) $y := \ln(x)$

Solution. (a) Write the integral $\displaystyle\int \sin(x)\, dx$

Evaluate symbolically. The result is: $-\cos(x)$

(b) Write the integral $\displaystyle\int x^n\, dx$

Evaluate symbolically. The result is: $\dfrac{x^{(n+1)}}{(n + 1)}$

(c) Write the integral $\displaystyle\int \ln(x)\, dx$

Evaluate symbolically. The result is: $x\cdot\ln(x) - x$

<u>Example 9.7.</u> Find the integral of the function $y := \dfrac{1}{2}\cdot x$

Solution. Write the integral $\displaystyle\int \dfrac{1}{2}\cdot x\,dx$

Evaluate symbolically. The result is: $\dfrac{1}{4}\cdot x^2$

(compare to Example 3.29, Chapter 3).

<u>Example 9.8.</u> Find the integral of the function $y := \sqrt{2\cdot x - 1}$

Solution. Write the integral $\displaystyle\int \sqrt{2\cdot x - 1}\,dx$

Evaluate symbolically. The result is: $\dfrac{1}{3}\cdot(2\cdot x - 1)^{\left(\frac{3}{2}\right)}$

(compare to Example 3.30, Chapter 3).

<u>Example 9.9.</u> Find the integral of the function $y := \sin(\phi)\cdot(\cos(\phi))^3$

Solution. Write the integral $\displaystyle\int \sin(\phi)\cdot(\cos(\phi))^3\,d\phi$

Evaluate symbolically. The result is: $\dfrac{-1}{4}\cdot\cos(\phi)^4$

(see Example 3.51, Chapter 3).

<u>Example 9.10</u> Find the integral of the function $y := (\cos(\theta))^5\cdot(\sin(\theta))^3$

Solution. Write the integral $\displaystyle\int (\cos(\theta))^5\cdot(\sin(\theta))^3\,d\theta$

Evaluate symbolically. The result is:
$$\frac{-1}{8}\cdot\sin(\theta)^2\cdot\cos(\theta)^6 - \frac{1}{24}\cdot\cos(\theta)^6$$

(see Example 3.53, Chapter 3).

The basic steps to evaluate the **definite integral** are as follows:

1. Choose the *Calculus* menu and click on the *Integral* sign $\int_{\blacksquare}^{\blacksquare} \blacksquare\ d\blacksquare$

2. In the placeholders write the function you want to integrate, the argument of integration, and the integral limits.

3. Put the sign = after the integral.

Example 9.11 Find the integral of the function $y := \sqrt{2\cdot x - 1}$ at the interval $(5,13)$

Solution. Write the integral and put a sign = at the end:

$$\int_5^{13} \sqrt{2\cdot x - 1}\ dx = 32.667$$

(compare to Example 3.33, Chapter 3).

Example 9.12 Find the integral of the function $y := \dfrac{\sin(x)^2}{2 + \cos(x)}$
at the interval $(0, \pi)$

Solution. Write the integral and put a sign = at the end:

$$\int_0^\pi \frac{\sin(x)^2}{2 + \cos(x)}\ dx = 0.842$$

9.4 How to Use the Computer to Evaluate Partial Derivatives, Double and Triple Integrals

The basic steps to evaluate the **partial derivatives** are the same as for the ordinary derivatives (see Section 9.2).

<u>Example 9.13</u> Find the partial derivatives of the first and second orders with respect to x and the mixed derivative for the function $z := x^3 \cdot y^2 + 2 \cdot x^2 \cdot y - 6$

Solution. Write the first derivative $\quad \dfrac{d}{dx}\left(x^3 \cdot y^2 + 2 \cdot x^2 \cdot y - 6\right)$

Evaluate symbolically. The result is: $\quad 3 \cdot x^2 \cdot y^2 + 4 \cdot x \cdot y$

Write the second derivative $\quad \dfrac{d^2}{dx^2}\left(x^3 \cdot y^2 + 2 \cdot x^2 \cdot y - 6\right)$

Evaluate symbolically. The result is: $\quad 6 \cdot x \cdot y^2 + 4 \cdot y$

Write the mixed derivative $\quad \dfrac{d}{dx}\dfrac{d}{dy}\left(x^3 \cdot y^2 + 2 \cdot x^2 \cdot y - 6\right)$ by clicking twice on $\dfrac{d}{d\blacksquare}\blacksquare$

Evaluate symbolically. The result is: $\quad 6 \cdot x^2 \cdot y + 4 \cdot x$

(compare to Example 3.48, Chapter 3).

The basic steps to evaluate the **double** and **triple integrals** are the same as for definite integrals (see section 9.3) by clicking two times (or three times) on

$$\int_{\blacksquare}^{\blacksquare} \blacksquare \, d\blacksquare \text{ sign}$$

<u>Example 9.14</u> Find the integral $I := \displaystyle\int_1^3 \int_2^5 \left(5 \cdot x^2 \cdot y - 2 \cdot y^3\right) dx \, dy$

Solution. Write the integral and put a sign = at the end:

$$\int_1^3 \int_2^5 \left(5 \cdot x^2 \cdot y - 2 \cdot y^3\right) dx \, dy = 660$$

(compare to Example 3.50, Chapter 3).

Example 9.15 Find the integral $I := \int_0^1 \int_2^4 \int_0^3 (x + y + z)\, dx\, dy\, dz$

Solution. Write the integral and put a sign $=$ at the end:

$$\int_0^1 \int_2^4 \int_0^3 (x + y + z)\, dx\, dy\, dz = 30$$

9.5 How to Use the Computer to Solve Differential Equations

In the numerical solution of differential equations, the user gets back a matrix containing the values of the function evaluated over a set of points. Thus, the following three entries have to be specified:

1. The differential equation itself;

2. The initial conditions;

3. A range of points over which the solution is to be evaluated.

To solve **differential equations**, MATHCAD uses the Runge-Kutta method specified by the function

rkfixed(y, x1, x2, npoints, D)

where y is a vector on n initial values; n is the order of the differential equation;

x1, x2 are the endpoints of the interval on which the solution of the differential equation is to be evaluated;

npoints is the number of points beyond the initial point at which the solution is to be found;

$D(x,y)$ is an n-element vector-valued function containing the first derivatives of the unknown functions.

The function *rkfixed* returns a two column matrix in which: the left-hand column contains the points at which the solution to the differential equation is evaluated (x); the right hand columns contains the corresponding values of the solution (y, dy/dx, ...).

Example 9.16 Find the solution of the first-order differential equation

$$\frac{d}{dx}y + 3y = 0$$

with initial conditions y(0) = 4 at 100 points between 0 and 4

Solution.

Step 1. Define initial value $y_0 := 4$

Step 2. Define the function for the first derivative. From the equation we can define:

$$\frac{d}{dx}y = -3y$$

Thus, $D(x,y) := -3 \cdot y_0$

Step 3. As given, x1 = 0, x2 = 4, npoints = 100. Thus, the solution is:

$$Y := \text{rkfixed}(y, 0, 4, 100, D)$$

The first 16 solutions are given below:

	0	1
0	0	4
1	0.04	3.548
2	0.08	3.147
3	0.12	2.791
4	0.16	2.475
5	0.2	2.195
6	0.24	1.947
7	0.28	1.727
8	0.32	1.532
9	0.36	1.358
10	0.4	1.205
11	0.44	1.069
12	0.48	0.948
13	0.52	0.841
14	0.56	0.745
15	0.6	0.661

Y =

Let us compare this solution to Example 3.60, Chapter 3, where the analytical solution of this equation was found:

$$y := 4 \cdot \exp(-3 \cdot x)$$

For the first three test points: $n := 0 .. 2$ $x_0 := 0$ $x_1 := 0.04$ $x_2 := 0.08$

and from the analytical solution we obtain:

$$y_n := 4 \cdot \exp\left(-3 \cdot x_n\right)$$

$$y = \begin{pmatrix} 4 \\ 3.548 \\ 3.147 \end{pmatrix}$$

which matches the results of the numerical solution.

<u>Example 9.17</u> Find the solution of the second-order differential equation

$$\frac{d^2}{dt^2} z = \frac{d}{dt} z + 2 \cdot z$$

with initial conditions z(0) = 1, dz/dt (0) = 3 at 200 points between 0 and 5.

Solution.

Step 1. Define the initial value $z_0 := 1$ $z_1 := 3$

Step 2. Define the function for the first and second derivatives. From original equation we can define:

$$D(t, z) := \begin{pmatrix} z_1 \\ -z_1 + 2 \cdot z_0 \end{pmatrix}$$

Step 3. As given, x1 = 0, x2 = 5, npoints = 400. Thus, the solution is:

$$Z := \mathrm{rkfixed}(z, 0, 5, 200, D)$$

The first 16 solutions are given below:

$$t \qquad z(t) \qquad \frac{d}{dt}z$$

$$Z = \begin{array}{|c|c|c|c|}
\hline
 & 0 & 1 & 2 \\
\hline
0 & 0 & 1 & 3 \\
\hline
1 & 0.025 & 1.075 & 2.977 \\
\hline
2 & 0.05 & 1.149 & 2.959 \\
\hline
3 & 0.075 & 1.223 & 2.944 \\
\hline
4 & 0.1 & 1.296 & 2.934 \\
\hline
5 & 0.125 & 1.369 & 2.927 \\
\hline
6 & 0.15 & 1.443 & 2.924 \\
\hline
7 & 0.175 & 1.516 & 2.925 \\
\hline
8 & 0.2 & 1.589 & 2.929 \\
\hline
9 & 0.225 & 1.662 & 2.937 \\
\hline
10 & 0.25 & 1.736 & 2.949 \\
\hline
11 & 0.275 & 1.81 & 2.963 \\
\hline
12 & 0.3 & 1.884 & 2.982 \\
\hline
13 & 0.325 & 1.959 & 3.003 \\
\hline
14 & 0.35 & 2.034 & 3.027 \\
\hline
15 & 0.375 & 2.11 & 3.055 \\
\hline
\end{array}$$

Example 9.18 Find the solution of the system of the second-order differential equations

$$\frac{d^2}{dp^2}u = 2 \cdot v$$

$$\frac{d^2}{dp^2}v = 4 \cdot v - 2 \cdot u$$

with initial conditions u(0) = 1.5, du/dp (0) = 1.5 u(0) = 1.0, dv/dp (0) = 1.0 at 100 points between 0 and 1.

Solution.

Step 1. Define the vector of the initial conditions:

V = [u(0), du/dp(0), v(0), dv/dp (0)]^T:

$$V_0 := 1.5 \qquad V_1 := 1.5 \qquad V_2 := 1 \qquad V_3 := 1$$

$$V = \begin{pmatrix} 1.5 \\ 1.5 \\ 1 \\ 1 \end{pmatrix}$$

Step 2. Define the function for the first and second derivatives. From the original equations we can define $D = [du/dp, d^2u/dp^2, dv/dp, dv^2/dp^2]^T$:

$$D(p, V) := \begin{pmatrix} V_1 \\ 2 \cdot V_2 \\ V_3 \\ 4 \cdot V_2 - 2 \cdot V_0 \end{pmatrix}$$

Step 3. As given, x1 = 0, x2 = 1, npoints = 100. Thus, the solution is:

$UV := rkfixed(V, 0, 1, 100, D)$ The first 16 solutions are given below:

		p	u	$\dfrac{d}{dp}u$	v	$\dfrac{d}{dp}v$
		0	1	2	3	4
	0	0	1.5	1.5	1	1
	1	0.01	1.515	1.52	1.01	1.01
	2	0.02	1.53	1.54	1.02	1.02
	3	0.03	1.546	1.561	1.03	1.03
	4	0.04	1.562	1.582	1.041	1.041
	5	0.05	1.578	1.603	1.051	1.051
	6	0.06	1.594	1.624	1.062	1.062
UV =	7	0.07	1.61	1.645	1.073	1.072
	8	0.08	1.627	1.667	1.083	1.083
	9	0.09	1.643	1.688	1.094	1.094
	10	0.1	1.66	1.71	1.105	1.105
	11	0.11	1.678	1.733	1.116	1.116
	12	0.12	1.695	1.755	1.127	1.127
	13	0.13	1.713	1.778	1.139	1.138
	14	0.14	1.731	1.801	1.15	1.15
	15	0.15	1.749	1.824	1.162	1.161

9.6 How to Use the Computer in Matrix Calculus

The basic steps to perform **arithmetic operations** with matrices are as follows:

1. Choose *Insert,* then *Matrix.*

2. Specify the matrices required divided by the sign of operation required (addition, subtraction, multiplication, division).

3. Use sign = to obtain the result.

<u>Example 9.19</u> Add two matrices as shown:

$$\begin{pmatrix} 1 & 2 & 3 \\ 4 & 5 & 6 \end{pmatrix} + \begin{pmatrix} 1 & -2 & 0 \\ 3 & -6 & -10 \end{pmatrix} = \begin{pmatrix} 2 & 0 & 3 \\ 7 & -1 & -4 \end{pmatrix}$$

(compare to Example 6.2, Chapter 6).

<u>Example 9.20</u> Multiply two matrices as shown:

$$\begin{pmatrix} 1 & 1 & 0 \\ -3 & -2 & 1 \\ 2 & 0 & 1 \end{pmatrix} \cdot \begin{pmatrix} 0 & 1 & 4 \\ 1 & 3 & 2 \\ 5 & 0 & 1 \end{pmatrix} = \begin{pmatrix} 1 & 4 & 6 \\ 3 & -9 & -15 \\ 5 & 2 & 9 \end{pmatrix}$$

(compare to Example 6.4, Chapter 6).

The basic steps to transpose a matrix are as follows:

1. Choose *Insert,* then *Matrix* and specify the elements of the matrix.

2. Select (highlight) the specified matrix.

3. From *Symbolics* menu choose *Matrix* and *Transpose.*

<u>Example 9.21</u> Find a transposed matrix for the matrix:

$$\begin{pmatrix} 2 & 4 & 5 \\ 3 & 2 & 8 \\ 7 & 1 & 1 \end{pmatrix}$$

Solution. Following the steps cited above, obtain the transposed matrix:

$$\begin{pmatrix} 2 & 3 & 7 \\ 4 & 2 & 1 \\ 5 & 8 & 1 \end{pmatrix}$$ (compare to Example 6.8, Chapter 6).

The basic steps to evaluate a determinant are as follows:

1. Choose *Insert*, then *Matrix*, and specify the elements of the matrix.

2. Select (highlight) the specified matrix.

3. From *Symbolics* menu, choose *Matrix* and *Determinant*.

Example 9.22 Find a determinant for the matrix:

$$\begin{pmatrix} 1 & 2 & 4 \\ -1 & 0 & 3 \\ 3 & 1 & -2 \end{pmatrix}$$

Solution. Following the steps cited above, obtain the determinant D = 7

(compare to Example 6.10, Chapter 6).

Certainly, this tool is the most useful when determinants of large matrices have to be found like in the following example.

Example 9.23 Find a determinant for the matrix:

$$\begin{pmatrix} 1 & 2 & 3 & 4 & 5 \\ 12 & 23 & 36 & 45 & 18 \\ 14 & 7 & 67 & 13 & 15 \\ 0 & 7 & 89 & 114 & 11 \\ 41 & 8 & 9 & 0 & 1 \end{pmatrix}$$

Solution. Following the steps cited above obtain the determinant

D = −15675098

The basic steps to obtain the **inverse matrix** are as follows:

1. Choose *Insert*, then *Matrix*, and specify the elements of the matrix.

2. Select (highlight) the specified matrix.

3. From the *Symbolics* menu choose *Matrix* and *Invert*.

<u>Example 9.24</u> Find the inverse matrix for:

$$\begin{pmatrix} 1 & 2 \\ 2 & 3 \end{pmatrix}$$

Solution. Following the steps above obtain the inverse matrix:

$$\begin{pmatrix} -3 & 2 \\ 2 & -1 \end{pmatrix}$$

(compare to Example 6.12, Chapter 6).

Certainly, this tool is the most useful when large inverse matrices have to be found, as in the following example.

<u>Example 9.25</u> Find the inverse matrix for:

$$\begin{pmatrix} 1 & 2 & 3 & 4 & 5 \\ 12 & 23 & 36 & 45 & 18 \\ 14 & 7 & 67 & 13 & 15 \\ 0 & 7 & 89 & 114 & 11 \\ 41 & 8 & 9 & 0 & 1 \end{pmatrix}$$

Solution. Following the steps above, obtain the inverse matrix:

$$\begin{pmatrix} \dfrac{382601}{7837549} & \dfrac{-112347}{7837549} & \dfrac{-33245}{7837549} & \dfrac{34714}{7837549} & \dfrac{226062}{7837549} \\[3mm] \dfrac{-1735032}{7837549} & \dfrac{587543}{7837549} & \dfrac{8247}{7837549} & \dfrac{-171987}{7837549} & \dfrac{-132462}{7837549} \\[3mm] \dfrac{-904393}{15675098} & \dfrac{11625}{7837549} & \dfrac{143243}{7837549} & \dfrac{-5057}{7837549} & \dfrac{-82571}{15675098} \\[3mm] \dfrac{241171}{7837549} & \dfrac{-25976}{7837549} & \dfrac{-113097}{7837549} & \dfrac{83439}{7837549} & \dfrac{40339}{7837549} \\[3mm] \dfrac{4526767}{15675098} & \dfrac{-198742}{7837549} & \dfrac{7882}{7837549} & \dfrac{-1865}{7837549} & \dfrac{545}{15675098} \end{pmatrix}$$

The basic steps to solve **the set of linear equations Ax = B** are as follows:

1. Specify the elements of the matrices **A** and **B,** based on the equations coefficients (see Section 6.1.13).

2. Obtain the inverse matrix A^{-1}.

3. Obtain the product $A^{-1} \cdot B$ to find the matrix **x.**

<u>Example 9.26</u> Solve the set of linear equations:

$$x + y + z = 6$$
$$x + 2y + 5z = 20$$
$$x + 4y + 9z = 36$$

Solution.

Step 1:

$$A := \begin{pmatrix} 1 & 1 & 1 \\ 1 & 2 & 5 \\ 1 & 4 & 9 \end{pmatrix} \qquad B := \begin{pmatrix} 6 \\ 20 \\ 36 \end{pmatrix}$$

Step 2:

$$A^{-1} = \begin{pmatrix} 0.5 & 1.25 & -0.75 \\ 1 & -2 & 1 \\ -0.5 & 0.75 & -0.25 \end{pmatrix}$$

Step 3:

$$x := \begin{pmatrix} 0.5 & 1.25 & -0.75 \\ 1 & -2 & 1 \\ -0.5 & 0.75 & -0.25 \end{pmatrix} \cdot \begin{pmatrix} 6 \\ 20 \\ 36 \end{pmatrix}$$

Thus, after multiplication:

$$x = \begin{pmatrix} 1 \\ 2 \\ 3 \end{pmatrix}$$

Answer: $x = 1, y = 2, z = 3$

9.7 How to Use the Computer in Stochastic Simulation

The basic steps to simulate a **random variable** with specified distribution are as follows:

1. Simulate the samples of a random variable with uniform distribution at (0,1) using the standard function rnd (1).
2. Transform the sample of a random variable with uniform distribution into the sample with specified distribution using transformation algorithms (see Appendix 21).

<u>Example 9.27</u> Find first three samples of the random variable with Gaussian distribution with parameters $m = 0$, $\sigma = 1$

Solution. To obtain the first sample, we generate two independent samples of the variable with uniform distribution:

$$k := 0..1 \quad u_k := rnd(1) \quad u_0 = 1.268 \times 10^{-3} \quad u_1 = 0.193$$

Since $m := 0$ and $\sigma := 1$ the first sample is:

$$zG := m + \sigma \cdot \sqrt{-2 \cdot \ln\left(u_0 + 10^{-10}\right)} \cdot \cos\left(2 \cdot \pi \cdot u_1\right) \quad zG = 1.273$$

Analogously, the subsequent two samples are:

$$k := 0..1 \quad u_k := rnd(1) \quad u_0 = 0.585 \quad u_1 = 0.35$$

$$zG := m + \sigma \cdot \sqrt{-2 \cdot \ln\left(u_0 + 10^{-10}\right)} \cdot \cos\left(2 \cdot \pi \cdot u_1\right) \quad zG = -0.61$$

$$k := 0..1 \quad u_k := rnd(1) \quad u_0 = 0.823 \quad u_1 = 0.174$$

$$zG := m + \sigma \cdot \sqrt{-2 \cdot \ln\left(u_0 + 10^{-10}\right)} \cdot \cos\left(2 \cdot \pi \cdot u_1\right) \quad zG = 0.287$$

Note. The factor 10^{-10} is useful to add in case the random generator rnd(1) returns 0

<u>Example 9.28</u> Find first two samples of the random variable with Rayleigh distribution with the parameter $\sigma = 10$.

Solution. To obtain the first sample we generate one independent sample of the variable with uniform distribution:

$$k := 0 \quad u_k := rnd(1) \quad u_0 = 0.71$$

Since $\sigma := 10$ the first sample is:

$$zR := \sigma \cdot \sqrt{-2 \cdot \ln\left(u_k + 10^{-10}\right)} \quad zR = 8.268$$

Analogously, the next sample is:

$$k := 0 \quad u_k := rnd(1) \quad u_0 = 0.304$$

$$zR := \sigma \cdot \sqrt{-2 \cdot \ln\left(u_k + 10^{-10}\right)} \qquad zR = 15.432$$

The basic steps to simulate an **uncorrelated random process** with specified distribution are as follows:

1. Define the length of the process T and sampling interval Δt which has to be greater than the correlation interval.
2. Define the number of samples to be simulated as $N = T/\Delta t$.
3. Simulate samples of a random variable with uniform distribution at $(0,1)$ using the standard function rnd (1).
4. Transform the sample of a random variable with uniform distribution into N sample with specified distribution using transformation algorithms (see Appendix 21).

Example 9.29 Simulate the uncorrelated stationary random process with Gaussian distribution with parameters: T = 10 s, Δt = 0.01 s, m = 10, σ = 5.

Solution. The number of samples is: $\qquad N := \dfrac{T}{\Delta t} \qquad N = 1000$

The sample range is: $\qquad n := 0 .. N - 1 \quad$ Parameters are: $\quad m := 10 \quad \sigma := 5$

The samples of the random variables with uniform distribution are:

$$k := 0 .. 1 \quad u_{n,k} := rnd(1)$$

The sample of the Gaussian process is (Figure 9.1):

$$zG_n := m + \sigma \cdot \sqrt{-2 \cdot \ln\left(u_{n,0} + 10^{-10}\right)} \cdot \cos\left(2 \cdot \pi \cdot u_{n,1}\right)$$

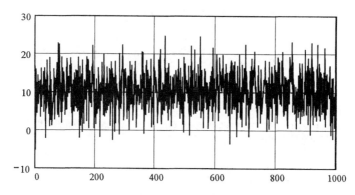

Figure 9.1 The sample of uncorrelated stationary Gaussian random process (amplitude vs. number of samples) with m = 10 and σ = 5.

<u>Example 9.30</u> Simulate the uncorrelated stationary random process with Rayleigh distribution with parameters:T = 100 s, Δt = 0.5 s, s = 100.

Solution. The number of samples is: $\quad N := \dfrac{T}{\Delta t} \qquad N = 200$

The samples range is: $\quad n := 0 .. N - 1 \quad$ The parameter is: $\quad \sigma := 100$

The samples of the random variables with uniform distribution are:

$z_n := \mathrm{rnd}(1)$

The sample of the Rayleigh process is (Figure 9.2):

$$zR_n := \sigma \cdot \sqrt{-2 \cdot \ln\!\left(z_n + 10^{-10}\right)}$$

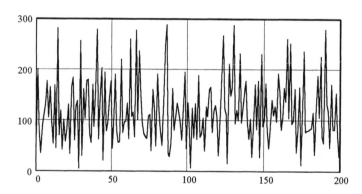

Figure 9.2 The sample of uncorrelated stationary Rayleigh random process (amplitude vs. number of samples) with σ = 100.

<u>Example 9.31</u> Simulate the uncorrelated nonstationary random process with Rayleigh distribution with parameters: T = 100 s, Dt = 0.5 s, $\sigma := n \cdot \Delta t$, where *n* is the number of the current sample.

Solution. The number of samples is: $\quad N := \dfrac{T}{\Delta t} \qquad N = 200$

The sample range is: $\quad n := 0 .. N - 1 \quad$ The parameter is: $\quad \sigma_n := n \cdot \Delta t$

The samples of the random variables with uniform distribution are:

$z_n := \mathrm{rnd}(1)$

The sample of the Rayleigh process is (Figure 9.3):

$$zR_n := \sigma_n \cdot \sqrt{-2 \cdot \ln(z_n + 10^{-10})}$$

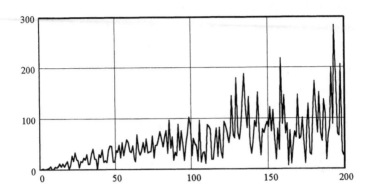

Figure 9.3 The sample of uncorrelated nonstationary Rayleigh random process (amplitude vs. number of samples) with $\sigma := n \cdot \Delta t$.

The basic steps to simulate a **correlated random process** with specified distribution and correlation function are as follows:

1. Based on specified correlation function define the correlation interval τ
2. Define the length of the process T and sampling interval Δt which must be less than correlation interval τ
3. Define number of samples to be simulated as $N = T/\Delta t$.
4. Simulate samples of a random process based on procedures outlined in Section 8.3.4 and algorithms cited in Appendix 21

Example 9.32 Simulate the correlated stationary random process with Gaussian distribution with parameters:$T = 10$ s, $\Delta t = 0.1$ s, $m = 0$, $\sigma = 5$,
$K(\tau) := \sigma \cdot \exp(-\alpha \cdot \tau)$, where $\alpha = 1$.

Solution. The number of samples is:　$N := \dfrac{T}{\Delta t}$　　$N = 1000$

The samples range is:　$n := 0 .. N - 1$

Parameters of the process are:　　$m := 0$　$\sigma := 5$　$\alpha := 1$

Simulation parameters are:　$a_1 := \exp(-\alpha \cdot \Delta t)$　　$b_1 := \sigma \cdot \sqrt{1 - \exp(-2 \cdot \alpha \cdot \Delta t)}$

The samples of the random variables with uniform distribution are:

$k := 0..1 \quad u_{n,k} := rnd(1)$

The samples of the standard Gaussian process are:

$$z0G_n := \sqrt{-2 \cdot \ln\left(u_{n,0} + 10^{-10}\right)} \cdot \cos\left(2 \cdot \pi \cdot u_{n,1}\right)$$

The first sample of samples of the correlated process is:

$z_0 := \sigma \cdot z0G_0$

The new sample range is: $\quad n := 1..N-1$

The sample of the correlated Gaussian process is (see Figure 9.4):

$z_n := a_1 \cdot z_{n-1} + b_1 \cdot z0G_n$

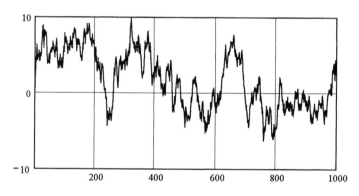

Figure 9.4 The sample of the correlated stationary Gaussian random process (amplitude vs. number of samples).

Selected Bibliography

Leonov, S., A., and A. I. Leonov, *Handbook of Computer Simulation in Radio Engineering, Communications and Radar*, Norwood, MA: Artech House, 2001.

MATHCAD 2000 Professional, User's Guide, Cambridge, MA: Mathsoft, 2000.

PART II

MATHEMATICAL ALGORITHMS TO SOLVE COMMON PROBLEMS IN ELECTRICAL ENGINEERING

Chapter 10

ELECTRICAL CIRCUITS AND DEVICES

Problem 10.1 Alternating Current and RCL Components. Write mathematical equations for the voltage at the resistor R, capacitance C, and inductance L when the alternating current i = Isinωt (I is the current amplitude, ω is a circular frequency, t is time) flows in a circuit. Find the voltages for the following parameters:

$$R := 50 \quad [\Omega] \qquad C := 10^{-4} \quad [F] \qquad L := 0.5 \quad [H]$$

$$I := 5 \quad [A] \qquad \omega := 50 \quad [Hz]$$

at the moment $t := 0$

Solution. The equations for the voltages at R, C, L elements are as follows:

$$V_R := I \cdot R \cdot \sin(\omega \cdot t) \qquad \qquad \text{current and voltage are in phase}$$

$$V_C := \frac{I}{\omega \cdot C} \cdot \sin\left(\omega \cdot t - \frac{\pi}{2}\right) \qquad \text{voltage lags behind current by } \frac{\pi}{2}$$

$$V_L := \omega \cdot L \cdot I \cdot \sin\left(\omega \cdot t + \frac{\pi}{2}\right) \qquad \text{voltage leads current by } \frac{\pi}{2}$$

For parameters specified, the voltage for t = 0 is:

$$V_R = 0 \qquad V_C = -1 \times 10^3 \qquad V_L = 125 \qquad [V]$$

Based on Euler's formula (see Section 4.1.9), equations for the voltage can also be written as:

$$V_R := \text{Im}(I \cdot R \cdot \exp(j \cdot \omega \cdot t)) \qquad \qquad V_R = 0$$

$$V_C := \text{Im}\left[\frac{I}{\omega \cdot C} \cdot \exp\left[j \cdot \left(\omega \cdot t - \frac{\pi}{2}\right)\right]\right] \qquad V_C = -1 \times 10^3$$

$$V_L := \mathrm{Im}\left[\,\omega \cdot L \cdot I \cdot \exp\left[\,j \cdot \left(\omega \cdot t + \frac{\pi}{2}\right)\right]\right] \qquad V_L = 125$$

Problem 10.2   Complex Impedance. Write mathematical equations for complex impedance, complex voltage, impedance, phase, and actual voltage of LCR circuit given in Figures 10.1 - 10.3. Find impedance, phase, and voltage for $I = 1$ A, $\omega = 100$ Hz, and $t = \pi/(2\omega)$ s.

Solution.

A. *Circuit 1* (Figure 10.1)

$$R := 200 \ [\Omega] \quad L := 50 \cdot 10^{-3} \ [\mathrm{H}] \quad C := 10^{-4} \ [\mathrm{F}] \quad \omega := 100 \ [\mathrm{Hz}] \quad I := 1 \ [\mathrm{A}]$$

$$t := \frac{\pi}{2 \cdot \omega} \quad [\mathrm{s}]$$

The complex impedances of R, C, L elements are given by the formulas:

$$Z_R := R \qquad Z_C := \frac{-j}{\omega \cdot C} \qquad Z_L := j \cdot \omega \cdot L$$

Complex impedance Z_{comp} of the series connection is:

$$Z_{comp} := Z_R + Z_C + Z_L \qquad Z_{comp} := R + j \cdot \left(\omega \cdot L - \frac{1}{\omega \cdot C}\right)$$

Complex voltage V_{comp} is: $V_{comp} := I \cdot Z_{comp}$

Impedance is $Z = |Z_{comp}|$:

$$Z := |Z_{comp}| \qquad\qquad Z = 221.416 \qquad [\Omega]$$

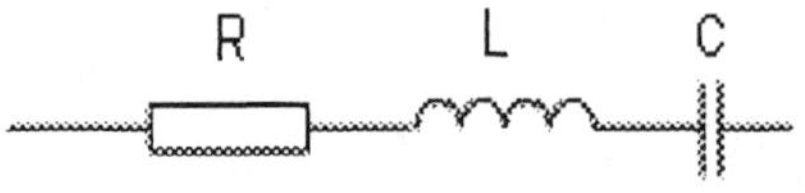

Figure 10.1 Connection of RCL elements in a series.

or in terms of R, C, L:

$$Z := \sqrt{R^2 + \left(\omega \cdot L - \frac{1}{\omega \cdot C}\right)^2}$$ $Z = 221.416$ [Ω]

The phase is:

$$\psi := \text{atan}\left(\frac{\text{Im}(Z_{comp})}{\text{Re}(Z_{comp})}\right)$$ $\psi = -0.443$ [rad]

or in terms of R, C, L:

$$\psi := \text{atan}\left(\frac{\omega \cdot L - \frac{1}{\omega \cdot C}}{R}\right)$$ $\psi = -0.443$ [rad]

The actual voltage is:

$$V := \text{Im}(V_{comp} \cdot \exp(j \cdot \omega \cdot t))$$ $V = 200$ [V]

Based on Euler's formula another representation for V is:

$$V := I \cdot Z \cdot \sin(\omega \cdot t + \psi)$$ $V = 200$ [V]

B. *Circuit 2* (Figure 10.2)

$$R := 100 \ [\Omega] \quad C := 10^{-4} \ [F] \quad \omega := 100 \ [Hz] \quad I := 1 \ [A] \quad t := \frac{\pi}{2 \cdot \omega}$$

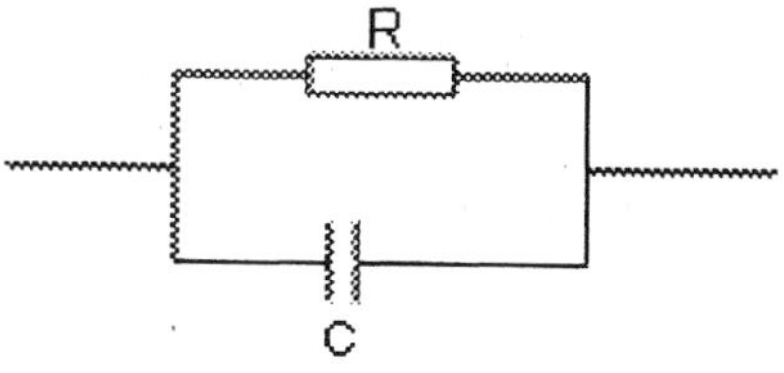

Figure 10.2 Parallel connection of RCL elements.

The complex impedances of R and C elements are given by the formulas:

$$Z_R := R \qquad Z_C := \frac{-j}{\omega \cdot C}$$

Complex impedance Z_{comp} of the parallel connection is:

$$\frac{1}{Z_{comp}} := \frac{1}{Z_R} + \frac{1}{Z_C}$$

Thus, $\qquad Z_{comp} := \dfrac{Z_C \cdot Z_R}{Z_R + Z_C} \qquad Z_{comp} := \dfrac{R \cdot \dfrac{-j}{\omega \cdot C}}{R - \dfrac{j}{\omega \cdot C}}$

Complex voltage V_{comp} is: $\qquad V_{comp} := I \cdot Z_{comp}$

Impedance is $Z = |Z_{comp}|$:

$$Z := |Z_{comp}| \qquad\qquad Z = 70.711 \qquad [\Omega]$$

The phase is:

$$\psi := atan\left(\frac{Im(Z_{comp})}{Re(Z_{comp})}\right) \qquad\qquad \psi = -0.785 \qquad [rad]$$

The actual voltage is:

$$V := Im\left(V_{comp} \cdot exp(j \cdot \omega \cdot t)\right) \qquad\qquad V = 50 \qquad\qquad [V]$$

Based on Euler's formula, another representation for V is:

$$V := I \cdot Z \cdot sin(\omega \cdot t + \psi) \qquad\qquad V = 50 \qquad\qquad [V]$$

C. *Circuit 3* (Figure 10.3)

$R := 40$ [Ω] $C := 10^{-6}$ [F] $L := 10^{-3}$ [H] $\omega := 100$ [Hz] $I := 1$ [A]

$$t := \frac{\pi}{2 \cdot \omega}$$

The complex impedances of R and C elements are given by the formulas:

$$Z_R := R \qquad Z_C := \frac{-j}{\omega \cdot C} \qquad Z_L := j \cdot \omega \cdot L$$

The problem can be solved by the following stages:

Stage 1. <u>Find impedance $Z1_{comp}$ of the inductance L.</u>

$$Z1_{comp} := Z_L$$

Stage 2. <u>Find impedance $Z2_{comp}$ of the parallel connection of the resistor R and capacitance C.</u>

$$\frac{1}{Z2_{comp}} := \frac{1}{Z_R} + \frac{1}{Z_C}$$

Thus, $$Z2_{comp} := \frac{Z_C + Z_R}{Z_R \cdot Z_C} \qquad Z2_{comp} := \frac{R \cdot \frac{-j}{\omega \cdot C}}{R - \frac{j}{\omega \cdot C}}$$

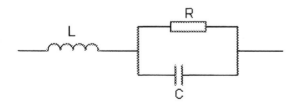

Figure 10.3 Composite connection of RCL elements.

Stage 3. Find impedance of the series connection of $Z1_{comp}$ and $Z2_{comp}$

$$Z_{comp} := Z1_{comp} + Z2_{comp}$$

Complex voltage V_{comp} is: $V_{comp} := I \cdot Z_{comp}$

Impedance is $Z = |Z_{comp}|$:

$$Z := |Z_{comp}|$$ $Z = 39.999$ [Ω]

The phase is:

$$\psi := atan\left(\frac{Im(Z_{comp})}{Re(Z_{comp})}\right)$$ $\psi = -1.5 \times 10^{-3}$ [rad]

The actual voltage is:

$$V := Im\left(V_{comp} \cdot exp(j \cdot \omega \cdot t)\right)$$ $V = 39.999$ [V]

Based on Euler's formula another representation for V is:

$$V := I \cdot Z \cdot sin(\omega \cdot t + \psi)$$ $V = 39.999$ [V]

Problem 10.3   Passive RF Device and Noise Figure. Calculate the noise figure of the power divider with port 3 terminated in a matched load (Figure 10.4). The dissipative insertion loss from port 1 to 2 or port 1 to port 3 is L_in = 0.1 dB. The physical temperature of the divider is T = 300 °K.

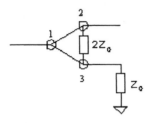

Figure 10.4 Three-port power divider.

Solution. Let us convert the insertion loss from decibels to ratio:

$$L\_in := 0.1$$

dB, insertion loss

$$L := 10^{\frac{L\_in}{10}} \qquad L = 1.023$$

ratio, insertion loss

The reference temperature T_0 and physical temperature T of the divider are:

$$T_0 := 290 \qquad T := 300$$

^{0}K, temperatures

The formula for the scattering matrix (see Section 6.2.4) of the divider is:

$$S := \frac{-j}{\sqrt{2 \cdot L}} \cdot \begin{pmatrix} 0 & 1 & 1 \\ 1 & 0 & 0 \\ 1 & 0 & 0 \end{pmatrix}$$

Thus, the elements of the matrix are (see Section 6.1.4):

$$S_{11} := 0 \qquad S_{12} := \frac{-j}{\sqrt{2 \cdot L}} \qquad S_{13} := \frac{-j}{\sqrt{2 \cdot L}}$$

$$S_{21} := \frac{-j}{\sqrt{2 \cdot L}} \qquad S_{22} := 0 \qquad S_{23} := 0$$

$$S_{31} := \frac{-j}{\sqrt{2 \cdot L}} \qquad S_{32} := 0 \qquad S_{33} := 0$$

Since we assume a matched source at port 1, then the reflection coefficient looking towards generator Γ_1:

$$\Gamma_1 := 0$$

The reflection coefficient Γ_2 looking towards port 2 can be calculated as:

$$\Gamma_2 := S_{22} - \frac{S_{12} \cdot S_{21} \cdot \Gamma_1}{1 - S_{11} \cdot \Gamma_1} \qquad \Gamma_2 = 0$$

The available gain G can be expressed in terms of the S-parameters of the device and port mismatches as:

$$G := \frac{\left(|S_{21}|\right)^2 \cdot \left[1 - \left(|\Gamma_1|\right)^2\right]}{\left(|1 - S_{11} \cdot \Gamma_1|\right)^2 \cdot \left[1 - \left(|\Gamma_2|\right)^2\right]} \qquad G = 0.489$$

The equivalent noise temperature of the device is:

$$T_e := \frac{1 - G}{G} \cdot T \qquad T_e = 313.976$$

The noise F figure is:

$$F := 1 + \frac{T_e}{T_0} \qquad F = 2.083 \qquad \text{ratio, noise figure}$$

$$F\_db := 10 \cdot \log(F) \qquad F\_db = 3.186 \qquad \text{dB, noise figure}$$

Problem 10.4 RF Receiver and Noise Figure. The superheterodyne RF receiver consists of four stages: (1) RF input circuit (front end), (2) RF amplifier and mixer, (3) IF amplifier, (4) analog-to-digital converter (ADC). Parameters of each stage of the receiver are given below.

RF Input Circuit

$LOSS_1 := 1.0$	dB, loss
$B_1 := 50$	MHz, bandwidth
$T_1 := 290$	^0K, physical noise temperature of input termination used in noise figure definition

RF Amplifier

$NF_2 := 2$	dB, noise figure
$GN_2 := 30$	dB, gain
$B_2 := 50$	MHz, bandwidth

Mixer and IF Amplifier

$NF_3 := 10$ dB, noise figure

$GN_3 := 50$ dB, gain

$B_3 := 5$ MHz, bandwidth

Analog-to-Digital Converter

$N := 7$
                                                                number of bits used for signal digitization

$V := 3$ Volt, full scale of amplitudes quantized by ADC

Find the noise figure of the receiver and contribution of each individual stage to the total noise figure.

Solution. The following stages can be introduced to solve the problem.

Stage 1. Define Constants

$k := 1.38054 \cdot 10^{-23}$ W*s/^{0}K, Boltzmann's constant

$dB(x) := 10 \cdot \log\left(\dfrac{x}{10^{-3}}\right)$ conversion equation

$ratio(x) := 10^{0.1 \cdot x}$ conversion equation

Stage 2. Noise Parameters for RF Input Circuit

A. *Parameters at the Input* (Point 1, Figure 10.5)

$p_1 := k \cdot T_1$ $p_1 \cdot 10^{21} = 4.004$ W/Hz, noise power density

$P_1 := p_1 \cdot B_1 \cdot 10^6$ $P_1 = 2.002 \times 10^{-13}$ W, noise power

$p_{1\_dBm} := dB(p_1)$ $p_{1\_dBm} = -173.976$ dB(mW/Hz), noise power density

$P_{1\_dBm} := dB(P_1)$ $P_{1\_dBm} = -96.986$ dBm, noise power

B. *Contribution of Elements Between Point 1 and Point 2 to the Equivalent Input Noise at Point 1*(Figure 10.5)

$L_1 := ratio(LOSS_1)$ $L_1 = 1.259$ input circuit loss

$p_2 := p_1 \cdot L_1$ $\qquad$ $p_2 \cdot 10^{21} = 5.04$ $\qquad$ W/Hz, noise power density

$P_2 := p_2 \cdot B_1 \cdot 10^6$ $\qquad$ $P_2 = 2.52 \times 10^{-13}$ $\qquad$ W, noise power

$p_{2\_dBm} := dB(p_2)$ $\qquad$ $p_{2\_dBm} = -172.976$ $\qquad$ dB(mW/Hz), noise power density

$P_{2\_dBm} := dB(P_2)$ $\qquad$ $P_{2\_dBm} = -95.986$ $\qquad$ dBm, noise power

C. *Noise Figure Contribution from Elements Between Point 1 and Point 2 , Referred to Point 1*

$NF_{12} := 10 \cdot \log(L_1)$ $\qquad$ $NF_{12} = 1$ $\qquad$ dB, noise figure for point 1 to point 2 path

$\Delta NF_{12} := NF_{12}$ $\qquad$ $\Delta NF_{12} = 1$ $\qquad$ dB, noise figure contribution by front end

Stage 3. Noise Parameters for RF Amplifier

A. *Contribution of Elements Between Point 1 and Point 3 to the Equivalent Input Noise at Point 1* (Figure 10.5)

$F_2 := \text{ratio}(NF_2)$ $\qquad$ $F_2 = 1.585$ $\qquad$ amplifier noise figure

$G_2 := \text{ratio}(GN_2)$ $\qquad$ $G_2 = 1 \times 10^3$ $\qquad$ amplifier gain

$F_{13} := L_1 \cdot F_2$ $\qquad$ $F_{13} = 1.995$ $\qquad$ noise figure contribution from elements between points 1 and 3

$G_{13} := \dfrac{G_2}{L_1}$ $\qquad$ $G_{13} = 794.328$ $\qquad$ gain between points 1 and 3

$p_3 := p_1 \cdot F_{13} \cdot G_{13}$ $\qquad$ $p_3 \cdot 10^{18} = 6.345$ $\qquad$ W/Hz, noise power density

$B_{13} := \begin{vmatrix} B_1 & \text{if } B_1 \le B_2 \\ B_2 & \text{otherwise} \end{vmatrix}$ $\qquad$ $B_{13} = 50$ $\qquad$ MHz, the narrowest bandwidth between points 1 and 3

$P_3 := p_3 \cdot B_{13} \cdot 10^6$ $\qquad$ $P_3 = 3.173 \times 10^{-10}$ $\qquad$ W, noise power

$p_{3\_dBm} := dB(p_3)$ $\qquad$ $p_{3\_dBm} = -141.976$ $\qquad$ dB(mW/Hz), noise power density

$P_{3\_dBm} := dB(P_3)$ $\qquad$ $P_{3\_dBm} = -64.986$ $\qquad$ dBm, noise power

B. *Noise Figure Contribution from Elements Between Point 2 and Point 3, Referred to Point 1*

$NF_{13} := 10 \cdot \log(F_{13})$ $NF_{13} = 3$ dB, noise figure for point 1 to point 3 path

$\Delta NF_{23} := NF_{13} - NF_{12}$ dB, noise figure contribution by amplifier

$\Delta NF_{23} = 2$

Stage 4. Noise Parameters for Mixer and IF Amplifier

A. *Contribution of Elements Between Point 1 and Point 4 to the Equivalent Input Noise at Point 1* (Figure 10.5)

$F_3 := \text{ratio}(NF_3)$ $F_3 = 10$ mixer and IF amplifier noise figure

$G_3 := \text{ratio}(GN_3)$ $G_3 = 1 \times 10^5$ mixer and IF amplifier gain

$$F_{14} := L_1 \cdot \left(F_2 + \frac{F_3 - 1}{G_{13}} \right)$$ noise figure contribution from elements between points 1 and 4

$F_{14} = 2.01$

$G_{14} := G_2 \cdot G_3$ $G_{14} = 1 \times 10^8$ gain between points 1 and 4

$p_4 := p_1 \cdot F_{14} \cdot G_{14}$ $p_4 \cdot 10^{14} = 80.453$ W/Hz, noise power density

$B_{14} := B_3$ $B_{14} = 5$ MHz, bandwidth

Note. Bandwidth B_{14} is equal to the bandwidth of the narrowest filter (typically, it is the last filter prior to analog-to-digital conversion).

$P_4 := p_4 \cdot B_{14} \cdot 10^6$ $P_4 = 4.023 \times 10^{-6}$ W, noise power

$p_{4\_dBm} := dB(p_4)$ $p_{4\_dBm} = -90.945$ dB(mW/Hz), noise power density

$P_{4\_dBm} := dB(P_4)$ $P_{4\_dBm} = -23.955$ dBm, noise power

B. *Noise Figure Contribution from Elements Between Point 3 and Point 4, Referred to Point 1*

$NF_{14} := 10 \cdot \log(F_{14})$ $NF_{14} = 3.031$ dB, noise figure for point 1 to point 4 path

$\Delta NF_{34} := NF_{14} - NF_{13}$ $\qquad$ dB, noise figure contribution by
$\qquad\qquad\qquad\qquad\qquad\qquad\qquad$ mixer and IF amplifier

$\Delta NF_{34} = 0.031$

Stage 5. Noise Parameters for Analog-to-Digital Converter

A. *Contribution of Elements Between Point 1 and Point 5 to the Equivalent Input Noise at Point 1* (Figure 10.5)

$$\Delta V := \frac{V}{2^{N-1}} \qquad \Delta V = 0.047 \qquad \text{V, quantization level}$$

$$P_{qn} := \frac{\Delta V^2}{12} \qquad P_{qn} = 1.831 \times 10^{-4} \qquad \text{W, quantization noise power}$$

$$P_5 := P_4 + P_{qn} \qquad P_5 = 1.871 \times 10^{-4} \qquad \text{W, noise power}$$

$$p_5 := \frac{P_5}{B_{14} \cdot 10^6} \qquad p_5 = 3.743 \times 10^{-11} \qquad \text{W/Hz, noise power density}$$

$$p_{5\_dBm} := dB(p_5) \qquad p_{5\_dBm} = -74.268 \qquad \text{dB(mW/Hz), noise power density}$$

$$P_{5\_dBm} := dB(P_5) \qquad P_{5\_dBm} = -7.279 \qquad \text{dBm, noise power}$$

B. *Noise Figure Contribution from Elements Between Point 4 and Point 5, Referred to Point 1*

$$F_4 := \frac{P_5}{P_4} \qquad F_4 = 46.519 \qquad\qquad \text{ADC noise figure}$$

$$NF_4 := 10 \cdot \log(F_4) \qquad NF_4 = 16.676 \qquad \text{dB, ADC noise figure}$$

$$F_{15} := L_1 \cdot \left(F_2 + \frac{F_3 - 1}{G_{13}} + \frac{F_4 - 1}{G_{14}} \right) \qquad \begin{array}{l} \text{noise figure contribution from elements} \\ \text{between points 1 and 5} \end{array}$$

$$F_{15} = 2.01$$

$$NF_{15} := 10 \cdot \log(F_{15}) \qquad NF_{15} = 3.031 \qquad \text{dB, noise figure for point 1 to point 5 path}$$

$\Delta NF_{45} := NF_{15} - NF_{14}$
dB, noise figure contribution by
analog-to-digital converter

$\Delta NF_{45} = 1.238 \times 10^{-6}$

Stage 6. Total Receiver Noise Figure

The total receiver noise figure and contribution of various stages are shown in Figure 10.5.

1 - RF input - 2 - amplifier - 3 -mixer & IF amplifier - 4 ----- ADC ----- 5

LOSS$_1$ = 1	NF$_2$ = 2	NF$_3$ = 10	NF$_4$ = 16.676	dB
	GN$_2$ = 30	GN$_3$ = 50		dB
B$_1$ = 50	B$_2$ = 50	B$_3$ = 5		MHz
			# of bits	
			N = 7	

Contribution to receiver noise figure by each stage, dB

$\Delta NF_{12} = 1 \qquad \Delta NF_{23} = 2 \qquad \Delta NF_{34} = 0.031 \qquad \Delta NF_{45} = 1.238 \times 10^{-6}$

Noise figure of receiver stages referred to the input of receiver, dB

$NF_{12} = 1 \qquad NF_{13} = 3 \qquad NF_{14} = 3.031 \qquad NF_{15} = 3.031$

Figure 10.5 Receiver noise figure calculation chart.

Problem 10.5 Binary Integrator. M signals are integrated (accumulated) in a radar binary integrator. The decision rule used is that the target is detected when at least N signals out of M exceed the detection threshold. The probability of detection p at the input of the integrator is 0.6. Find what the probability of detection will be at the output of the binary integrator for N = 2, M = 4.

Solution. As given the parameters are as follows:

$p := 0.6 \qquad N := 2 \qquad M := 4$

The probability P of detection (or false alarm) at the output of the N-out-of-M integrator as the function of the input probability of detection (or false alarm) p is given by Newton's binomial (binomial expansion, see Section 1.4.4):

$$P := \sum_{n=N}^{M} \frac{M!}{n! \cdot (M-n)!} \cdot p^n \cdot (1-p)^{M-n} \qquad P = 0.821$$

Problem 10.6 Interference Rejection Filters. In a radar system with an antenna rotating 360 degrees in azimuth, the interference caused by reflections from the earth's surface (clutter) passes through the set of rejection filters (three filters). The clutter has Gaussian spectrum, and rejection is based on the difference of a target return (desired signal) and a clutter return (interference signal) in doppler frequency. Describe mathematically the process of interference rejection, and find the clutter attenuation CA after filtering, assuming a target return falls in the middle filter (#2) for parameters given below:

frequency := 1000	MHz, radar frequency
rpm := 20	rpm, antenna rotation rate
$\Theta := 1$	deg, antenna beamwidth
$f_p := 1000$	Hz, pulse repetition frequency
N := 3	number of pulses for coherent integration
K := 3	number of filters in a doppler bank

$$c := \begin{pmatrix} 0.85 & 0.85 + 0.85j & 0.85 \\ -0.85 \cdot j & -3 - 3 \cdot j & 0.85 \cdot j \\ -0.85 & 3 + 3 \cdot j & -0.85 \\ 0.85 \cdot j & -0.85 - 0.85 \cdot j & -0.85 \cdot j \end{pmatrix} \qquad \begin{array}{l} \text{rejection filters} \\ \text{coefficients} \end{array}$$

I_lim := 50	dB, equipment limitations for clutter attenuation due to instability of system circuits
$v_{rc} := 0$	m/s, clutter radial velocity
$\sigma_v := 5$	m/s, rms velocity spread due to clutter motion

Solution. The process of interference rejection mathematically can be described in the following stages:

Stage 1. Define Constants, Sampling Rates, and Intervals

$f\_s := 5$ MHz, digital sampling rate

$Vmax := 1000$ knots, maximum velocity to be plotted

$M := 2^{10}$ number of discrete velocity samples in response

$T_p := \dfrac{1}{f_p}$ $T_p = 1 \times 10^{-3}$ s, pulse repetition interval

$\lambda := \dfrac{2.997925 \cdot 10^8}{\text{frequency} \cdot 10^6}$ $\lambda = 0.3$ m, radar wavelength

$m := 0 .. M - 1$ cycle

$v_m := Vmax \cdot \dfrac{m}{M}$ knots, current velocity sample

$f_m := 2 \cdot \dfrac{v_m}{\lambda} \cdot \dfrac{1852}{3600}$ Hz, current doppler frequency sample

$f\_max := 2 \cdot \dfrac{Vmax}{\lambda} \cdot \dfrac{1852}{3600}$ Hz, maximum doppler frequency

$f\_max = 3.432 \times 10^3$

$p := \text{floor}\left(\dfrac{f\_max}{f_p} \right)$

$filter\_peaks := p + 1$ $filter\_peaks = 4$ number of filter response maxima occurring at 0 - V_{max}

Stage 2. Describe Interference Spectrum and Power

$f0 := \dfrac{2 \cdot v_{rc}}{\lambda}$ $f0 = 0$ Hz, clutter central frequency

$t_0 := \dfrac{\Theta}{rpm \cdot \dfrac{360}{60}}$ $t_0 = 8.333 \times 10^{-3}$ s, time-on-target

$$\sigma_{c\_sm} := \frac{1.178}{\sqrt{2} \cdot \pi \cdot t_0}$$

Hz, spectrum spread by scanning modulation due to antenna rotation

$$\sigma_{c\_sm} = 31.817$$

$$\sigma_{v\_sm} := \frac{\sigma_{c\_sm} \cdot \lambda}{2}$$

m/s, rms velocity spread by scanning modulation due to antenna rotation

$$\sigma_{v\_sm} = 4.769$$

$$\sigma_v = 5$$

m/s, rms velocity spread by interference motion

$$\sigma_{v\Xi} := \sqrt{\sigma_v^2 + \left(\sigma_{v\_sm}\right)^2}$$

m/s, resultant rms velocity spread

$$\sigma_{v\Xi} = 6.91$$

$$\sigma_c := \frac{2 \cdot \sigma_{v\Xi}}{\lambda} \qquad \sigma_c = 46.098$$

Hz, resultant clutter spectrum spread

$$u := 0..p$$

cycle

$$S\_clutter_m := \sum_u \exp\left[-\frac{\left[f_m - \left(f0 + u \cdot f_p\right)\right]^2}{2 \cdot \left(\sigma_c\right)^2}\right]$$

discrete Gaussian clutter spectrum equation

$$S\_clutter\_db_m := 10 \cdot \log\left(S\_clutter_m + 10^{-10}\right)$$

dB, discrete spectrum

$$z := 0..1$$

cycle

$$S\_clutter\_d_m := \sum_z \exp\left[-\frac{\left[f_m - \left(f0 + z \cdot f_p\right)\right]^2}{2 \cdot \left(\sigma_c\right)^2}\right]$$

discrete clutter spectrum function

$$S\_clutter(f) := \sum_z \exp\left[-\frac{\left[f - \left(f0 + z \cdot f_p\right)\right]^2}{2 \cdot \left(\sigma_c\right)^2}\right]$$

continuous clutter spectrum function

$$P\_c := \int_{0}^{f_p} \left(S\_clutter(f) + 10^{-10} \right) df \qquad \text{clutter power}$$

$$P\_c = 115.549$$

$$P\_c\_db := 10 \cdot \log(P\_c) \qquad P\_c\_db = 20.628 \qquad \text{dB, clutter power}$$

Stage 3. Frequency Response of Rejection Filters Bank

$$n := 0 .. N - 1 \qquad k := 0 .. K - 1 \qquad\qquad \text{cycles}$$

$$c\_rms_{n,k} := \frac{c_{n,k}}{\sqrt{\displaystyle\sum_{n} \left(\left| c_{n,k} + 10^{-10} \right| \right)^2}} \qquad \text{filter coefficients normalized to noise bandwidth}$$

$$c\_max_{n,k} := \frac{c_{n,k}}{\displaystyle\sum_{n} \left(\left| c_{n,k} + 10^{-10} \right| \right)} \qquad \text{filter coefficients normalized to maximum}$$

Filter # 1 responses (discrete and continuous)

$$R1\_rms_m := \sum_{n} \left(c\_rms_{n,0} \cdot \exp\left(j \cdot 2 \cdot \pi \cdot n \cdot T_p \cdot f_m \right) + 10^{-10} \right)$$

$$R1\_max_m := \sum_{n} \left(c\_max_{n,0} \cdot \exp\left(j \cdot 2 \cdot \pi \cdot n \cdot T_p \cdot f_m \right) + 10^{-10} \right)$$

$$R1(f) := \sum_{n} \left(c\_max_{n,0} \cdot \exp\left(j \cdot 2 \cdot \pi \cdot n \cdot T_p \cdot f \right) \right)$$

Filter # 2 responses (discrete and continuous)

$$R2\_rms_m := \sum_{n} \left(c\_rms_{n,1} \cdot \exp\left(j \cdot 2 \cdot \pi \cdot n \cdot T_p \cdot f_m \right) + 10^{-10} \right)$$

$$R2\_max_m := \sum_{n} \left(c\_max_{n,1} \cdot \exp\left(j \cdot 2 \cdot \pi \cdot n \cdot T_p \cdot f_m \right) + 10^{-10} \right)$$

$$R2(f) := \sum_{n} \left(c\_max_{n,1} \cdot \exp\left(j \cdot 2 \cdot \pi \cdot n \cdot T_p \cdot f\right) \right)$$

Filter # 3 responses (discrete and continuous)

$$R3\_rms_m := \sum_{n} \left(c\_rms_{n,2} \cdot \exp\left(j \cdot 2 \cdot \pi \cdot n \cdot T_p \cdot f_m\right) + 10^{-10} \right)$$

$$R3\_max_m := \sum_{n} \left(c\_max_{n,2} \cdot \exp\left(j \cdot 2 \cdot \pi \cdot n \cdot T_p \cdot f_m\right) + 10^{-10} \right)$$

$$R3(f) := \sum_{n} \left(c\_max_{n,2} \cdot \exp\left(j \cdot 2 \cdot \pi \cdot n \cdot T_p \cdot f\right) \right)$$

Filter gains normalized to noise bandwidth

$$G1\_rms_m := 20 \cdot \log\left(\left| R1\_rms_m \right| \right) \qquad\qquad \text{dB, filter \#1 gain}$$

$$G2\_rms_m := 20 \cdot \log\left(\left| R2\_rms_m \right| \right) \qquad\qquad \text{dB, filter \#2 gain}$$

$$G3\_rms_m := 20 \cdot \log\left(\left| R3\_rms_m \right| \right) \qquad\qquad \text{dB, filter \#3 gain}$$

Filter gains normalized to maximum

$$G1\_max_m := 20 \cdot \log\left(\left| R1\_max_m \right| \right) \qquad\qquad \text{dB, filter \#1 gain}$$

$$G2\_max_m := 20 \cdot \log\left(\left| R2\_max_m \right| \right) \qquad\qquad \text{dB, filter \#2 gain}$$

$$G3\_max_m := 20 \cdot \log\left(\left| R3\_max_m \right| \right) \qquad\qquad \text{dB, filter \#3 gain}$$

$$G\_rms_m := \begin{array}{|l} par\_in \leftarrow -1000 \\ par_0 \leftarrow G1\_rms_m \\ par_1 \leftarrow G2\_rms_m \\ par_2 \leftarrow G3\_rms_m \\ \text{for } ind \in 0 .. K - 1 \\ \quad par\_in \leftarrow par_{ind} \text{ if } par_{ind} \geq par\_in \\ MAX \leftarrow par\_in \\ MAX \end{array}$$

$$G\_av := \frac{1}{M-1} \cdot \sum_m G\_rms_m \qquad G\_av = 2.903 \qquad dB, \text{ average gain}$$

$$Gav_m := G\_av$$

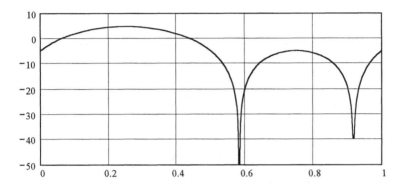

Figure 10.6 Frequency response of filter # 1 normalized to noise bandwidth (dB) vs. doppler frequency relative to PRF.

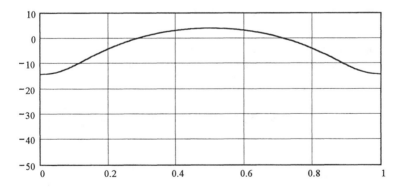

Figure 10.7 Frequency response of filter # 2 normalized to noise bandwidth (dB) vs. doppler frequency relative to PRF.

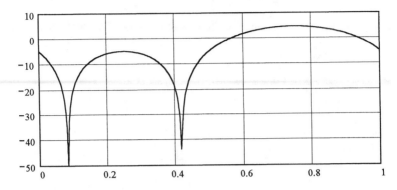

Figure 10.8 Frequency response of filter # 3 normalized to noise bandwidth (dB) vs. doppler frequency relative to PRF.

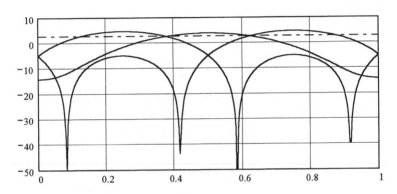

Figure 10.9 Filter bank responses normalized to noise bandwidth (dB) vs. doppler frequency relative to PRF (dashed line is an average gain).

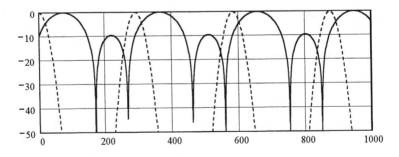

Figure 10.10 Filter # 1 response (solid) overlaid on the clutter spectrum (dashed) vs. velocity in knots.

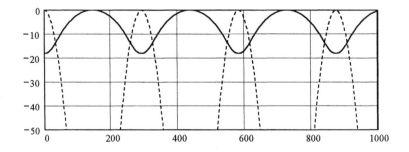

Figure 10.11 Filter # 2 response (solid) overlaid on the clutter spectrum (dashed) vs. velocity in knots.

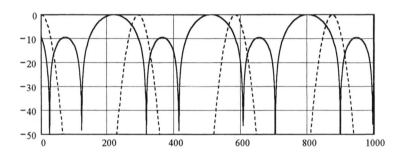

Figure 10.12 Filter # 3 response (solid) overlaid on the clutter spectrum (dashed) vs. velocity in knots.

Stage 4. Clutter Power in Filter # 2 (Notch Filter)

$$\text{F2\_res\_d}_m := \left(\left| \text{R2\_max}_m \right| \right)^2 \cdot \text{S\_clutter\_d}_m + 10^{-10}$$

$$\text{res\_F2}_m := 10 \cdot \log\left(\text{F2\_res\_d}_m \right)$$
dB, discrete clutter residue in filter # 2

$$\text{F2\_res(f)} := \left(\left| \text{R2(f)} \right| \right)^2 \cdot \text{S\_clutter(f)}$$
clutter residue (continuous function)

$$\text{P\_c\_F2} := \int_0^{f_p} \text{F2\_res(f) } df$$
clutter residue power

$$\text{P\_c\_F2} = 2.176$$

$$\text{clutter\_res\_F2\_db} := 10 \cdot \log\left(\frac{P\_c\_F2}{P\_c}\right) \qquad \text{dB, clutter residue power}$$

$$\text{clutter\_res} := -\text{clutter\_res\_F2\_db} \qquad \text{a sign change}$$

$$\text{clutter\_res} = 17.251 \qquad \text{dB, clutter residue power}$$

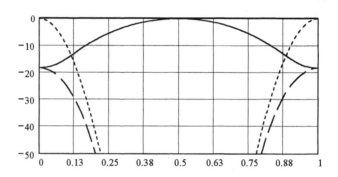

Figure 10.13 Clutter residue (dashed) in filter #2 (solid line is filter #2 frequency response, dotted line is the clutter spectrum) vs. doppler frequency relative to pulse repetition frequency.

Stage 5. Calculate Clutter Attenuation

$$CA := \frac{1}{\dfrac{1}{10^{\frac{I\_lim}{10}}} + \dfrac{1}{10^{\frac{\text{clutter\_res}}{10}}}} \qquad \text{clutter attenuation}$$

$$CA\_db := 10 \cdot \log(CA)$$

$$CA\_db = 17.249 \qquad \text{dB, clutter attenuation}$$

Problem 10.7 A Sensor on a Moving Platform: Orientation Errors.
The angular position of a missile (target) is measured by a shipborne infrared sensor. Find the angle measurement errors in azimuth and elevation due to the errors in knowledge of the ship (platform) orientation for the parameters specified below:

$\psi := 25$ deg, azimuth of the target

$\theta := 80$ deg, elevation of the target

$\Delta\alpha_1 := 1$ ang. min, the error of platform yaw angle measurement

$\Delta\alpha_2 := 1$ ang. min, the error of platform pitch angle measurement

$\Delta\alpha_3 := 1$ ang. min, the error of platform roll angle measurement

Platform orientation errors positive directions are as given in Figure 10.14.

Solution.

$$z := \frac{\pi}{180}$$ coefficient to transform degrees to radians

$\Theta := 90 - \theta$ deg, an angle complement to the target elevation angle

$\Theta = 10$

Coordinate transformation matrix elements are (see Appendix 6):

$$c_1 := \cos\left(z \cdot \frac{\Delta\alpha_1}{60}\right) \qquad s_1 := \sin\left(z \cdot \frac{\Delta\alpha_1}{60}\right) \qquad c_2 := \cos\left(z \cdot \frac{\Delta\alpha_2}{60}\right)$$

$$s_2 := \sin\left(z \cdot \frac{\Delta\alpha_2}{60}\right) \qquad c_3 := \cos\left(z \cdot \frac{\Delta\alpha_3}{60}\right) \qquad s_3 := \sin\left(z \cdot \frac{\Delta\alpha_3}{60}\right)$$

$$A_0 := \cos(z \cdot \psi) \cdot \sin(z \cdot \Theta) \cdot c_1 \cdot c_2 \qquad A_1 := \sin(z \cdot \psi) \cdot \sin(z \cdot \Theta) \cdot (s_1 \cdot c_3 - c_1 \cdot s_2 \cdot s_3)$$

$$A_2 := \cos(z \cdot \Theta) \cdot (s_1 \cdot s_3 + c_1 \cdot c_3 \cdot s_2) \qquad B_0 := \cos(z \cdot \psi) \cdot \sin(z \cdot \Theta) \cdot - s_1 \cdot c_2$$

$$B_1 := \sin(z \cdot \psi) \cdot \sin(z \cdot \Theta) \cdot (s_1 \cdot s_2 \cdot s_3 + c_1 \cdot c_3) \quad B_2 := \cos(z \cdot \Theta) \cdot (c_1 \cdot s_3 - s_1 \cdot c_3 \cdot s_2)$$

$$C_0 := \cos(z \cdot \psi) \cdot \sin(z \cdot \Theta) \cdot - s_2 \quad C_1 := \sin(z \cdot \psi) \cdot \sin(z \cdot \Theta) \cdot - s_3 \cdot c_2$$

$$C_2 := \cos(z \cdot \Theta) \cdot c_2 \cdot c_3$$

The equations to find the azimuth $\Delta\psi$ and elevation $\Delta\theta$ measurement errors are as follows:

$k := 0 .. 2$ cycle

$$U := \frac{\displaystyle\sum_k A_k}{\sqrt{\left(\displaystyle\sum_k A_k\right)^2 + \left(\displaystyle\sum_k B_k\right)^2}}$$

$$V := \frac{\displaystyle\sum_k C_k}{\sqrt{\left(\displaystyle\sum_k A_k\right)^2 + \left(\displaystyle\sum_k B_k\right)^2 + \left(\displaystyle\sum_k C_k\right)^2}}$$

$$\Delta\psi := \begin{vmatrix} \dfrac{acos(U) - z \cdot \psi}{z} \cdot 60 & \text{if } 0 \le z \cdot \psi \le \pi \\[2ex] \dfrac{2 \cdot \pi - acos(U) - z \cdot \psi}{z} \cdot 60 & \text{otherwise} \end{vmatrix}$$

$\Delta\psi = 1.737$ ang. min., the error of azimuth measurement

$$\Delta\theta := -\left(\frac{acos(V) - z \cdot \Theta}{z} \cdot 60\right)$$

$\Delta\theta = -1.329$ ang. min., the error of elevation measurement

Problem 10.8   A Sensor on a Moving Platform: Position Errors. The range to a spacecraft (target) is measured by a laser radar mounted on the satellite. Find the range and angle measurement errors due to the errors in knowledge of the satellite (platform) position for the parameters specified below:

$R := 10$ nm, target range

$\psi := 25$ deg, azimuth of the target

$\theta := 85$ deg, elevation of the target

$\Delta x_1 := 10$ m, the error of the platform center of gravity position measurement along the X1 axis

$\Delta x_2 := 10$ m, the error of the platform center of gravity position measurement along the X2 axis

$\Delta x_3 := 10$ m, the error of the platform center of gravity position measurement along the X3 axis

Platform position errors positive directions are as given in Figure 10.14.

$z := \dfrac{\pi}{180}$ coefficient to transform degrees to radians

$\Theta := 90 - \theta$ deg, an angle complement to the target elevation angle

$\Theta = 5$

$R_m := R \cdot 1852$ $R_m = 1.852 \times 10^4$ m, target range

Transformation of the Platform Position Error Components to Polar Coordinates:

$\eta_1 := \sqrt{\Delta x_1{}^2 + \Delta x_2{}^2 + \Delta x_3{}^2}$ $\eta_1 = 17.321$ m, position error vector modulus

$\eta_2 := \operatorname{atan}\left(\dfrac{\Delta x_2}{\Delta x_1 + 10^{-10}} \right)$ rad, azimuth of position error vector

$\dfrac{\eta_2}{z} = 45$ deg, azimuth of position error vector

$\eta_3 := \operatorname{atan}\left(\dfrac{\sqrt{\Delta x_1{}^2 + \Delta x_2{}^2}}{\Delta x_3 + 10^{-10}} \right)$ rad, angle of position error vector complement to elevation of this vector ($\pi/2$ - elevation)

$\dfrac{\eta_3}{z} = 54.736$ deg, angle of position error vector complement to elevation of this vector ($\pi/2$ - elevation)

Coordinate Transformation Matrix Elements

$$D_0 := \cos(z \cdot \psi) \cdot \sin(z \cdot \Theta) - \dfrac{\eta_1}{R_m} \cdot \cos(\eta_2) \cdot \sin(\eta_3)$$

$$D_1 := \sin(z \cdot \psi) \cdot \sin(z \cdot \Theta) - \dfrac{\eta_1}{R_m} \cdot \sin(\eta_2) \cdot \sin(\eta_3)$$

$$D_2 := \cos(z \cdot \Theta) - \frac{\eta_1}{R_m} \cdot \cos(\eta_3)$$

$$U := \frac{D_0}{\sqrt{\sum_{k=0}^{1} (D_k)^2}} \qquad V := \frac{D_2}{\sqrt{\sum_{k=0}^{2} (D_k)^2}}$$

$$\Delta\psi 2 := \begin{vmatrix} \dfrac{\mathrm{acos}(U) - z \cdot \psi}{z} \cdot 60 & \text{if } 0 \le z \cdot \psi \le \pi \\[2mm] \dfrac{2 \cdot \pi - \mathrm{acos}(U) - z \cdot \psi}{z} \cdot 60 & \text{otherwise} \end{vmatrix}$$

$\Delta\psi 2 = -10.387$ ang. min., the error of azimuth measurement

$$\Delta\theta 2 := -\left(\frac{\mathrm{acos}(V) - z \cdot \Theta}{z} \cdot 60 \right)$$

$\Delta\theta 2 = 2.296$ ang. min., the error of elevation measurement

$$a_R := \sin(z \cdot \Theta) \cdot \cos(z \cdot \psi - \eta_2) \cdot \sin(\eta_3) + \cos(z \cdot \Theta) \cdot \cos(\eta_3)$$

$$\delta_R := \sqrt{1 - 2 \cdot \frac{\eta_1}{R_m} \cdot a_R + \left(\frac{\eta_1}{R_m} \right)^2}$$

$\Delta R := R_m \cdot (\delta_R - 1) \qquad \Delta R = -11.115$ m, range measurement error

Note. The actual value of a coordinate $C = (R, \psi, \theta)$, the measured value C_m and measurement error ΔC are related as $C = C_m + \Delta C$. If the error has a minus sign, then it means that actual coordinate is less than a measured value. The equations may lead to discontinuous functions when azimuth ψ and elevation θ are close to 0, π or 2π, so exercise caution when calculating errors for those angles.

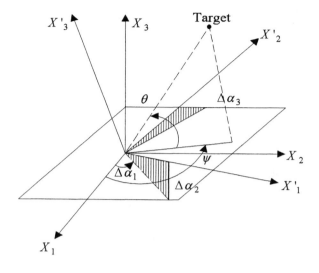

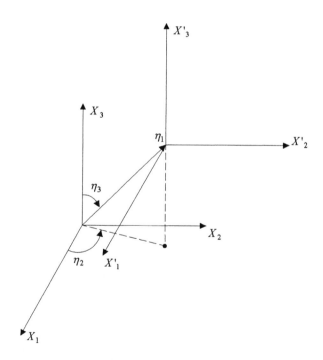

Figure 10.14 Parameters of a sensor platform position errors.

Chapter 11

ANTENNAS AND PROPAGATION

Problem 11.1 Antenna Patterns and Gain. Write mathematical equations for the following common antenna patterns: (1) omnidirectional, (2) sin(x)/x type, Gaussian, (4) cosecant-squared. Plot the patterns for the following parameters: angle of maximum radiation θ_{max} = 10^0 and 3-dB beamwidth BW = 5^0. Calculate power gain for the antenna with parameters: wavelength λ = 10 cm, antenna physical aperture A = 20 m^2, and aperture efficiency η_{ap} = 0.95.

Solution. The following parameters are given:

$$BW := 5 \quad \theta_{max} := 10 \quad [deg] \quad \lambda := 0.1 \quad [m] \quad A := 20 \quad [m2] \quad \eta_{ap} := 0.95$$

The following constants can be introduced:

$$z := \frac{\pi}{180} \qquad \qquad \text{coefficient to transform degrees to radians}$$

$$c1 := \frac{1.39157}{\sin\left(\frac{z \cdot BW}{2}\right)} \qquad c1 = 31.9 \qquad \text{normalization coefficient for sin(x)/x pattern}$$

$$c2 := \frac{\sqrt{2 \cdot \ln(2)}}{\sin\left(\frac{z \cdot BW}{2}\right)} \qquad c2 = 27 \qquad \text{normalization coefficient for Gaussian pattern}$$

The equations for normalized voltage patterns for the specified antenna types can be written as:

$$f_1(\theta) := 1 \qquad \qquad \text{omnidirectional pattern}$$

$$f_2(\theta) := \frac{\sin\left[c1 \cdot z \cdot \left(\theta - \theta_{max}\right)\right]}{c1 \cdot z \cdot \left(\theta - \theta_{max}\right)}$$

sin(x)/x pattern

$$f_3(\theta) := \exp\left[\left[-c2^2 \cdot \frac{\left[z \cdot \left(\theta - \theta_{max}\right)\right]^2}{4}\right]\right]$$

Gaussian pattern

$$f_4(\theta) := \begin{vmatrix} \left[\exp\left[\left[-c2^2 \cdot \dfrac{\left[z \cdot \left(\theta - \theta_{max}\right)\right]^2}{4}\right]\right]\right] & \text{if } \theta \le \theta_{max} + \dfrac{BW}{2} \\[3em] \dfrac{1}{\sqrt{2}} \cdot \dfrac{\sin\left[z \cdot \left(\theta_{max} + \dfrac{BW}{2}\right)\right]}{\sin(z \cdot \theta)} & \text{if } \theta > \theta_{max} + \dfrac{BW}{2} \end{vmatrix}$$

cosecant-squared pattern

The plots of the normalized voltage patterns versus angle are given in Figures 11.1 to 11.4.

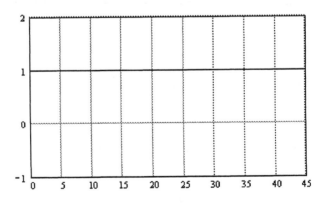

Figure 11.1 Normalized voltage gain vs. angle (degrees) for omnidirectional pattern.

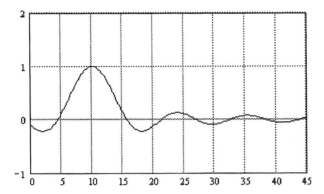

Figure 11.2 Normalized voltage gain vs. angle (degrees) for sin(x)/x pattern.

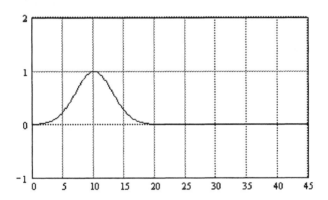

Figure 11.3 Normalized voltage gain vs. angle (degrees) for Gaussian pattern.

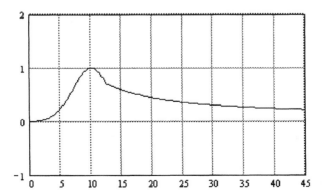

Figure 11.4 Normalized voltage gain vs. angle (degrees) for cosecant-squared pattern.

Antenna gain is given by the formula:

$$G := \frac{4 \cdot \pi \cdot A \cdot \eta_{ap}}{\lambda^2} \qquad\qquad G = 2.388 \times 10^4 \qquad\qquad \text{antenna power gain}$$

$$G\_db := 10 \cdot \log(G) \qquad\qquad G\_db = 43.78 \qquad\qquad \text{dB, antenna power gain}$$

Problem 11.2 Linear Phased Array. Write a mathematical equation for the linear phased array radiation pattern. Find the values of the pattern and power gain for angles 0^0, 3^0, and 10^0 from the normal to array for the following parameters: wavelength = 10 cm, number of elements N = 100, distance between array elements d = 1 cm, phase shift between adjacent elements = 0.36^0, uniform distribution of the field across array, array radiation efficiency η_{ar} = 1. Plot the patterns for the following parameters: a) number of elements in the array N = 100, spacing between elements d = 1 cm; b) N = 10, d = 1 cm; c) N = 100, d = 10 cm.

Solution: As given, the parameters of the array are as follows:

$$\lambda := 0.1 \quad [\text{m}] \quad N := 100 \quad d := 0.01 \quad [\text{m}] \quad \Delta\psi := 0.36 \quad [\text{deg}] \quad \eta_{ar} := 1$$

A linear array can be described by the set of following equations:

$$n := 0 .. N - 1 \qquad\qquad \text{number of the current element}$$

$$x_n := (n + 0.5) \cdot d \qquad\qquad \begin{array}{l} \text{m, } x \text{ coordinate of the } n\text{th element for} \\ \text{uniformly spaced array} \end{array}$$

$$L := (N - 1) \cdot d \qquad L = 0.99 \qquad\qquad \text{m, length of the array}$$

Amplitude distribution of the illuminating field of the array nth element is given by some function. Common assumptions are uniform distribution (distribution = 1) or cosine squared on a pedestal with parameters a_0, b_0, $a_0 + b_0 = 1$ (distribution = 2). As given:

distribution := 1

$$A(n) := \begin{vmatrix} 1 \quad \text{if distribution} = 1 \\[2ex] a_0 + b_0 \cdot \cos\left[\pi \cdot \left(\frac{x_n}{L} + 0.5 \right) \right]^2 \quad \text{if distribution} = 2 \end{vmatrix}$$

$\phi(n) := -n \cdot z \cdot \Delta \psi$ rad, phase distribution of the illuminating field for the array nth element

$I(n) := A(n) \cdot e^{j \cdot \phi(n)}$ amplitude-phase distribution of the illuminating field for the array nth element

Denoting the wave number as:

$$k := \frac{2 \cdot \pi}{\lambda}$$

the equation for the linear array factor can be written as (see Appendix 7):

$$f_a(\theta) := \frac{\left| \sum_n I(n) \cdot \exp(j \cdot k \cdot x_n \cdot \sin(z \cdot \theta)) \right|}{\sum_n A(n)}$$

The array factor is modified by the individual element pattern. Let us assume that element pattern is given by the equation (c is the matching correction coefficient):

$$f_e(\theta) := (\cos(z \cdot \theta))^{\frac{c}{2}}$$ Typical value of the coefficient $c := \frac{3}{2}$

Then the normalized linear phased array pattern is:

$$f_{array}(\theta) := f_a(\theta) \cdot f_e(\theta)$$

For angles specified the normalized array pattern values are:

$\theta := 0$ $f_a(\theta) = 0.984$ $f_e(\theta) = 1$ $f_{array}(\theta) = 0.984$

$\theta := 3$ $f_a(\theta) = 0.73$ $f_e(\theta) = 0.999$ $f_{array}(\theta) = 0.729$

$\theta := 10$ $f_a(\theta) = 0.177$ $f_e(\theta) = 0.989$ $f_{array}(\theta) = 0.175$

Array power gain is given by equation:

$$G := \pi \cdot \eta_{ar} \cdot N \cdot \cos(z \cdot \theta)^c$$

Thus, for angles specified array power gain in dB:

$\theta := 0$ $G := \pi \cdot \eta_{ar} \cdot N \cdot \cos(z \cdot \theta)^c$ $G\_db := 10 \cdot \log(G)$ $G\_db = 24.971$

$\theta := 3$ $G := \pi \cdot \eta_{ar} \cdot N \cdot \cos(z \cdot \theta)^c$ $G\_db := 10 \cdot \log(G)$ $G\_db = 24.963$

$\theta := 10$ $G := \pi \cdot \eta_{ar} \cdot N \cdot \cos(z \cdot \theta)^c$ $G\_db := 10 \cdot \log(G)$ $G\_db = 24.872$

The plots of the normalized voltage patterns for different combinations of parameters N and d are given in Figures 11.5 through 11.7.

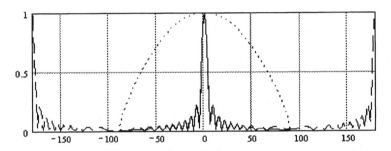

Figure 11.5 Normalized voltage gain vs. angle (deg) for the linear array of N = 100 elements with 1-cm spacing between the elements and uniform amplitude distribution (dashed line is array factor, dotted line is an element pattern, solid line is array pattern).

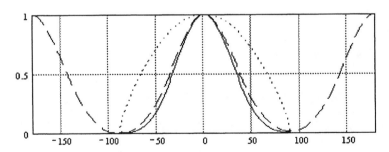

Figure 11.6 Normalized voltage gain vs. angle (deg) for the linear array of N = 10 elements with 1-cm spacing between the elements and uniform amplitude distribution (dashed line is array factor, dotted line is an element pattern, solid line is array pattern).

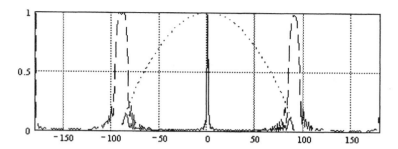

Figure 11.7 Normalized voltage gain vs. angle (deg) for the linear array of
N = 100 elements with 10-cm spacing between the elements and
uniform amplitude distribution (dashed line is array factor, dotted line
is an element pattern, solid line is array pattern).

Problem 11.3 Planar Phased Array. Write a mathematical equation for the
normalized array pattern (array factor) of the planar phased array of 10,000
elements (M = 100 elements in a row, number of rows is N = 100). Find the power
gain for radiation normal to the plane of array (υ = 0) for the following parameters:
wavelength = 10 cm, distance between array elements d = 1 cm, increments of
phase control algorithm $\Delta 1 = \pi/8$, $\Delta 2 = \pi/4$, uniform distribution of the field
across array, array radiation efficiency $\eta_{ar} = 1$. Angles are defined as follows: υ is
the angle from normal to array, and θ is the angle in the plane of the array between
the X axis and projection of an incident wave.

Solution: As given the parameters of the array are as follows:

$$\lambda := 0.1 \ [m] \quad N := 100 \quad M := 100 \quad d := 0.01 \ [m] \quad \eta_{ar} := 1$$

$$\Delta 1 := \frac{\pi}{8} \quad \Delta 2 := \frac{\pi}{4} \quad [rad]$$

Planar array can be described by the set of following equations:.

$$n := 0 .. N - 1$$

$$m := 0 .. M - 1 \qquad \qquad \text{numbers of the current element}$$

$$u_0 := \cos(z \cdot \upsilon) \cdot \sin(z \cdot \theta)$$

$$v_0 := \sin(z \cdot \upsilon) \cdot \sin(z \cdot \theta) \qquad \begin{array}{l} \text{sine-space coordinates for a planar array (see} \\ \underline{\text{Appendix 7}}) \end{array}$$

$$x_n := n \cdot d \qquad \qquad \text{m, } x \text{ coordinate of the } nm\text{th element}$$

$$y_m := m \cdot d \qquad \qquad \text{m, } y \text{ coordinate of the } nm\text{th element}$$

Amplitude distribution of the illuminating field of the array nth element

$$A(n,m) := \begin{vmatrix} 1 & \text{if distribution} = 1 \\[3ex] a_0 + b_0 \cdot \cos\left[\pi \cdot \dfrac{\sqrt{(x_n)^2 + (y_m)^2}}{d \cdot \sqrt{N^2 + M^2}}\right]^2 & \text{if distribution} = 2 \end{vmatrix}$$

Denoting the wave number as:

$$k := \frac{2 \cdot \pi}{\lambda}$$

$\phi(n) := \Delta 1 \cdot \text{floor}\big(k \cdot u_0 \cdot x_n + 0.5\big)$ phase distribution for x coordinate

$\phi(m) := \Delta 1 \cdot \text{floor}\big(k \cdot v_0 \cdot y_m + 0.5\big)$ phase distribution for y coordinate

Phase distribution of the illuminating field of the array nmth element is:

$$\phi 0(n,m) := \Delta 2 \cdot \text{floor}\left(\frac{\phi(n) + \phi(m)}{\Delta 2} + 0.5\right)$$

Amplitude-phase distribution of the illuminating field of the array nmth element is:

$$I(n,m) := A(n,m) \cdot e^{j \cdot \phi 0(n,m)}$$

Normalized planar array pattern is:

$$f_a(\theta, \upsilon) := \frac{\left|\displaystyle\sum_n \sum_m I(n,m) \cdot \exp\big[j \cdot k \cdot (x_n \cdot u_0 + y_m \cdot v_0)\big]\right|}{\displaystyle\sum_n \sum_m A(n,m)}$$

Array power gain is given by equation:

$$G := \pi \cdot \eta_{ar} \cdot N \cdot M \cdot \cos(z \cdot \upsilon)^c$$

Thus, for angle specified array power gain in dB:

$$\upsilon := 0 \quad G_2 := \pi \cdot \eta_{ar} \cdot N \cdot M \cdot \cos(z \cdot \upsilon)^c \quad G\_db := 10 \cdot \log(G) \quad G\_db = 44.971$$

Problem 11.4   Free-Space Propagation Loss. The signal with frequency f = 1000 MHz is transmitted by RF source and propagates in free-space (vacuum). Find the reduction in signal energy L_p due to propagation (free-space propagation loss) at the distance R = 100 meters from the source. Plot the dependence of the propagation loss versus distance to the source.

Solution. As given, the input parameters to calculate propagation loss are as follows:

$f := 1000$ [MHz] $R := 100$ [m]

The speed of light is (see <u>Appendix 1</u>):

$c := 2.997925 \cdot 10^8$ [m]

The wavelength is:

$\lambda := \dfrac{c}{f \cdot 10^6}$ $\lambda = 0.3$ [m]

Free-space propagation loss is given by the formula:

$L_p := 10 \cdot \log\left[\left(4 \cdot \pi \cdot \dfrac{R}{\lambda}\right)^2\right]$ $L_p = 72.448$ [dB]

Dependence of the free-space propagation loss versus distance to the RF source is given in Figure 11.8.

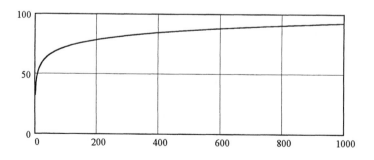

Figure 11.8 Dependence of the free-space propagation loss (dB) vs. distance to the RF source (m) for frequency f = 1000 MHz.

Problem 11.5   Pattern-Propagation Factor. The signal with frequency f = 1000 MHz is transmitted by a radar with Gaussian antenna pattern (antenna height h_a = 25 m, polarization is horizontal, 3-dB beamwidth BW = 2^0, beam axis elevation angle θ_{max} = 5^0, elevation pointing angle θ = 0.5^0), and propagates over the earth's surface. Find the reduction in signal amplitude F due to the combined effects of antenna pattern and propagation (pattern-propagation factor PPF) at the target point for propagation over dry soil with no vegetation. Distance to the target R_t = 25 km from antenna, target height h_t = 10 m.

Solution. As given, the input parameters to calculate pattern-propagation loss are as follows:

f := 1000   [MHz] R_t := 25000 [m] h_t := 10   [m] h_a := 25   [m]

BW := 2   [deg] θ_{max} := 5.0   [deg] θ := 0.5   [deg]

Let us write the mathematical equations in general form for different types of antenna polarization, different antenna patterns, and different kinds of underlying surfaces specified as:

polarization := 1

$\begin{array}{l} 1 \rightarrow \text{horizontal} \\ 2 \rightarrow \text{vertical} \end{array}$

antenna_type := 3

$\begin{array}{l} 1 \rightarrow \text{omnidirectional} \\ 2 \rightarrow \sin(x)/x \text{ type} \\ 3 \rightarrow \text{Gaussian} \end{array}$

Underlying Surface

surface_type := 3

$0 \rightarrow$ absolutely refractive surface (ρ_0 = 1)
$1 \rightarrow$ good soil (wet)
$2 \rightarrow$ average soil
$3 \rightarrow$ bad soil (dry)
$4 \rightarrow$ salt water
$5 \rightarrow$ fresh water
$6 \rightarrow$ snow or ice

σ_h := 1   m, rms roughness of the surface

vegetation := 0

$\begin{array}{l} 0 \rightarrow \text{no vegetation} \\ 1 \rightarrow \text{thin grass} \\ 2 \rightarrow \text{dense weeds or brush} \\ 3 \rightarrow \text{dense trees} \end{array}$

The problem can be solved by breaking it into the following stages.

Stage 1. Define Constant

$z := \dfrac{\pi}{180}$ coefficient to transform degrees to radians

$a_e := 8493333$ m, effective earth radius (4/3 approximation)

$c := 2.997925 \cdot 10^8$ m/s, velocity of light

$c1 := \dfrac{1.39157}{\sin\left(\dfrac{z \cdot BW}{2}\right)}$ $c1 = 79.735$ normalization coefficient for sin(x)/x pattern

$c2 := \dfrac{\sqrt{2 \cdot \ln(2)}}{\sin\left(\dfrac{z \cdot BW}{2}\right)}$ $c2 = 67.464$ normalization coefficient for Gaussian pattern

$\lambda := \dfrac{c}{f \cdot 10^6}$ $\lambda = 0.3$ m, wavelength

Stage 2. Calculate Path Difference for Direct and Reflected Rays and Propagation Region

Parameters to calculate the path difference are:

$$p := \sqrt{\dfrac{4 \cdot a_e \cdot (h_a + h_t) + R_t^2}{3}}$$ $p = 2.459 \times 10^4$

$$\Phi := \operatorname{acos}\left[\dfrac{2 \cdot a_e \cdot (h_t - h_a) \cdot R_t}{p^3}\right]$$ $\Phi = 2.014$

Distance from the radar to the point of specular reflection is:

$$d_1 := \dfrac{R_t}{2} - p \cdot \cos\left(\dfrac{\Phi + \pi}{3}\right)$$ $d_1 = 1.612 \times 10^4$ [m]

Height of radar antenna above the plane tangent at the point of specular reflection is:

$$H_1 := h_a - \frac{d_1^2}{2 \cdot a_e} \qquad\qquad H_1 = 9.711 \qquad\qquad [m]$$

Height of target above the plane tangent at point of specular reflection is:

$$H_2 := \frac{(R_t - d_1) \cdot H_1}{d_1} \qquad\qquad H_2 = 5.353 \qquad\qquad [m]$$

Path difference is the extra one-way distance traveled by the specular reflected ray, as compared to the direct path to the target (meters):

$$\delta 0 := \frac{2 \cdot H_1 \cdot H_2}{R_t} \qquad\qquad \delta 0 = 4.159 \times 10^{-3} \qquad\qquad [m]$$

Threshold to compare against the path difference to define the propagation region is:

$$Z := \frac{\lambda}{8} \qquad\qquad Z = 0.037 \qquad\qquad [m]$$

The propagation regions are defined as follows:

region = *1* is the interference (optical) region where the lobbing structure of electromagnetic field exists;

region = *2* is the transition (intermediate) region where the lobbing field structure is corrected by diffraction;

region = *3* is the diffraction region where no lobbing structure is present (attenuation only) for smooth-sphere or knife-edge diffraction cases.

Stage 3. Equations for the Interference Region (region = 1)

A. *Pattern Factors*

$$\psi := \theta \qquad\qquad \psi = 0.5 \qquad\qquad\qquad \text{deg, grazing angle}$$

$\theta_{p\_d} := \theta - \theta_{max}$ $\theta_{p\_d} = -4.5$ deg, angle of direct ray (target) relative to beam axis

$\theta_{p\_r} := -\theta - \theta_{max}$ $\theta_{p\_r} = -5.5$ deg, angle of reflected ray relative to beam axis

Normalized voltage pattern factor for the direct beam is:

$$f\_d := \begin{vmatrix} 1 & \text{if antenna\_type} = 1 \\ \text{if antenna\_type} = 2 \\ \quad \begin{vmatrix} f \leftarrow 1 & \text{if } \theta_{p\_d} = 0 \\ \dfrac{\sin\left[c1 \cdot z \cdot \left(\theta_{p\_d}\right)\right]}{c1 \cdot z \cdot \left(\theta_{p\_d}\right)} & \text{otherwise} \end{vmatrix} \\ f \leftarrow \exp\left[-c2^2 \cdot \dfrac{\left[z \cdot \left(\theta_{p\_d}\right)\right]^2}{4}\right] & \text{if antenna\_type} = 3 \end{vmatrix}$$

$f\_d = 8.949 \times 10^{-4}$

Normalized voltage pattern factor for the reflected beam is:

$$f\_r := \begin{vmatrix} 1 & \text{if antenna\_type} = 1 \\ \text{if antenna\_type} = 2 \\ \quad \begin{vmatrix} f \leftarrow 1 & \text{if } \theta_{p\_r} = 0 \\ \dfrac{\sin\left[c1 \cdot z \cdot \left(\theta_{p\_r}\right)\right]}{c1 \cdot z \cdot \left(\theta_{p\_r}\right)} & \text{otherwise} \end{vmatrix} \\ f \leftarrow \exp\left[\left[-c2^2 \cdot \dfrac{\left[z \cdot \left(\theta_{p\_r}\right)\right]^2}{4}\right]\right] & \text{if antenna\_type} = 3 \end{vmatrix}$$

$f\_r = 2.795 \times 10^{-5}$

B. *Fresnel Reflection Coefficient*

$\varepsilon_r :=$ | 0 if surface_type = 0
| 25 if surface_type = 1
| 15 if surface_type = 2
| 3 if surface_type = 3
| if surface_type = 4
| | 80 if $f \le 1500$
| | $80 - 0.00733 \cdot (f - 1500)$ if $1500 < f \le 3000$
| | $69 - 0.00243 \cdot (f - 3000)$ if $3000 < f \le 10000$
| if surface_type = 5
| | 80 if $f \le 1500$
| | $80 - 0.00733 \cdot (f - 1500)$ if $1500 < f \le 3000$
| | $69 - 0.00243 \cdot (f - 3000)$ if $3000 < f \le 10000$
| if surface_type = 6
| | 3.2 if $f \le 2000$
| | 3.2 if $2000 < f \le 10000$

relative dielectric constant of the surface material

$\sigma_e :=$ | 0 if surface_type = 0
| 0.02 if surface_type = 1
| 0.005 if surface_type = 2
| 0.001 if surface_type = 3
| if surface_type = 4
| | 4.3 if $f \le 1500$
| | $4.3 + 0.00148 \cdot (f - 1500)$ if $1500 < f \le 3000$
| | $6.52 + 0.001314 \cdot (f - 3000)$ if $3000 < f \le 10000$
| if surface_type = 5
| | 1 if $f \le 1500$
| | $1 + 0.00106 \cdot (f - 1500)$ if $1500 < f \le 3000$
| | $1 + 0.00106 \cdot (f - 3000)$ if $3000 < f \le 10000$
| if surface_type = 6
| | 0.000057 if $f \le 2000$
| | $0.000057 + 6.79 \cdot 10^{-8} \cdot (f - 2000)$ if $2000 < f \le 10000$

mho/m, conductivity of the surface material

$$\varepsilon_c := \varepsilon_r - j \cdot 60 \cdot \frac{c}{f \cdot 10^6} \cdot \sigma_e \qquad \text{complex dielectric constant}$$

Complex Fresnel reflection coefficient for horizontal polarization is:

$$\Gamma_H := \begin{vmatrix} 1 & \text{if surface\_type} = 0 \\ \dfrac{\sin(z \cdot \psi) - \sqrt{\varepsilon_c - \cos(z \cdot \psi)^2}}{\sin(z \cdot \psi) + \sqrt{\varepsilon_c - \cos(z \cdot \psi)^2}} & \text{otherwise} \end{vmatrix}$$

Complex Fresnel reflection coefficient for vertical polarization is:

$$\Gamma_V := \begin{vmatrix} 1 & \text{if surface\_type} = 0 \\ \dfrac{\varepsilon_c \cdot \sin(z \cdot \psi) - \sqrt{\varepsilon_c - \cos(z \cdot \psi)^2}}{\varepsilon_c \cdot \sin(z \cdot \psi) + \sqrt{\varepsilon_c - \cos(z \cdot \psi)^2}} & \text{otherwise} \end{vmatrix}$$

$$\rho_0 := \begin{vmatrix} |\Gamma_H| & \text{if polarization} = 1 \\ |\Gamma_V| & \text{if polarization} = 2 \end{vmatrix} \qquad \text{magnitude of Fresnel reflection coefficient}$$

$$\rho_0 = 0.988$$

$$\phi := \begin{vmatrix} 180 & \text{if surface\_type} = 0 \\ \text{otherwise} \\ \quad \begin{vmatrix} \dfrac{\arg(\Gamma_H)}{z} & \text{if polarization} = 1 \\ \dfrac{\arg(\Gamma_V)}{z} & \text{if polarization} = 2 \end{vmatrix} \end{vmatrix} \qquad \begin{array}{l} \text{phase of Fresnel reflection} \\ \text{coefficient, deg} \end{array}$$

$$\phi = 179.997$$

C. *Specular Scattering Coefficient*

$$\rho_s := \exp\left[-2 \cdot \left(2 \cdot \pi \cdot \frac{\sigma_h}{\lambda} \cdot \sin(z \cdot \psi) \right)^2 \right] \qquad \rho_s = 0.935$$

D. *Vegetation Coefficient*

$$K := \begin{vmatrix} 1 & \text{if vegetation} = 0 \\ 1 & \text{if vegetation} = 1 \\ 3 & \text{if vegetation} = 2 \\ 10 & \text{if vegetation} = 3 \end{vmatrix}$$

multiplication factor depending
on vegetation type

$$\rho_v := \begin{vmatrix} 1 & \text{if vegetation} = 0 \\ \exp\left(-\frac{K}{\lambda}\cdot\sin(z\cdot\psi)\right) & \text{otherwise} \end{vmatrix}$$

$\rho_v = 1$ vegetation coefficient

E. *Divergence Factor*

$$u := \sqrt{\frac{a_e}{2\cdot h_a}}\cdot\tan(z\cdot\theta)$$

parameter

$$D := \sqrt{\frac{1}{3}\cdot\left(1 + 2\cdot\frac{u}{\sqrt{u^2 + 3}}\right)}$$ $D = 0.966$

divergence factor accounting for
sphericity of the earth

F. *Reflection PPF*

$$\delta := 2\cdot h_a\cdot\sin(z\cdot\theta)$$ $\delta = 0.436$

m, path difference between direct
and reflected beams

$$\alpha := \frac{2\cdot\pi}{\lambda}\cdot\delta + z\cdot\phi$$ $\alpha = 12.286$

rad, PPF phase angle

PPF in interference region

$$F_r := f\_d\cdot\left|1 + \frac{f\_r}{f\_d}\cdot D\cdot\rho_0\cdot\rho_s\cdot\rho_v\cdot e^{-j\cdot\alpha}\right|$$ $F_r = 9.189 \times 10^{-4}$

or equivalent expression

$$F_r := f\_d\cdot\sqrt{1 + \left(\frac{f\_r}{f\_d}\cdot D\cdot\rho_0\cdot\rho_s\cdot\rho_v\right)^2 + 2\cdot\left(\frac{f\_r}{f\_d}\cdot D\cdot\rho_0\cdot\rho_s\cdot\rho_v\right)\cdot\cos(\alpha)}$$

$F_r = 9.189 \times 10^{-4}$ pattern-propagation factor in region = 1

$F_{r\_db} := 20 \cdot \log(F_r)$ dB, pattern-propagation factor in region = 1

$F_{r\_db} = -60.735$

Stage 4. Equations for Diffraction Region (region = 3)

A. *Pattern Factors*

$R_h := \sqrt{2 \cdot a_e} \cdot \left(\sqrt{h_a} + \sqrt{h_t} \right)$ $R_h = 3.364 \times 10^4$ m, horizon range

$\theta_0 := -\theta_{max}$ $\theta_0 = -5$ deg, angle of antenna pattern with respect to maximum radiation angle at the horizon

Normalized voltage pattern factor

$f\_0 := \begin{cases} 1 & \text{if antenna\_type} = 1 \\ \quad \text{if antenna\_type} = 2 \\ \quad \begin{cases} f \leftarrow 1 & \text{if } \theta_0 = 0 \\ \dfrac{\sin\left[c1 \cdot z \cdot (\theta_0) \right]}{c1 \cdot z \cdot (\theta_0)} & \text{otherwise} \end{cases} \\ f \leftarrow \exp\left[\left[-c2^2 \cdot \dfrac{\left[z \cdot (\theta_0) \right]^2}{4} \right] \right] & \text{if antenna\_type} = 3 \end{cases}$

$f\_0 = 1.725 \times 10^{-4}$

B. *Smooth-Sphere Diffraction*

Height-gain factors

$b := \begin{cases} 0 & \text{if polarization} = 1 \\ 1 & \text{if polarization} = 2 \end{cases}$ $b = 0$

$h_{min} := \dfrac{\lambda}{2 \cdot \pi} \cdot \dfrac{\left[\varepsilon_r^2 + \left(60 \cdot \sigma_e \cdot \lambda \right)^2 \right]^{\frac{b}{2}}}{\left[\left(\varepsilon_r^2 - 1 \right)^2 + \left(60 \cdot \sigma_e \cdot \lambda \right)^2 \right]^{\frac{1}{4}}}$ $h_{min} = 0.017$

$$h1 := \begin{vmatrix} h_{min} & \text{if } h_a < h_{min} \\ h_a & \text{otherwise} \end{vmatrix}$$

$$h1 = 25$$

$$h2 := \begin{vmatrix} h_{min} & \text{if } h_t < h_{min} \\ h_t & \text{otherwise} \end{vmatrix}$$

$$h2 = 10$$

$$h_c := 30 \cdot \lambda^{\frac{2}{3}}$$

$$h_c = 13.438$$

$$g1 := \begin{vmatrix} 1 & \text{if } h_a \le h_c \\ 0.1356 \cdot \left(\dfrac{h_a}{h_c}\right)^{-0.904} \cdot 10^{0.948 \cdot \sqrt{\frac{h_a}{2 \cdot h_c}}} & \text{otherwise} \end{vmatrix}$$

$$g1 = 0.635$$

$$g2 := \begin{vmatrix} 1 & \text{if } h_t \le h_c \\ 0.1356 \cdot \left(\dfrac{h_t}{h_c}\right)^{-0.904} \cdot 10^{0.948 \cdot \sqrt{\frac{h_t}{2 \cdot h_c}}} & \text{otherwise} \end{vmatrix}$$

$$g2 = 1$$

Smooth-sphere diffraction propagation factor

$$F_{sm} := 9.29 \cdot 10^{-6} \cdot \lambda^{-\frac{3}{2}} \cdot R_t^{\frac{1}{2}} \cdot \exp\left(-7.12 \cdot 10^{-5} \cdot R_t \cdot \lambda^{-\frac{1}{3}}\right) \cdot g1 \cdot h1 \cdot g2 \cdot h2$$

$$F_{sm} = 0.099$$

C. *Knife-Edge Diffraction*

$$a := \frac{2}{\lambda} \cdot \left(\frac{1}{R_h} + \frac{1}{R_t - R_h}\right)$$

$$a = -5.738 \times 10^{-4} \qquad \text{parameter}$$

$$par := \frac{R_h}{R_t} \cdot \left[\frac{(R_t - R_h)^2}{2 \cdot a_e} + \sigma_h\right] \cdot \sqrt{a}$$

$$par = 0.174i \qquad \text{parameter}$$

$$p := \begin{vmatrix} par & \text{if } a > 0 \\ 1 & \text{otherwise} \end{vmatrix}$$

$$p = 1$$

$$F_{ke\_db} := \begin{cases} -(6 + 8 \cdot p) & \text{if } p \le 1 \\ -\left(6.4 + 20 \cdot \log\left(\sqrt{p^2 + 1} + p\right)\right) & \text{otherwise} \end{cases}$$

dB, knife-edge diffraction propagation factor

$$F_{ke\_db} = -14$$

$$F_{ke} := 10^{\frac{F_{ke\_db}}{10}}$$

knife-edge diffraction propagation factor

$$F_{ke} = 0.04$$

D. *Diffraction PPF*

$$\sigma_h = 1$$

m, rms roughness of the surface specified

$$\sigma_{h\_cr} := \frac{\sqrt{\lambda \cdot R_h}}{2} \qquad \sigma_{h\_cr} = 50.213$$

m, the required rms roughness to support knife-edge diffraction

$$F := \begin{cases} F_{sm} & \text{if } \sigma_h \le \sigma_{h\_cr} \\ F_{ke} & \text{otherwise} \end{cases}$$

diffraction propagation factor

$$F_d := f\_0 \cdot F$$

pattern-propagation factor for region = 3

$$F_d = 1.715 \times 10^{-5}$$

$$F_{d\_db} := 20 \cdot \log(F_d)$$

dB, pattern-propagation factor for region = 3

$$F_{d\_db} = -95.315$$

Stage 5. Equations for Transition Region (region = 2)

$$F_{t\_db} := F_{r\_db} \cdot \frac{\delta 0}{Z} + F_{d\_db} \cdot \left(1 - \frac{\delta 0}{Z}\right)$$

dB, pattern-propagation factor for region = 2

$$F_{t\_db} = -91.477$$

Stage 6. Pattern-Propagation Factor for Specified Conditions

The pattern-propagation region is defined by the formula:

region :=

$\left|\begin{array}{l} \text{if } H_1 > 0 \\ \quad \left|\begin{array}{l} 1 \quad \text{if } \delta 0 > Z \\ 2 \quad \text{if } \delta 0 \le Z \\ \end{array}\right. \\ 3 \quad \text{if } \delta 0 \le Z \text{ if } H_1 \le 0 \end{array}\right.$

region = 2   Thus, in this problem the target is in the transition region.

The pattern-propagation factor is:

F_db := $\left|\begin{array}{ll} F_{r\_db} & \text{if region} = 1 \\ F_{t\_db} & \text{if region} = 2 \\ F_{d\_db} & \text{if region} = 3 \end{array}\right.$

F_db = −91.477   dB, PPF

Problem 11.6 Propagation Loss Due to Attenuation in Atmosphere.

The signal with frequency $f = 1000$ MHz is transmitted by an RF source and propagates in the standard atmosphere. Antenna height $h_a = 70$ m and elevation angle of the beam $\theta = 1$ deg. Find the loss due to attenuation in atmosphere at the distance $R = 100$ nautical miles from the source.

Solution. As given, the input parameters to calculate atmospheric attenuation loss L_a are as follows:

$f := 1000$ [MHz] $h_a := 70$ [m] $\theta := 1$ [deg] $R := 100$ [nm]

The predominant source of attenuation for this frequency is absorption of RF energy by atmospheric gas (molecules of oxygen). The problem can be solved by breaking it into the following stages.

Stage 1. Define Constant and Parameters of the Atmosphere

$N := 0.000313$	atmosphere refractivity
$c_e := -0.1439$	1/km, decay constant
$n0 := 1.000313$	refractive index
$r_{km} := 6370$	km, Earth radius
$r_m := 6370000$	m, Earth radius
$a_e := 8493333$	m, Earth effective radius taking into account refraction in atmosphere
$p0 := 1013.25$	mbar, atmospheric pressure constant

$\alpha1 := 5.2561222$

$\alpha2 := 0.034164794$ troposphere model constants

$\alpha3 := 11.388265$

$T0 := 300$ ^0K, standard temperature

$C := 2.0058$ the absorption coefficient constants

$z := \dfrac{\pi}{180}$ coefficient to transform degrees to radians

$K := 100$ number of points to calculate integral numerically

Oxygen Resonance Frequencies (GHz)

$f\_N\_plus_0 := 56.2648$	$f\_N\_minus_0 := 118.7505$	$Z_0 := 0$
$f\_N\_plus_1 := 56.2648$	$f\_N\_minus_1 := 118.7505$	$Z_1 := 1$
$f\_N\_plus_2 := 58.4466$	$f\_N\_minus_2 := 62.4862$	$Z_2 := 0$
$f\_N\_plus_3 := 58.4466$	$f\_N\_minus_3 := 62.4862$	$Z_3 := 1$
$f\_N\_plus_4 := 59.5910$	$f\_N\_minus_4 := 60.3061$	$Z_4 := 0$
$f\_N\_plus_5 := 59.5910$	$f\_N\_minus_5 := 60.3061$	$Z_5 := 1$
$f\_N\_plus_6 := 60.4348$	$f\_N\_minus_6 := 59.1642$	$Z_6 := 0$
$f\_N\_plus_7 := 60.4348$	$f\_N\_minus_7 := 59.1642$	$Z_7 := 1$
$f\_N\_plus_8 := 61.1506$	$f\_N\_minus_8 := 58.3239$	$Z_8 := 0$
$f\_N\_plus_9 := 61.1506$	$f\_N\_minus_9 := 58.3239$	$Z_9 := 1$
$f\_N\_plus_{10} := 61.8002$	$f\_N\_minus_{10} := 57.6125$	$Z_{10} := 0$
$f\_N\_plus_{11} := 61.8002$	$f\_N\_minus_{11} := 57.6125$	$Z_{11} := 1$
$f\_N\_plus_{12} := 62.4212$	$f\_N\_minus_{12} := 56.9682$	$Z_{12} := 0$
$f\_N\_plus_{13} := 62.4212$	$f\_N\_minus_{13} := 56.9682$	$Z_{13} := 1$
$f\_N\_plus_{14} := 62.9980$	$f\_N\_minus_{14} := 56.3634$	$Z_{14} := 0$
$f\_N\_plus_{15} := 62.9980$	$f\_N\_minus_{15} := 56.3634$	$Z_{15} := 1$
$f\_N\_plus_{16} := 63.5685$	$f\_N\_minus_{16} := 55.7839$	$Z_{16} := 0$
$f\_N\_plus_{17} := 63.5685$	$f\_N\_minus_{17} := 55.7839$	$Z_{17} := 1$
$f\_N\_plus_{18} := 64.1272$	$f\_N\_minus_{18} := 55.2214$	$Z_{18} := 0$

$f\_N\_plus_{19} := 64.1272$	$f\_N\_minus_{19} := 55.2214$	$Z_{19} := 1$
$f\_N\_plus_{20} := 64.6779$	$f\_N\_minus_{20} := 54.6728$	$Z_{20} := 0$
$f\_N\_plus_{21} := 64.6779$	$f\_N\_minus_{21} := 54.6728$	$Z_{21} := 1$
$f\_N\_plus_{22} := 65.2240$	$f\_N\_minus_{22} := 54.1294$	$Z_{22} := 0$
$f\_N\_plus_{23} := 65.2240$	$f\_N\_minus_{23} := 54.1294$	$Z_{23} := 1$
$f\_N\_plus_{24} := 65.7626$	$f\_N\_minus_{24} := 53.5960$	$Z_{24} := 0$
$f\_N\_plus_{25} := 65.7626$	$f\_N\_minus_{25} := 53.5960$	$Z_{25} := 1$
$f\_N\_plus_{26} := 66.2978$	$f\_N\_minus_{26} := 53.0695$	$Z_{26} := 0$
$f\_N\_plus_{27} := 66.2978$	$f\_N\_minus_{27} := 53.0695$	$Z_{27} := 1$
$f\_N\_plus_{28} := 66.8313$	$f\_N\_minus_{28} := 52.5458$	$Z_{28} := 0$
$f\_N\_plus_{29} := 66.8313$	$f\_N\_minus_{29} := 52.5458$	$Z_{29} := 1$
$f\_N\_plus_{30} := 67.3627$	$f\_N\_minus_{30} := 52.0259$	$Z_{30} := 0$
$f\_N\_plus_{31} := 67.3627$	$f\_N\_minus_{31} := 52.0259$	$Z_{31} := 1$
$f\_N\_plus_{32} := 67.8923$	$f\_N\_minus_{32} := 51.5091$	$Z_{32} := 0$
$f\_N\_plus_{33} := 67.8923$	$f\_N\_minus_{33} := 51.5091$	$Z_{33} := 1$
$f\_N\_plus_{34} := 68.4205$	$f\_N\_minus_{34} := 50.9949$	$Z_{34} := 0$
$f\_N\_plus_{35} := 68.4205$	$f\_N\_minus_{35} := 50.9949$	$Z_{35} := 1$
$f\_N\_plus_{36} := 68.9478$	$f\_N\_minus_{36} := 50.4830$	$Z_{36} := 0$
$f\_N\_plus_{37} := 68.9478$	$f\_N\_minus_{37} := 50.4830$	$Z_{37} := 1$
$f\_N\_plus_{38} := 69.4741$	$f\_N\_minus_{38} := 49.9730$	$Z_{38} := 0$
$f\_N\_plus_{39} := 69.4741$	$f\_N\_minus_{39} := 49.9730$	$Z_{39} := 1$
$f\_N\_plus_{40} := 70.0000$	$f\_N\_minus_{40} := 49.4648$	$Z_{40} := 0$
$f\_N\_plus_{41} := 70.0000$	$f\_N\_minus_{41} := 49.4648$	$Z_{41} := 1$
$f\_N\_plus_{42} := 70.5249$	$f\_N\_minus_{42} := 48.9582$	$Z_{42} := 0$
$f\_N\_plus_{43} := 70.5249$	$f\_N\_minus_{43} := 48.9582$	$Z_{43} := 1$
$f\_N\_plus_{44} := 71.0497$	$f\_N\_minus_{44} := 48.4530$	$Z_{44} := 0$
$f\_N\_plus_{45} := 71.0497$	$f\_N\_minus_{45} := 48.4530$	$Z_{45} := 1$

Stage 2. Calculation of Absorption Coefficient in Atmosphere

$R_{km} := 1.852 \cdot R$ $R_{km} = 185.2$ km, target range

$R_m := 1852 \cdot R$ $R_m = 1.852 \times 10^5$ m, target range

$$h_m := R_m \cdot \sin(z \cdot \theta) + \frac{\left(R_m \cdot \cos(z \cdot \theta)\right)^2}{2 \cdot a_e}$$ m, target height above antenna

$h_m = 5.251 \times 10^3$

$h_{km} := \dfrac{h_m}{1000}$ $h_{km} = 5.251$ km, target height

$\Delta h := \dfrac{h_{km}}{K}$ $\Delta h = 0.053$ km, target height increment

$k := 0 .. K$ cycle

$h_k := \dfrac{h_a}{1000} + k \cdot \Delta h$ km, current height

$h_{g\_m_k} := \dfrac{r_m \cdot h_k \cdot 1000}{r_m + h_k \cdot 1000}$ m, geopotential altitude

$h_{g\_km_k} := \dfrac{h_{g\_m_k}}{1000}$ $h_{g\_km_{100}} = 5.316$ km, geopotential altitude

^{0}K, temperature of the atmosphere

$$T_k := \begin{vmatrix} 288.16 - 0.0065 \cdot h_k \cdot 1000 & \text{if } h_{g\_km_k} \le 11 \\ 216.66 & \text{if } 11 < h_{g\_km_k} < 25 \\ 216.66 + 0.003 \cdot \left(h_k - 25\right) \cdot 1000 & \text{otherwise} \end{vmatrix}$$

Millibars, pressure of the atmosphere

$$p_k := \begin{vmatrix} p0 \cdot \left(\dfrac{T_k}{288.16} \right)^{\alpha 1} & \text{if } h_{g\_km_k} \le 11 \\[2em] \dfrac{226.32}{T_k} \cdot e^{-\alpha 2 \cdot (h_k - 11) \cdot 1000} & \text{if } 11 < h_{g\_km_k} < 25 \\[2em] 24.886 \cdot \left(\dfrac{216.66}{T_k} \right)^{\alpha 3} & \text{otherwise} \end{vmatrix}$$

$$f\_GHz := \frac{f}{1000} \qquad\qquad f\_GHz = 1 \qquad\qquad \text{GHz, RF source frequency}$$

The line-breadth constant parameter

$$g_k := \begin{vmatrix} 0.640 & \text{if } h_k \le 8 \\ 0.640 + 0.04218 \cdot (h_k - 8) & \text{if } 8 < h_k \le 25 \\ 1.357 & \text{otherwise} \end{vmatrix}$$

$$\Delta f_k := g_k \cdot \frac{p_k}{p0} \cdot \frac{T0}{T_k} \qquad\qquad\qquad \text{the line-breadth constant}$$

$$F0_k := \frac{\Delta f_k}{\left(\Delta f_k \right)^2 + f\_GHz^2} \qquad\qquad \text{the nonresonant contribution}$$

$$n := 1 .. 45 \qquad\qquad\qquad\qquad\qquad\qquad\qquad \text{cycle}$$

Parameters to calculate the absorption coefficient

$$\mu\_plus_n := \frac{n \cdot (2 \cdot n + 3)}{n + 1}$$

$$\mu\_minus_n := \frac{(n + 1) \cdot (2 \cdot n - 1)}{n}$$

$$\mu\_0_n := \frac{2 \cdot \left(n^2 + n + 1 \right) \cdot (2 \cdot n + 1)}{n \cdot (n + 1)}$$

$$\Sigma 1\_plus_{k,n} := \frac{\Delta f_k}{\left(f\_N\_plus_n - f\_GHz\right)^2 + \left(\Delta f_k\right)^2}$$

$$\Sigma 2\_plus_{k,n} := \frac{\Delta f_k}{\left(f\_N\_plus_n + f\_GHz\right)^2 + \left(\Delta f_k\right)^2}$$

$$\Sigma 1\_minus_{k,n} := \frac{\Delta f_k}{\left(f\_N\_minus_n - f\_GHz\right)^2 + \left(\Delta f_k\right)^2}$$

$$\Sigma 2\_minus_{k,n} := \frac{\Delta f_k}{\left(f\_N\_minus_n + f\_GHz\right)^2 + \left(\Delta f_k\right)^2}$$

$$F\_N\_plus_{k,n} := \Sigma 1\_plus_{k,n} + \Sigma 2\_plus_{k,n}$$

$$F\_N\_minus_{k,n} := \Sigma 1\_minus_{k,n} + \Sigma 2\_minus_{k,n}$$

$$A1_{k,n} := \mu\_plus_n \cdot F\_N\_plus_{k,n}$$

$$A2_{k,n} := \mu\_minus_n \cdot F\_N\_minus_{k,n}$$

$$E_n := 2.06844 \cdot n \cdot (n + 1)$$

$$\Sigma_k := \sum_{n=1}^{45} Z_n \cdot \left[\left(A1_{k,n} + A2_{k,n} + \mu\_0_n \cdot F0_k\right) \cdot \exp\left(-\frac{E_n}{T_k}\right)\right]$$

$$\gamma_k := C \cdot p_k \cdot \left(T_k\right)^{-3} \cdot f\_GHz^2 \cdot \Sigma_k \qquad\qquad \text{dB/km, absorption coefficient}$$

Stage 3. Calculation of Propagation Loss Based on the Absorption Coefficient

Refractive index of a standard atmosphere is:

$$m_k := 1 + N \cdot \exp\left(c_e \cdot h_k\right)$$

Absorption loss as the function of the absorption coefficient γ is given by the integral:

$$L_a := \int_0^h \frac{\gamma}{\sqrt{1 - \left[\dfrac{n0 \cdot \cos(z \cdot \theta)}{m \cdot \left(1 + \dfrac{h}{r_{km}}\right)}\right]^2}} \, dh \qquad \text{dB, absorption loss}$$

It can be evaluated with the aid of a computer (see Chapter 9), but often it is time-consuming to compute the integral above as is. In order to obtain much quicker results with acceptable accuracy this integral can be represented in discrete form and calculated approximately based on *trapezoid rule* (see Section 3.2.10):

$$y_0 := \frac{\gamma_0}{\sqrt{1 - \left[\dfrac{n0 \cdot \cos(z \cdot \theta)}{m_0 \cdot \left(1 + \dfrac{h_0}{r_{km}}\right)}\right]^2}} \qquad y_K := \frac{\gamma_K}{\sqrt{1 - \left[\dfrac{n0 \cdot \cos(z \cdot \theta)}{m_K \cdot \left(1 + \dfrac{h_K}{r_{km}}\right)}\right]^2}}$$

$$S := \sum_{k=1}^{K-1} \frac{\gamma_k}{\sqrt{1 - \left[\dfrac{n0 \cdot \cos(z \cdot \theta)}{m_k \cdot \left(1 + \dfrac{h_k}{r_{km}}\right)}\right]^2}}$$

$$L_\alpha := (y_0 + y_K + 2 \cdot S) \cdot \frac{\Delta h}{2}$$

$$L_\alpha = 0.786 \qquad \text{dB, atmospheric attenuation loss}$$

Problem 11.7 Scattering Cross Section. Find the scattering cross section (or radar cross section RCS, or echo area) of the conductive sphere with radius R = 1 m illuminated by a radar with frequency f = 1000 MHz. Plot the dependence of RCS versus sphere radius.

Solution. As given, the parameters are as follows:

$R := 1$ [m] $\quad$ $f := 1000$ [MHz]

The following constants and parameters can be introduced:

$c := 2.997925 \cdot 10^8$ $\qquad\qquad\qquad$ m/s, velocity of light

$\lambda := \dfrac{c}{f \cdot 10^6}$ $\quad \lambda = 0.3$ $\qquad\qquad\qquad$ m, wavelength

$k := \dfrac{2 \cdot \pi}{\lambda}$ $\quad k = 20.958$ $\qquad\qquad\qquad$ wave number

$\rho := k \cdot R$ $\quad \rho = 20.958$ $\quad$ the ratio of sphere circumference to wavelength

Coefficients in series for RCS calculation, where js and ys are spherical Bessel functions of the first and second kind of order n, are given by the equations:

$$a(n,\rho) := \frac{js(n,\rho)}{js(n,\rho) - j \cdot ys(n,\rho)}$$

$$b(n,\rho) := \frac{-\left[\dfrac{d}{d\rho}(\rho \cdot js(n,\rho))\right]}{\dfrac{d}{d\rho}(\rho \cdot js(n,\rho) - j \cdot \rho \cdot ys(n,\rho))}$$

RCS of a sphere [m²] is given by the formula:

$$RCS := \pi \cdot R^2 \cdot \left[\frac{1}{\rho^2} \cdot \left[\left|\sum_{n=1}^{2 \cdot \text{ceil}(\rho)} (-1)^n \cdot (2 \cdot n + 1) \cdot (a(n,\rho) + b(n,\rho))\right|\right]^2\right]$$

Note. The summation should actually extend to n = ∞, but it is sufficient to set the upper limit to n > ρ with some margin.

$RCS = 3.027$ $\qquad\qquad\qquad\qquad$ m², RCS

The dependence of the sphere RCS versus its normalized radius is given in Figure 11.9.

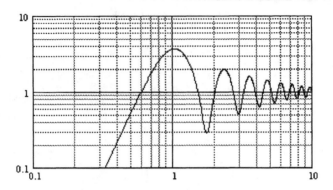

Figure 11.9 The RCS of a conducting sphere normalized to its projected area πR^2 vs. the ratio of sphere circumference to wavelength ρ.

Chapter 12

WAVEFORMS AND SIGNAL PROCESSING

Problem 12.1 Harmonic Oscillation. Write the equation for the signal which is the harmonic oscillation of duration 1 s with unity amplitude, zero initial phase, and carrier frequency 10 Hz. Find the expression for the instantaneous frequency f(t). Plot time-domain and frequency-domain representations of the waveform. Calculate the average power and energy of the signal.

Solution. The mathematical equation for harmonic oscillation with amplitude A(t), initial phase ϕ, and carrier frequency f_0 is:

$$V(t) := A(t) \cdot \cos(\Phi(t)) \tag{12.1}$$

where $\Phi(t) := 2 \cdot \pi \cdot f_0 \cdot t + \phi$ is the complete phase.

The complex waveform can be written as:

$$WF(t) := A(t) \cdot \exp(-j \cdot \Phi(t)) \quad \text{or in the equivalent form:}$$

$$WF(t) := A(t) \cdot \exp(-j \cdot \phi) \cdot \exp(-j \cdot 2 \cdot \pi \cdot f_0 \cdot t)$$

Based on Euler's formula (see Section 4.1.9) the equation of the harmonic oscillation can be represented as:

$$V(t) := Re(WF(t)) \quad \text{or in the equivalent form:}$$

$$V(t) := Re(A(t) \cdot \exp(-j \cdot \Phi(t))) \quad \text{or in the equivalent form:}$$

$$V(t) := Re(A(t) \cdot \cos(2 \cdot \pi \cdot f_0 \cdot t + \phi) - j \cdot \sin(2 \cdot \pi \cdot f_0 \cdot t + \phi))$$

Since Re{X} is the operation of extraction of the real part of the expression {X}

$$V(t) := A(t) \cdot \cos(2 \cdot \pi \cdot f_0 \cdot t + \phi) \quad \text{that matches (12.1).}$$

The instantaneous circular frequency $\omega(t)$ is the derivative of the complete phase $\Phi(t)$:

327

$$\omega(t) := \frac{d}{dt}\Phi(t)$$

and the linear instantaneous frequency: $f(t) := \dfrac{\omega(t)}{2 \cdot \pi}$

Thus: $\omega(t) := \dfrac{d}{dt}\left(2 \cdot \pi \cdot f_0 \cdot t + \phi\right)$

that results in $\omega(t) := 2 \cdot \pi \cdot f_0$ or $f(t) := f_0$

Time-Domain Representation

In order to graph the waveform, we must represent it in discrete form and specify:

$f\_s := 10^3$	Hz, sampling rate
$\tau := 1$	s, duration
$f_0 := 10$	Hz, carrier frequency
$\phi := 0$	deg, initial phase

This results in the following parameters:

$\Delta t := \dfrac{1}{f\_s}$ $\Delta t = 1 \times 10^{-3}$ s, sampling interval

$N := \text{floor}\left(\dfrac{\tau}{\Delta t}\right)$ $N = 1 \times 10^3$ number of samples in the waveform

                                    sampling cycle

$n := 0 .. N - 1$

Since amplitude is equal to unity and constant: $A_n := 1$

The expression for the complex waveform $WF(t_n) = WF_n$ is:

$$WF_n := A_n \cdot \exp\left[-j\left(2 \cdot \pi \cdot f_0 \cdot \Delta t \cdot n + \phi\right)\right]$$

Thus, the real waveform is: $V := \text{Re}(WF)$

The graph of the waveform $V(t_n) = V_n$ is given in Figure 12.1.

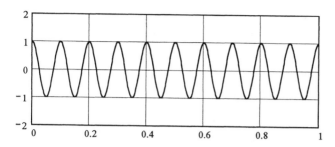

Figure 12.1 Amplitude (V) of the harmonic oscillation with $f_0 = 10$ Hz vs. time (s).

The instantaneous frequency $f(t_n) = f_n$ of the harmonic oscillation is constant (Figure 12.2):

$$f_n := f_0 \qquad f_0 = 10 \qquad [Hz]$$

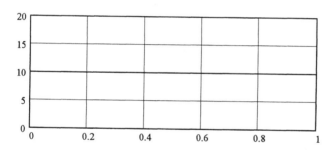

Figure 12.2 The instantaneous frequency (Hz) of the harmonic oscillation with $f_0 = 10$ Hz vs. time (s).

Power and Energy

Average power of the signal and its energy can be calculated as:

$$POWER(WF) := \left[\sum_n \left(|WF_n| \right)^2 \right] \cdot \frac{1}{N} \qquad [W], \text{ average power of the signal}$$

$$POWER(WF) = 1 \qquad [W]$$

$$\text{ENERGY(WF)} := \left[\sum_n \left(|WF_n| \right)^2 \right] \cdot \Delta t \qquad \text{[W*s], energy of the signal}$$

$$\text{ENERGY(WF)} = 1 \qquad \text{[W*s]}$$

Frequency-Domain Representation

In order to graph the spectrum of the waveform, we must also specify:

$y\_th := -3$ $\qquad\qquad\qquad$ dB, level to calculate spectrum width

$I := 2^{12}$ $\qquad I = 4.096 \times 10^3$ $\qquad$ number of samples to calculate spectrum

The waveform must be extended in order to have a sufficient number of samples to calculate the spectrum.

$i\_s := \text{ceil}\left(\dfrac{I - N}{2} \right)$ $\qquad i\_s = 1.548 \times 10^3$ $\qquad$ first nonzero sample

$i\_f := i\_s + N - 1$ $\qquad i\_f = 2.547 \times 10^3$ $\qquad$ last nonzero sample

$i := 0 .. I - 1$ $\qquad\qquad\qquad\qquad\qquad\qquad$ cycle

$EWF_i := 0$

$EWF_{i\_s+n} := WF_n$ $\qquad\qquad$ extended waveform at the frequency f0

$EV_i := \text{Re}\left(EWF_i \right)$ $\qquad\qquad$ extended waveform voltage

The spectrum of the waveform is the Fourier transform of its time-domain representation (see Appendix 12).

$SP\_EWF := \text{cfft(EWF)}$ $\qquad\qquad$ extended waveform spectrum

$I := \text{last(SP\_EWF)}$ $\quad I = 4.095 \times 10^3$ $\qquad$ last sample in the waveform spectrum

$\Delta f := \dfrac{f\_s}{I}$ $\qquad \Delta f = 0.244$ $\qquad$ Hz, frequency sampling interval

$i\_0 := \dfrac{I + 1}{2}$ $\qquad i\_0 = 2.048 \times 10^3$ $\qquad\qquad$ central sample

$f_i := (i - i\_0) \cdot \Delta f$ $\qquad\qquad\qquad\qquad$ Hz, current frequency

$f_0 = 10$ Hz, carrier frequency

$f_0 = -500.122$

$f_I = 499.878$ Hz, frequency range

Note. The carrier frequency f0 must be within f_0 - f_M range. To increase this range, increase sampling rate f_s.

The procedure to center spectrum around zero frequency is:

i1_s := 0   il_s = 0

i1_f := i_0 − 1   il_f = 2.047×10^3

i2_s := 0   i2_s = 0

i2_f := i_0 − 1   i2_f = 2.047×10^3

i1 := il_s .. il_f   i2 := i2_s .. i2_f   cycles i1 and i2

$SP0\_EWF_{i1} := SP\_EWF_{i\_0+i1}$

$SP0\_EWF_{i2+i\_0} := SP\_EWF_{i2}$ spectrum centered around zero frequency

$SP_i := 20 \cdot \log\left(\left|SP0\_EWF_i + 10^{-10}\right|\right)$ dB, waveform spectrum

$SP\_MAX := \max(SP)$ spectrum maximum

$SPN := SP - SP\_MAX$ dB, normalized waveform spectrum

The spectrum width calculation is as follows:

Hz, frequency where the spectrum curve crosses the specified level upward

$$f\_b(SPC) := \begin{array}{|l} \text{for } i \in 1 .. I - 2 \\ \quad \text{break if } SPC_{i-1} \leq y\_th \text{ if } SPC_i > y\_th \\ \quad df\_b \leftarrow \dfrac{SPC_i - y\_th}{SPC_i - SPC_{i-1}} \cdot \Delta f \\ \quad f\_b \leftarrow f_i - df\_b \\ \quad f\_b \end{array}$$

Hz, frequency where the spectrum curve crosses the specified level downward

$$
\text{f\_e(SPC)} := \left|
\begin{array}{l}
\text{for } i \in 1 .. I - 1 \\
\quad \text{break if } SPC_{i-1} > y\_th \text{ if } SPC_i \leq y\_th \\
\quad df\_e \leftarrow \dfrac{SPC_{i-1} - y\_th}{SPC_{i-1} - SPC_i} \cdot \Delta f \\
\quad f\_e \leftarrow f_i + df\_e \\
\quad f\_e
\end{array}
\right.
$$

dSP := f_e(SPN) − f_b(SPN)   dSP = 1.1   Hz, waveform spectrum width
at   y_th = −3   dB level

Note. The spectrum width calculation procedure works only for monotonic spectrum functions at levels higher than y_th dB. Be cautious when using this for oscillating spectrum shapes with peak-to-valley spans more than y_th dB.

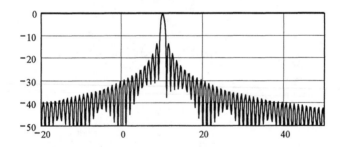

Figure 12.3 Normalized spectrum (dB) of the harmonic oscillation with $f_0 = 10$ Hz vs. frequency (Hz).

Problem 12.2 Amplitude-Modulated Signal. Write the equation for the amplitude-modulated harmonic oscillation of duration 1 second with cosine amplitude modulation with amplitude 5 V and modulation frequency 1 Hz, zero initial phase, and carrier frequency 10 Hz. Find the expression for the instantaneous frequency f(t). Plot time-domain and frequency-domain representations of the waveform. Calculate the average power and energy of the signal.

Solution. The mathematical equation for the modulated harmonic oscillation is the same as (12.1):

$$V(t) := A(t) \cdot \cos\big(\Phi(t)\big)$$

Only now the amplitude, instead of being constant over the time A = 1, varies as:

$A_0 := 5$ $\qquad$ $f_M := 1$ $\qquad\qquad$ $A(t) := A_0 \cdot \cos(t)$

Time-Domain Representation

Thus, the expression for the complex waveform $WF(t_n) = WF_n$ is:

$$WF_n := A_0 \cdot \cos\left(2 \cdot \pi \cdot f_M \cdot \Delta t \cdot n\right) \cdot \exp\left[-j \cdot \left(2 \cdot \pi \cdot f_0 \cdot \Delta t \cdot n + \phi\right)\right]$$

The real waveform is: $V := \mathrm{Re}(WF)$

The graph of the waveform $V(t_n) = V_n$ is given in Figure 12.4.

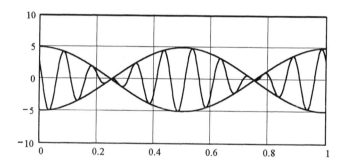

Figure 12.4 Amplitude (V) of harmonic oscillation with $f_0 = 10$ Hz of duration
1 s with amplitude modulation with amplitude 5 V and modulation
frequency 1 Hz vs. time (s).

The instantaneous frequency $f(t_n) = f_n$ of the amplitude-modulated harmonic
oscillation is also constant (Figure 12.5):

$f_n := f_0$ $\qquad$ $f_0 = 10$ $\qquad$ [Hz]

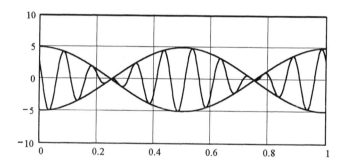

Figure 12.5 The instantaneous frequency (Hz) of the amplitude-modulated
harmonic oscillation with $f_0 = 10$ Hz vs. time (s).

Power and Energy

Average power of the signal and its energy can be calculated as:

$$\text{POWER(WF)} := \left[\sum_n \left(|WF_n| \right)^2 \right] \cdot \frac{1}{N} \qquad \text{[W], average power of the signal}$$

$$\text{POWER(WF)} = 12.5 \quad \text{[W]}$$

$$\text{ENERGY(WF)} := \left[\sum_n \left(|WF_n| \right)^2 \right] \cdot \Delta t \qquad \text{[W*s], energy of the signal}$$

$$\text{ENERGY(WF)} = 12.5 \quad \text{[W*s]}$$

Frequency-Domain Representation

Spectrum of the waveform can be found using the algorithm cited in Problem 12.1:

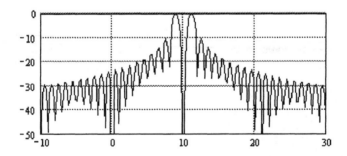

Figure 12.6 Normalized spectrum (dB) of harmonic oscillation $f_0 = 10$ Hz of
duration 1 s with amplitude modulation with modulation frequency
1 Hz vs. frequency (Hz).

Problem 12.3 Linear Frequency-Modulated Signal. Write the equation
for the linear frequency-modulated (LFM) oscillation of amplitude 1 V, carrier
frequency 5 MHz, duration $\tau = 50$ μs with modulation law $\psi(t) = - k_f \pi t^2$ and index
of modulation $k_f = 0.7$. Find the expression for the instantaneous frequency $f(t)$.
Plot time-domain and frequency-domain representations of the waveform.
Calculate the average power and energy of the signal.

Solution. The equation for the frequency-modulated oscillation with amplitude A(t) = A_0, initial phase ϕ, and carrier frequency f(t) is given by (12.1):

$$V(t) := A_0 \cdot \cos\big(\Phi(t)\big)$$

where $\Phi(t) := \omega_0 \cdot t + \psi(t) + \phi$ is the complete phase,

$\psi(t)$ is the factor describing phase variation in time.

Based on this representation of the complete phase, the complex waveform can be written as:

$$WF(t) := A_0 \cdot \exp\big(-j \cdot \psi(t)\big) \cdot \exp\Big[-\big(j \cdot 2 \cdot \pi \cdot f_0 \cdot t + \phi\big)\Big]$$

The factor $WF\_b(t) := A_0 \exp\big(-j \cdot \psi(t)\big)$ is called the baseband waveform

and the formula for the complex waveform at freqeuncy f_0 can be written as:

$$WF(t) := WF\_b(t) \cdot \exp\Big[-\big(j \cdot 2 \cdot \pi \cdot f_0 \cdot t + \phi\big)\Big]$$

The instantaneous circular frequency $\omega(t)$ is the derivative of the complete phase $\Phi(t)$:

$$\omega(t) := \frac{d}{dt}\Phi(t) \qquad \omega(t) := \frac{d}{dt}\big(\omega_0 \cdot t + \psi(t) + \phi\big) \qquad \omega(t) := \omega_0 + \frac{d}{dt}\psi(t)$$

Since the modulation law is given as:

$$k_f := 0.7 \quad \psi(t) := -k_f \cdot \tau \cdot t^2 \quad \text{then} \quad \omega(t) := \omega_0 - 2 \cdot k_f \cdot \tau \cdot t$$

and the linear instantaneous frequency $f(t) := \dfrac{\omega(t)}{2 \cdot \pi}$

Thus: $f(t) := f_0 - \dfrac{1}{\pi} \cdot k_f \cdot \tau \cdot t$

Time-Domain Representation

In order to graph the waveform, we have to represent it in discrete form and specify:

$f\_s := 20$	MHz, sampling rate
$\tau := 50$	μs, duration
$f_0 := 5$	MHz, carrier frequency

$\phi := 0$ deg, initial phase

This results in the following parameters:

$$\Delta t := \frac{1}{f\_s} \qquad \Delta t = 0.05 \qquad \qquad \text{μs, sampling interval}$$

$$N := \text{floor}\left(\frac{\tau}{\Delta t}\right) \qquad N = 1 \times 10^3 \qquad \text{number of samples in the waveform}$$

$$n := 0 .. N - 1 \qquad\qquad\qquad\qquad \text{sampling cycle}$$

To normalize and center the time sample, let us introduce the discrete time t_n (μs):

$$T_n := \frac{n}{f\_s} - \frac{1}{2 \cdot f\_s} \qquad t_n := \frac{T_n - \dfrac{\tau}{2}}{\left(\dfrac{\tau}{2}\right)}$$

Phase-modulation law for a linear frequency modulation is:

$$\psi_n := -k_f \cdot \tau \cdot (t_n)^2$$

The complex baseband waveform is: $WF\_b_n := A_0 \cdot \exp(-j \cdot \psi_n)$

The real part of the baseband waveform is: $V\_b := \text{Re}(WF\_b)$

The complex waveform at frequency f_0 is:

$$WF_n := WF\_b_n \cdot \exp\left[-\left(j \cdot 2 \cdot \pi \cdot f_0 \cdot \Delta t \cdot n + \phi\right)\right]$$

The real waveform is: $V := \text{Re}(WF)$

The graphs of the baseband waveform and waveform at frequency f_0 are given in Figures 12.7 and 12.8, correspondingly.

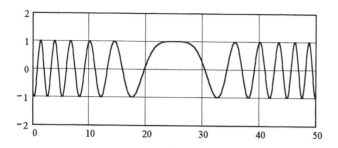

Figure 12.7 Amplitude (V) of the baseband LFM oscillation vs. time (μs).

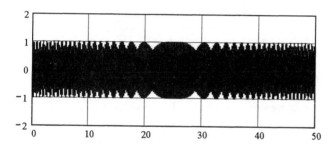

Figure 12.8 Amplitude (V) of the LFM oscillation with f_0 = 5 MHz vs. time (μs).

The instantaneous frequency $f(t_n) = f_n$ of the LFM oscillation (Figure12.9) is given by the equation:

$$f_n := f_0 - \frac{1}{\pi} \cdot k_f \cdot \tau \cdot t_n$$

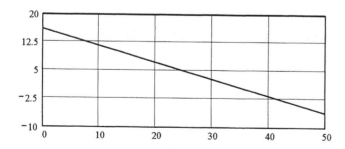

Figure 12.9 The instantaneous frequency (MHz) of the LFM oscillation with f_0 = 5 MHz vs. time (μs).

Power and Energy

Average power of the signal and its energy can be calculated as:

$$\text{POWER(WF)} := \left[\sum_n \left(|WF_n| \right)^2 \right] \cdot \frac{1}{N} \qquad \text{[W], average power of the signal}$$

$$\text{POWER(WF)} = 1 \qquad \text{[W]}$$

$$\text{ENERGY(WF)} := \left[\sum_n \left(|WF_n| \right)^2 \right] \cdot \Delta t \qquad \text{[W*µs], energy of the signal}$$

$$\text{ENERGY(WF)} = 50 \qquad \text{[W*µs]}$$

Frequency-Domain Representation

Spectrum of the waveform can be found using the algorithm cited in Problem 12.1.

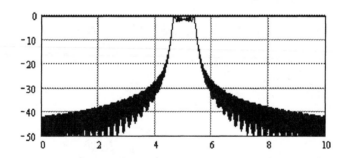

Figure 12.10 Normalized spectrum (dB) of the LFM oscillation $f_0 = 5$ MHz of duration 50 µs vs. frequency (MHz).

Problem 12.4 Nonlinear Frequency-Modulated Signal. Write the equation for the nonlinear frequency-modulated (NFM) oscillation with parameters, as in the previous problem and modulation law:

$$\psi(t) := k_f \cdot \tau \cdot \left[(-1)^{index} \cdot (t)^2 + \left[a\cos\left[(t)^2 \right] \right]^{\frac{1}{index+1}} \right] \qquad index := 3$$

Find the expression for the instantaneous frequency f(t). Plot time-domain and frequency-domain representations of the waveform. Calculate the average power and energy of the signal.

Solution. The equation for the nonlinear frequency-modulated complex waveform is the same as in the previous problem:

$$WF(t) := WF\_b(t) \cdot exp\left[-\left(j \cdot 2 \cdot \pi \cdot f_0 \cdot t + \phi \right) \right] \qquad WF\_b(t) := A_0 \, exp\left(-j \cdot \psi(t) \right)$$

The instantaneous circular frequency $\omega(t)$ is the derivative of the complete phase $\Phi(t)$:

$$\omega(t) := \frac{d}{dt}\Phi(t) \quad \omega(t) := \frac{d}{dt}\left(\omega_0 \cdot t + \psi(t) + \phi \right) \qquad \omega(t) := \omega_0 + \frac{d}{dt}\psi(t)$$

The derivative of the function $\dfrac{d}{dt}\psi(t)$ can be found with the aid of a computer (see Chapter 9):

$$\frac{d}{dt}\psi(t) := k_f \cdot \tau \cdot \left[-2 \cdot t - \frac{1}{\left[2 \cdot acos\left(t^2\right)^{\left(\frac{3}{4}\right)} \right]} \cdot \frac{t}{\left(1 - t^4\right)^{\left(\frac{1}{2}\right)}} \right]$$

and the linear instantaneous frequency: $\quad f(t) := \dfrac{\omega(t)}{2 \cdot \pi}$

Thus: $\quad f(t) := f_0 + \dfrac{1}{2 \cdot \pi} \cdot \left[k_f \cdot \tau \cdot \left[-2 \cdot t - \dfrac{1}{\left[2 \cdot acos\left(t^2\right)^{\left(\frac{3}{4}\right)} \right]} \cdot \dfrac{t}{\left(1 - t^4\right)^{\left(\frac{1}{2}\right)}} \right] \right]$

Time-Domain Representation

f_s := 1000   MHz, sampling rate

Note. Since the interpulse modulation is essentially nonlinear, the sampling rate must be fairly high in order to obtain the right results in the calculation of the power and energy.

$\tau := 50$ μs, duration

$f_0 := 5$ MHz, carrier frequency

$\phi := 0$ deg, initial phase

$\Delta t := \dfrac{1}{f\_s}$ $\Delta t = 1 \times 10^{-3}$ μs, sampling interval

$N := \text{floor}\left(\dfrac{\tau}{\Delta t}\right)$ $N = 5 \times 10^4$ number of samples in the waveform

$n := 0 .. N - 1$ sampling cycle

$T_n := \dfrac{n}{f\_s} - \dfrac{1}{2 \cdot f\_s}$ $t_n := \dfrac{T_n - \dfrac{\tau}{2}}{\left(\dfrac{\tau}{2}\right)}$ μs, time samples

Phase-modulation law for a nonlinear frequency modulation is:

$$\psi_n := k_f \cdot \tau \cdot \left[(-1)^{\text{index}} \cdot \left(t_n\right)^2 + \left[a\cos\left[\left(t_n\right)^2\right]^{\overline{\dfrac{1}{\text{index}+1}}}\right]\right]$$

The complex baseband waveform: $WF\_b_n := A_0 \cdot \exp\left(-j \cdot \psi_n\right)$

The real part of the baseband waveform is: $V\_b := \text{Re}(WF\_b)$

The complex waveform at frequency f_0 is:

$$WF_n := WF\_b_n \cdot \exp\left[-\left(j \cdot 2 \cdot \pi \cdot f_0 \cdot \Delta t \cdot n + \phi\right)\right]$$

The real waveform is: $V := \text{Re}(WF)$

The graphs of the baseband waveform and waveform at frequency f_0 are given in Figures 12.11 and 12.12, correspondingly.

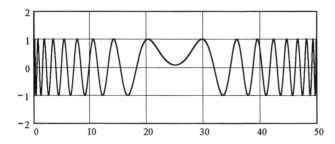

Figure 12.11 Amplitude (V) of the baseband NFM oscillation vs. time (μs).

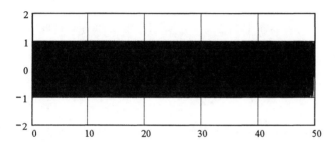

Figure 12.12 Amplitude (V) of the NFM oscillation with f_0 = 5 MHz vs. time (μs).

The instantaneous frequency $f(t_n) = f_n$ of the NFM oscillation (Figure 12.13) is given by the equation:

$$f_n := f_0 + \frac{1}{2 \cdot \pi} \cdot \left[k_f \cdot \tau \cdot \left[-2 \cdot t_n - \frac{1}{\left[2 \cdot acos\left[\left(t_n\right)^2 \right]^{\left(\frac{3}{4}\right)} \right]} \cdot \frac{t_n}{\left[1 - \left(t_n\right)^4 \right]^{\left(\frac{1}{2}\right)}} \right] \right]$$

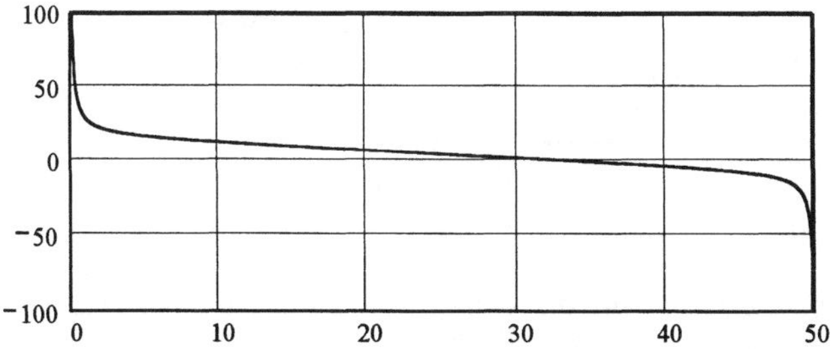

Figure 12.13 The instantaneous frequency (MHz) of the NFM oscillation
with f_0 = 5 MHz vs. time (μs).

Power and Energy

Average power of the signal and its energy can be calculated as:

$$\text{POWER(WF)} := \left[\sum_{n} \left(|WF_n| \right)^2 \right] \cdot \frac{1}{N} \qquad \text{[W], average power of the signal}$$

$$\text{POWER(WF)} = 1.076 \qquad \text{[W]}$$

$$\text{ENERGY(WF)} := \left[\sum_{n} \left(|WF_n| \right)^2 \right] \cdot \Delta t \qquad \text{[W*}\mu\text{s], energy of the signal}$$

$$\text{ENERGY(WF)} = 53.781 \qquad \text{[W*}\mu\text{s]}$$

Note. Theoretically average power $\to 1$ W and energy $\to 50$ W*μs when $\Delta t \to 0$.

Frequency-Domain Representation

Spectrum of the waveform can be found using the algorithm cited in Problem 12.1.

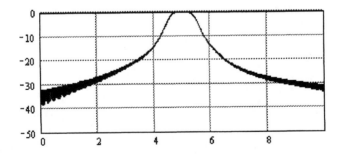

Figure 12.14 Normalized spectrum (dB) of the NFM oscillation $f_0 = 5$ MHz of duration 50 μs vs. frequency (MHz).

Problem 12.5 Phase-Modulated Signal. Write the equation for the phase-modulated (PHM) oscillation with parameters as in the Problem 12.3 and modulation law given by Barker code of length = 13. Plot time-domain and frequency-domain representations of the waveform. Calculate the average power and energy of the signal.

Solution. The equation for the phase-modulated complex waveform is the same as in the previous problem:

$$WF(t) := WF\_b(t)\cdot exp\left[-\left(j\cdot 2\cdot\pi\cdot f_0\cdot t + \phi\right)\right] \qquad WF\_b(t) := A_0 exp\left(-j\cdot\psi(t)\right)$$

Time-Domain Representation

As in the Problem 12.3:

$f\_s := 20$ MHz, sampling rate

$\tau := 50$ μs, duration

$f_0 := 5$ MHz, carrier frequency

$\phi := 0$ deg, initial phase

$\Delta t := \dfrac{1}{f\_s}$ $\Delta t = 0.05$ μs, sampling interval

$N := floor\left(\dfrac{\tau}{\Delta t}\right)$ $N = 1\times 10^3$ number of samples in the waveform

$n := 0..N-1$ sampling cycle

$T_n := \dfrac{n}{f\_s} - \dfrac{1}{2\cdot f\_s}$ $t_n := \dfrac{T_n - \dfrac{\tau}{2}}{\left(\dfrac{\tau}{2}\right)}$ μs, time samples

The following parameters can also be specified:

$LENGTH := 13$

$\Delta\tau := \dfrac{\tau}{LENGTH}$ $\Delta\tau = 3.846$ μs, phase modulation increment

$K := floor\left(\dfrac{\Delta\tau}{\Delta t}\right)$ $K = 76$ number of samples within modulation increment

The phase-modulation law for a modulation with Barker code is:

$$\psi_n := \begin{vmatrix} 0 & \text{if } n \le K \\ 0 & \text{if } K < n \le 2 \cdot K \\ 0 & \text{if } 2 \cdot K < n \le 3 \cdot K \\ 0 & \text{if } 3 \cdot K < n \le 4 \cdot K \\ 0 & \text{if } 4 \cdot K < n \le 5 \cdot K \\ \pi & \text{if } 5 \cdot K < n \le 6 \cdot K \\ \pi & \text{if } 6 \cdot K < n \le 7 \cdot K \\ 0 & \text{if } 7 \cdot K < n \le 8 \cdot K \\ 0 & \text{if } 8 \cdot K < n \le 9 \cdot K \\ \pi & \text{if } 9 \cdot K < n \le 10 \cdot K \\ 0 & \text{if } 10 \cdot K < n \le 11 \cdot K \\ \pi & \text{if } 11 \cdot K < n \le 12 \cdot K \\ 0 & \text{if } n > 12 \cdot K \end{vmatrix}$$

The complex baseband waveform: $WF\_b_n := A_0 \cdot \exp\left(-j \cdot \psi_n\right)$

The real part of the baseband waveform is: $V\_b := \text{Re}(WF\_b)$

The complex waveform at frequency f_0 is:

$$WF_n := WF\_b_n \cdot \exp\left[-\left(j \cdot 2 \cdot \pi \cdot f_0 \cdot \Delta t \cdot n + \phi\right)\right]$$

The real waveform is: $V := \text{Re}(WF)$

The graphs of the baseband waveform and waveform at frequency f_0 are given in Figures 12.15 and 12.16, correspondingly.

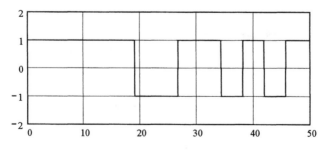

Figure 12.15 Amplitude (V) of the baseband PHM oscillation vs. time (μs).

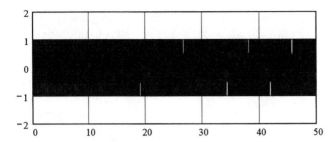

Figure 12.16 Amplitude (V) of the PHM oscillation with $f_0 = 5$ MHz vs. time (μs).

Power and Energy

Average power of the signal and its energy can be calculated as:

$$\text{POWER(WF)} := \left[\sum_n \left(|\text{WF}_n| \right)^2 \right] \cdot \frac{1}{N} \qquad \text{[W], average power of the signal}$$

$$\text{POWER(WF)} = 1 \qquad \text{[W]}$$

$$\text{ENERGY(WF)} := \left[\sum_n \left(|\text{WF}_n| \right)^2 \right] \cdot \Delta t \qquad \text{[W*}\mu\text{s], energy of the signal}$$

$$\text{ENERGY(WF)} = 50 \qquad \text{[W*}\mu\text{s]}$$

Frequency-Domain Representation

The spectrum of the waveform can be found using the algorithm cited in Problem 12.1.

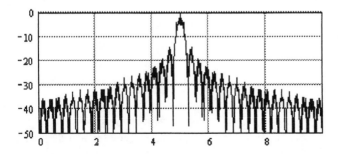

Figure 12.17 Normalized spectrum (dB) of the PHM oscillation $f_0 = 5$ MHz of duration 50 μs vs. frequency (MHz).

Problem 12.6 Digital Pulse Compression. Signal at the input of a radar pulse compressor is the one generated in Problem 12.4. Write the mathematical equations for the basic stages of the pulse compression with the following parameters:

$K := 251$ number of complex pulse compression coefficients

decimation $:= 10$ decimation ratio

$BW := 1$ MHz, bandwidth of the narrowest filter
                                          (Butterworth filter)

$\alpha := 1.0$ pulse compression weighting parameter
                                          (Kaiser weighting)

$f\_s := 20$ MHz, sampling rate

Plot the compressed waveform and time-domain / frequency-domain representations of the waveform through the basic stages of the pulse compression.

Solution. The following basic stages of the pulse compression algorithm can be set.

Stage 1. Define Constants, Sampling Rates, and Intervals

$y\_th := -3$ dB, level to calculate spectrum width

$M := 2^{12}$ $M = 4.096 \times 10^3$ number of waveform samples for spectrum
                                                  calculations

$N := 1000$ number of samples in the input waveform

$\Delta t := \dfrac{1}{f\_s}$ $\Delta t = 0.05$ µs, sampling interval before decimation

$f\_d := \dfrac{f\_s}{\text{decimation}}$ $f\_d = 2$ MHz, sampling rate after decimation

$\Delta t\_d := \dfrac{1}{f\_d}$ $\Delta t\_d = 0.5$ µs, sampling interval after decimation

$M1 := \text{floor}\left(\dfrac{M}{\text{decimation}} - 1 \right)$ number of samples after decimation

M1 = 408

n := 0 .. N − 1   cycle n

m := 0 .. M − 1   cycle m

m1 := 0 .. M1 − 1   cycle m1

dim(a, b) := b − a + 1   number of sample calculations:
                                a = first sample; b = last sample

Stage 2. Uncompressed Waveform Before Filtering and Decimation (N Samples)

WF1_UNC := S   complex baseband uncompressed waveform

V1 := Re(WF1_UNC)   waveform voltage

$\tau_{unc} := N \cdot \Delta t$ $\tau_{unc} = 50$ μs, uncompressed waveform duration

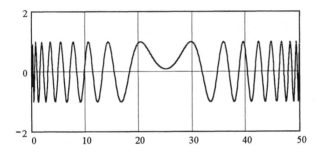

Figure 12.18 Baseband waveform before filtering and decimation vs. time (μs).

Stage 3. Uncompressed Waveform Before Filtering and Decimation Expanded
    for Better Transformation into Frequency Domain (M Samples)

$n\_s := \text{ceil}\left(\dfrac{M - N}{2}\right)$ $n\_s = 1.548 \times 10^3$ first nonzero sample (start)

$n\_f := n\_s + N - 1$ $n\_f = 2.547 \times 10^3$ last nonzero sample (finish)

$n\_0 := \dfrac{M}{2}$ $n\_0 = 2.048 \times 10^3$ central sample

$\dim(n\_s, n\_f) = 1 \times 10^3$ number of samples

$EWF1\_UNC_m := 0$ sets all values to zero

$EWF1\_UNC_{n\_s+n} := WF1\_UNC_n$ expanded complex baseband waveform

$EV1 := Re(EWF1\_UNC)$ waveform voltage

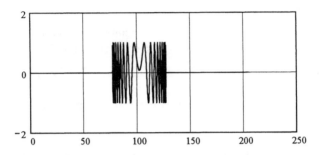

Figure 12.19 Expanded baseband waveform before filtering and decimation vs. time (μs.)

Stage 4. Spectrum of the Uncompressed Waveform Before Filtering

$SP\_F\_in := cfft(EWF1\_UNC)$ complex spectrum of the waveform at the filter input

$\Delta f := \dfrac{f\_s}{M}$ $\Delta f = 4.883 \times 10^{-3}$ MHz, frequency sampling interval

$f_m := (m - n\_0)\cdot\Delta f$ MHz, frequency counts

$f_{n\_0} = 0$ MHz, central frequency

$f_0 = -10$ $f_{M-1} = 9.995$ MHz, frequency limits

$p1 := 0..n\_0 - 1$ $p2 := 0..n\_0 - 1$ cycles $p1$ and $p2$

$SP0\_F\_in_{p1} := SP\_F\_in_{n\_0+p1}$

$SP0\_F\_in_{p2+n\_0} := \overset{\cdot}{SP}\_F\_in_{p2}$ complex spectrum of EWF1_UNC: centered around zero frequency

$SP0_m := 20\cdot\log\left(\left|SP0\_F\_in_m + 10^{-10}\right|\right)$ dB, waveform spectrum at the filter input

$SP0\_MAX := max(SP0)$ spectrum maximum

SP0N1 := SP0 − SP0_MAX   dB, normalized waveform spectrum at
                                                                           the filter input

Stage 5. Filter Frequency Response

H0_FIL := READPRN("filter5.prn") read the filter frequency response from the
                                                                file (see problem 12.7)

$H_m := 20 \cdot \log\left(\left| H0\_FIL_m + 10^{-10} \right| \right)$ dB, filter frequency response

HN := H − max(H)   dB, normalized filter frequency response

Stage 6. Spectrum of the Uncompressed Waveform After Filtering

$$\text{filtering}(SPC, H) := \left| \begin{array}{l} \text{spec} \leftarrow \overrightarrow{SPC \cdot H} \\ \text{spec} \end{array} \right.$$ filtering algorithm

SP0_F_out := filtering(SP0_F_in, H0_FIL)   spectrum of the filtered waveform

$SP0_m := 20 \cdot \log\left(\left| SP0\_F\_out_m + 10^{-10} \right| \right)$ dB, spectrum of the filtered
                                                                       waveform

SP0N2 := SP0 − max(SP0)   dB, normalized spectrum of the filtered waveform

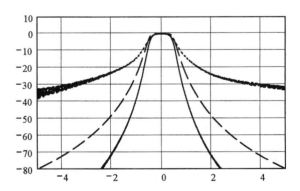

Figure 12.20 Normalized spectra (dB) of the filter input waveform (dotted line),
              filter frequency response (dashed line), and filter output waveform
              (solid line) vs. frequency (MHz).

Stage 7. Expanded Uncompressed Waveform at the Filter Output (M Samples)

$SP\_F\_out_{p1} := SP0\_F\_out_{n\_0+p1}$ transformation of the spectrum centered
around zero frequency
$SP\_F\_out_{n\_0+p2} := SP0\_F\_out_{p2}$

Note. MATHCAD Fourier transform functions ***icfft*** and ***cfft*** work with the spectra denoted SP but not with the spectra denoted SP0 that are centered around zero frequency.

$EWF2\_UNC := icfft(SP\_F\_out)$ expanded waveform at the filter output

$EV2 := Re(EWF2\_UNC)$ expanded waveform voltage

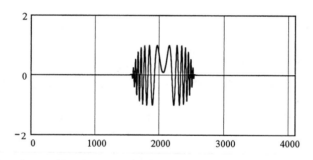

Figure 12.21 Expanded uncompressed waveform at the filter output vs. number of samples.

Stage 8. Filtered and Decimated Waveform (K Samples)

$K\_half := floor\left(\dfrac{K}{2}\right)$ $K\_half = 125$ the middle sample

$nT\_s := n\_0 - K\_half \cdot decimation$ $nT\_s = 798$ the first nonzero count

$nT\_f := n\_0 + K\_half \cdot decimation$ $nT\_f = 3.298 \times 10^3$ the last nonzero count

$k := 0 .. K - 1$ cycle

$EWF3\_UNC\_K_k := EWF2\_UNC_{nT\_s+k \cdot decimation}$

$EV3 := Re(EWF3\_UNC\_K)$ waveform voltage

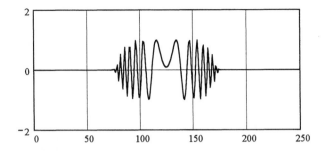

Figure 12.22 Filtered and decimated uncompressed waveform vs. number of samples.

Stage 9. Expanded Filtered and Decimated Waveform (M1 Samples)

$$\text{m1\_0} := \text{floor}\left(\frac{M1}{2}\right) \qquad \text{m1\_0} = 204 \qquad \text{the middle sample}$$

$$\text{mT\_s} := \text{m1\_0} - \text{K\_half} \qquad \text{mT\_s} = 79 \qquad \text{the first nonzero count}$$

$$\text{mT\_f} := \text{m1\_0} + \text{K\_half} \qquad \text{mT\_f} = 329 \qquad \text{the last nonzero count}$$

$$\text{EWF3\_UNC\_M1}_{m1} := 0 \qquad\qquad \text{sets all values to zero}$$

$$\text{EWF3\_UNC\_M1}_{mT\_s+k} := \text{EWF3\_UNC\_K}_k \qquad \text{expanded waveform}$$

Stage 10. Spectrum of the Filtered and Decimated Waveform (M1 Samples)

$$\text{SP\_EWF3\_UNC\_M1} := \text{cfft(EWF3\_UNC\_M1)} \qquad \text{waveform spectrum}$$

$$\Delta\text{fl\_d} := \frac{f\_d}{M1} \qquad \Delta\text{fl\_d} = 4.902 \times 10^{-3} \qquad \text{MHz, frequency sampling interval}$$

$$\text{fl}_{m1} := (\text{m1} - \text{m1\_0})\cdot\Delta\text{fl\_d} \qquad\qquad \text{MHz, frequency counts}$$

$$\text{fl}_{m1\_0} = 0 \qquad\qquad \text{MHz, central frequency}$$

$$\text{fl}_0 = -1 \qquad \text{fl}_{M1-1} = 0.995 \qquad\qquad \text{MHz, frequency limits}$$

$$\text{rl} := 0 .. \text{m1\_0} - 1 \qquad\qquad \text{cycle rl}$$

$$\text{SP0\_EWF3\_UNC\_M1}_{rl} := \text{SP\_EWF3\_UNC\_M1}_{m1\_0+rl}$$

spectrum centered around zero frequency

$$\text{SP0\_EWF3\_UNC\_M1}_{m1\_0+rl} := \text{SP\_EWF3\_UNC\_M1}_{rl}$$

$$\text{SP0}_{m1} := 20 \cdot \log\left(\left|\text{SP0\_EWF3\_UNC\_M1}_{m1} + 10^{-10}\right|\right)$$ dB, spectrum

$$\text{SP0N3} := \text{SP0} - \max(\text{SP0})$$ dB, normalized spectrum

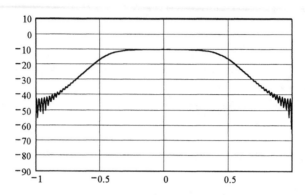

Figure 12.23   Spectrum of the filtered and decimated waveform (dB) vs. frequency (MHz).

Stage 11. Uncompressed Waveform at Pulse Compressor Input (M1 Samples)

$\text{mT\_s} = 79$ the first nonzero sample

$\text{mT\_f} = 329$ the last nonzero sample

$\text{t1} := 0 .. \text{mT\_s} - 1 \quad \text{t2} := \text{mT\_f} + 1 .. \text{M1} - 1$ cycles t1 and t2

$\text{EWF3\_UNC\_M1}_{t1} := 0$
$\text{EWF3\_UNC\_M1}_{t2} := 0$ setting waveform to zero beyond mTs and mT_f

$\text{WF\_UNC} := \text{EWF3\_UNC\_M1}$ uncompressed waveform

$\text{EV4} := \text{Re}(\text{WF\_UNC})$ waveform voltage

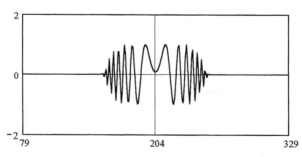

Figure 12.24   The uncompressed waveform at pulse compressor input vs. number of samples.

Stage 12. Unweighted Pulse Compression Coefficients (M1 Samples)

Zero-centered spectrum of the pulse compression coefficients

$SP0\_PC\_COEF\_UNW\_M1 := SP0\_EWF3\_UNC\_M1$

Transformation of the spectrum for the *icfft* function

$SP\_PC\_COEF\_UNW\_M1_{r1} := SP0\_PC\_COEF\_UNW\_M1_{m1\_0+r1}$

$SP\_PC\_COEF\_UNW\_M1_{m1\_0+r1} := SP0\_PC\_COEF\_UNW\_M1_{r1}$

Pulse compression coefficients before weighting

$PC\_COEF\_UNW\_M1 := icfft(SP\_PC\_COEF\_UNW\_M1)$

$PC\_r := Re(PC\_COEF\_UNW\_M1)$ real coefficients

$PC\_i := Im(PC\_COEF\_UNW\_M1)$ imaginary coefficients

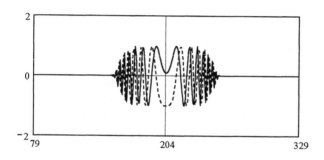

Figure 12.25 The unweighted pulse compression coefficients (solid line is a real part, dotted line is an imaginary part).

Stage 13. Frequency Weighting Function (M1 Samples)

$$f\_w_{m1} := \frac{m1 - m1\_0}{M1} \cdot f\_d$$ MHz, frequency counts

$$I_0(z) := \frac{1}{2 \cdot \pi} \cdot \int_0^{2 \cdot \pi} exp(z \cdot cos(u)) \, du$$ Bessel function

$$WF_{m1} := \frac{I_0\left[\pi \cdot \alpha \cdot \sqrt{1 - \left(\frac{f\_w_{m1}}{f\_w_0}\right)^2}\right]}{I_0(\pi \cdot \alpha)}$$ weighting function

$$WC_{m1} := 20 \cdot log\left[WF_{m1} \cdot \left(WF_{m1\_0}\right)^{-1} \right]$$ dB, weighting coefficients

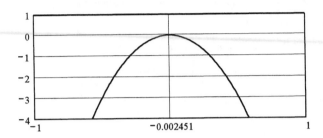

Figure 12.26 The weighting coefficients (dB) vs. frequency (MHz).

Stage 14. Spectrum of the Weighted Pulse Compression Coefficients (M1 Samples)

Zero-centered spectrum of the weighted pulse compression coefficients

$$SP0\_PC\_COEF\_W\_M1_{m1} := SP0\_PC\_COEF\_UNW\_M1_{m1} \cdot WF_{m1}$$

$$SP0_{m1} := 20 \cdot log\left(\left|SP0\_PC\_COEF\_W\_M1_{m1} + 10^{-10}\right|\right)$$ dB, spectrum

$$SP0N4 := SP0 - max(SP0)$$ dB, normalized spectrum

Spectrum of the weighted pulse compression coefficients

$$SP\_PC\_COEF\_W\_M1_{m1} := 0$$

$$SP\_PC\_COEF\_W\_M1_{r1} := SP0\_PC\_COEF\_W\_M1_{m1\_0+r1}$$

$$SP\_PC\_COEF\_W\_M1_{m1\_0+r1} := SP0\_PC\_COEF\_W\_M1_{r1}$$

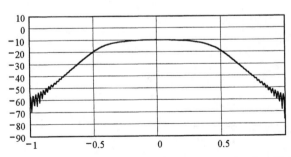

Figure 12.27 Spectrum of the weighted pulse compression coefficients (dB) vs. frequency (MHz).

Stage 15. Weighted Pulse Compression Coefficients (K Samples)

$PC\_COEF\_W\_M1 := icfft(SP\_PC\_COEF\_W\_M1)$ coefficients, M1 samples

$PC\_COEF\_W\_M1_{t1} := 0$

$PC\_COEF\_W\_M1_{t2} := 0$ setting coefficients to zero beyond mTs and mT_f

$PC\_COEF\_W\_K_k := PC\_COEF\_W\_M1_{mT\_s+k}$ coefficients, K samples

Stage 16. Pulse Compression Simulation

$PC\_COEF\_M1 := PC\_COEF\_W\_M1$ M1 samples

$PC\_COEF\_K_k := PC\_COEF\_M1_{mT\_s+k}$ K samples

$pc := Re(PC\_COEF\_K)$ real coefficients

Pulse compression simulation algorithm

$WF\_comp(WF\_unc, PC\_coef) :=$
|
$SP\_WF\_unc \leftarrow cfft(WF\_unc)$

$SP\_PC\_coef \leftarrow cfft(PC\_coef)$

$SP\_WF\_comp \leftarrow \overrightarrow{\overline{SP\_WF\_unc} \cdot SP\_PC\_coef}$

$WF\_compr \leftarrow icfft(SP\_WF\_comp)$

for $ind \in 0 .. m1\_0 - 1$
|
$WF0\_compr_{ind} \leftarrow WF\_compr_{m1\_0+ind}$

$WF0\_compr_{m1\_0+ind} \leftarrow WF\_compr_{ind}$

$WF0\_compr$

Stage 17. Compressed Waveform (M1 Samples)

$WF\_COMP := WF\_comp(WF\_UNC, PC\_COEF\_M1)$ compressed waveform

$L := rows(WF\_COMP)$ $L = 408$ number of samples in the waveform

Magnitude of the compressed waveform, dB

$$WF\_COMP\_db_{m1} := 20 \cdot log\left(\overrightarrow{\left|\left(WF\_COMP_{m1}\right) + 10^{-9}\right|}\right)$$

Normalized magnitude of the compressed waveform, dB

$$\text{WF\_COMP\_norm}_{m1} := \overrightarrow{\text{WF\_COMP\_db}_{m1} - \max(\text{WF\_COMP\_db})}$$

$$\text{WF}_{m1} := \text{WF\_COMP\_norm}_{m1}$$

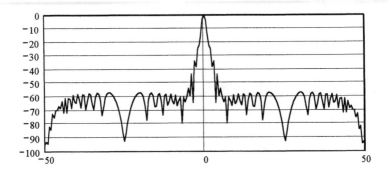

Figure 12.28 The magnitude of the compressed waveform (dB) vs. time (μs).

Stage 18. Spectrum of the Compressed Waveform (M1 Samples)

$$\text{SP\_WF\_COMP} := \text{cfft}(\text{WF\_COMP}) \qquad \text{spectrum of the compressed waveform}$$

$$\text{SP0\_WF\_COMP}_{r1} := \text{SP\_WF\_COMP}_{m1\_0+r1} \qquad \text{spectrum of the compressed}$$

$$\text{SP0\_WF\_COMP}_{m1\_0+r1} := \text{SP\_WF\_COMP}_{r1} \qquad \text{waveform centered around zero}$$

$$\text{SP0}_{m1} := 20 \cdot \log\left(\left|\text{SP0\_WF\_COMP}_{m1} + 10^{-10}\right|\right) \qquad \text{dB, spectrum}$$

$$\text{SP0\_MAX} := \max(\text{SP0})$$

$$\text{SP0N5} := \text{SP0} - \max(\text{SP0})$$

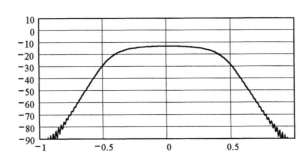

Figure 12.29 Spectrum of the compressed waveform (dB) vs. frequency (MHz).

Time Domain   *Frequency Domain (Spectrum)*

Waveform before filtering and decimation

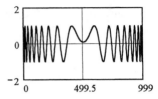 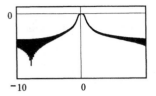

Waveform after filtering and decimation

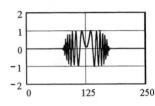 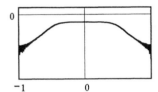

Weighting function

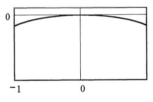

Pulse compression coefficients

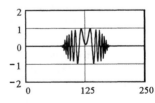 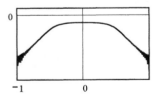

Compressed waveform

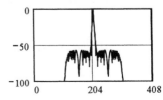 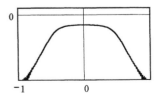

Figure 12.30 Pulse compression stages.

Problem 12.7 Ideal Rectangular Filter. Write the equations for the basic stages of filtering of the pulsed waveform given in Problem 12.1 with the ideal rectangular frequency-selective bandpass filter with central frequency $f_{0\_filter} = 10$ Hz and 3 dB bandwidth BW = 1 Hz. Plot the time-domain and frequency-domain representation of the signal at the filter input and output.

Solution. As given, the parameters of the filter:

$$BW := 1 \qquad f_{0\_filter} := 10$$

Stage 1. Define Constants, Sampling Interval, and Number of Samples

$f\_s := 10^3$ Hz, digital sampling rate

$y\_th := -3$ dB, level to calculate spectrum width

$M := 2^{12}$ $M = 4.096 \times 10^3$ number of waveform samples for spectrum calculations

$\Delta t := \dfrac{1}{f\_s}$ $\Delta t = 1 \times 10^{-3}$ s, sampling interval

$N := 1000$ number of samples in the input signal

$\tau := N \cdot \Delta t$ $\tau = 1$ s, signal duration

$n := 0 .. N - 1 \qquad m := 0 .. M - 1$ cycles n and m

$n\_s := \text{ceil}\left(\dfrac{M - N}{2}\right)$ $n\_s = 1.548 \times 10^3$ the first nonzero sample in the expanded signal (start)

$n\_0 := \dfrac{M}{2}$ $n\_0 = 2.048 \times 10^3$ the central sample

$n\_f := n\_s + N - 1$ $n\_f = 2.547 \times 10^3$ the last nonzero sample in the expanded signal (finish)

$\Delta f := \dfrac{f\_s}{M}$ $\Delta f = 0.244$ Hz, frequency sampling interval

$f_m := (m - n\_0) \cdot \Delta f$ Hz, frequency counts

$f_{0\_filter} = 10$ Hz, the central frequency of the filter

$f_0 = -500$

$f_{M-1} = 499.756$

Hz, frequency limits

Note. The central filter frequency must be within f_0 - f_{M-1} range. To increase this range, increase sampling rate f_s.

$\dim(a, b) := b - a + 1$ generic definition for the number of samples: a - first sample, b - last sample

$$\text{filtering(SPC,H)} := \begin{vmatrix} \text{spec} \leftarrow \overrightarrow{\text{SPC·H}} \\ \text{spec} \end{vmatrix}$$

filtering of the signal with the spectrum SPC by the filter with the frequency response H

Stage 2. The Signal at the Filter Input (N Samples)

$V := \text{Re}(S\_\text{in})$ input signal voltage (see Problem 12.1)

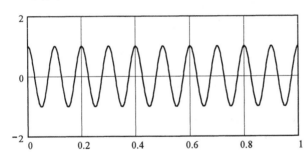

Figure 12.31 The signal voltage (V) vs. time (s) at the filter input.

Stage 3. Expanded Input Signal for Better Transformation into Frequency Domain (M Samples)

$\dim(n\_s, n\_f) = 1 \times 10^3$ number of samples

$ES\_\text{in}_m := 0$ sets all values to zero

$ES\_\text{in}_{n\_s+n} := S\_\text{in}_n$ sets nonzero values only for specified counts

$EV_m := \text{Re}(ES\_\text{in}_m)$ expanded signal voltage

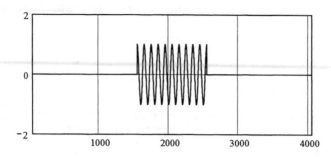

Figure 12.32 Expanded signal voltage (V) vs. number of samples at the filter input.

Stage 4. Spectrum of the Signal at the Filter Input (M Samples)

SP_S_in := cfft(ES_in)	input signal spectrum
$p := 0 .. n\_0 - 1$	cycle p

$$SP0\_S\_in_p := SP\_S\_in_{p+n\_0}$$

spectrum centered around zero frequency

$$SP0\_S\_in_{p+n\_0} := SP\_S\_in_p$$

$$SP0\_in_m := 20 \cdot log\left(\left|SP0\_S\_in_m + 10^{-10}\right|\right)$$ dB, spectrum

$$SPN\_in := SP0\_in - max(SP0\_in)$$ dB, normalized spectrum

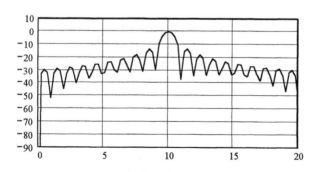

Figure 12.33 The spectrum (dB) of the signal at the filter input vs. frequency (Hz).

Stage 5. Filter Frequency Response

$$H\_FIL_m := \begin{vmatrix} 1 & \text{if} \quad \dfrac{-BW}{2} \le \left(f_m - f_{0\_filter}\right) \le \dfrac{BW}{2} \\[2mm] 0 & \text{otherwise} \end{vmatrix}$$

$$HF_m := 20 \cdot \log\left(\left|H\_FIL_m + 10^{-10}\right|\right) \qquad\qquad \text{dB, filter frequency response}$$

$$HFdb := HF - \max(HF) \qquad\qquad \text{dB, normalized filter frequency response}$$

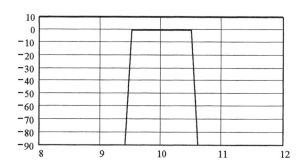

Figure 12.34 Normalized filter frequency response (dB) vs. frequency (Hz) for ideal rectangular filter.

Stage 6. Spectrum of the Signal at the Filter Output (M Samples)

$$SP0\_S\_out := filtering(SP0\_S\_in, H\_FIL) \qquad\qquad \text{spectrum of the signal} \\ \text{at the filter output}$$

$$SP0\_out_m := 20 \cdot \log\left(\left|SP0\_S\_out_m + 10^{-10}\right|\right) \qquad\qquad \text{dB, spectrum}$$

$$SPN\_out := SP0\_out - \max(SP0\_out) \qquad\qquad \text{dB, normalized spectrum}$$

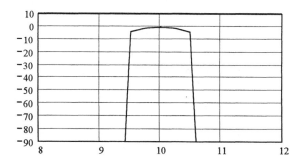

Figure 12.35 The spectrum (dB) of the signal at the filter output vs. frequency (Hz).

Stage 7. <u>Expanded Signal at the Filter Output (M Samples)</u>

Transformation of the spectrum for the *icfft* function

$$\text{SP\_S\_out}_p := \text{SP0\_S\_out}_{n\_0+p} \qquad \text{SP\_S\_out}_{n\_0+p} := \text{SP0\_S\_out}_p$$

$$\text{ES\_out} := \text{icfft}(\text{SP\_S\_out}) \qquad\qquad\qquad \text{expanded signal}$$

$$\text{EV}_m := \text{Re}\big(\text{ES\_out}_m\big) \qquad\qquad\qquad \text{expanded signal voltage}$$

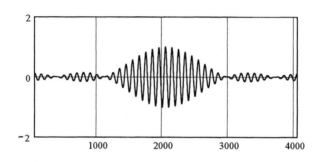

Figure 12.36 Expanded signal voltage (V) vs. number of samples at the filter output.

Stage 8. <u>Input and Output Signals Representation (K samples)</u>

$$C := 2 \quad\ K := C \cdot 2 \cdot N - 1 \qquad\qquad\qquad \text{expands time scale}$$

$$k := 0 .. K \qquad\qquad\qquad\qquad\qquad\qquad \text{cycle}$$

input signal	output signal	
$\text{S\_in}_k := \text{ES\_in}_{n\_0-C\cdot N+k}$	$\text{S\_out}_k := \text{ES\_out}_{n\_0-C\cdot N+k}$	complex signal
$\text{V1} := \text{Re}(\text{S\_in})$	$\text{V2} := \text{Re}(\text{S\_out})$	real signal
$A_k := \big|\text{S\_in}_k\big|$	$C_k := \big|\text{S\_out}_k\big|$	
$B_k := -\big|\text{S\_in}_k\big|$	$D_k := -\big|\text{S\_out}_k\big|$	signal envelope

Signal vs. time (s)   *Signal spectrum vs. frequency (Hz)*

Input signal

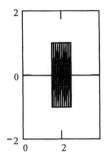 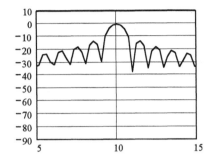

Filter frequency response

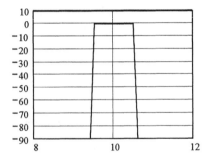

Output signal

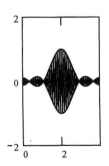 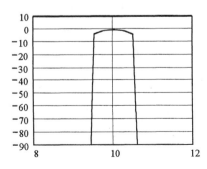

Figure 12.37 Frequency-selective filtering stages for ideal rectangular filter.

Problem 12.8 Gaussian Filter. Solve the Problem 12.7 for a Gaussian filter.

Solution. The only difference with respect to the Problem 12.7 is in the mathematical description of the frequency response of the filter (Stage 5).

Stage 5. Filter Frequency Response

$$x_m := 2 \cdot \left(f_m - f_{0\_filter} \right) \qquad \text{Hz, parameter}$$

$$H\_FIL_m := \begin{vmatrix} 0 \ \text{if} \ \dfrac{3}{10} \cdot \left(\dfrac{x_m}{BW} \right)^2 \ge 300 \\[4mm] \dfrac{1}{\sqrt{10^{\frac{3}{10} \cdot \left(\frac{x_m}{BW} \right)^2}}} \ \ \text{otherwise} \end{vmatrix} \qquad \text{filter frequency response}$$

$$HF_m := 20 \cdot \log\left(\left| H\_FIL_m + 10^{-10} \right| \right) \qquad \text{dB, filter frequency response}$$

$$HFdb := HF - \max(HF) \qquad \text{dB, normalized filter frequency response}$$

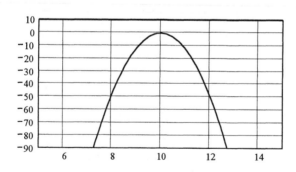

Figure 12.38 Normalized filter frequency response (dB) vs. frequency (Hz) for Gaussian filter.

With this change, the plots of the input and output waveforms are as follows:

Signal vs. time (s)   *Signal spectrum vs. frequency (Hz)*

Input signal

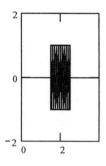 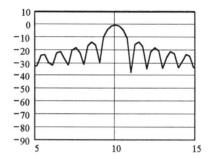

Filter frequency response

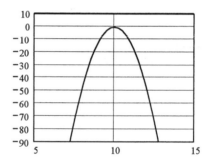

Output signal

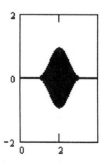 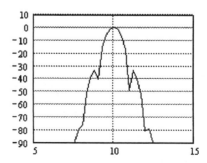

Figure 12.39 Frequency-selective filtering stages for the Gaussian filter.

Problem 12.9 Chebyshev Filter. Solve Problem 12.7 for a Chebyshev 4-poles filter with peak-to-valley parameter - 3 dB.

Solution. The only difference with respect to Problem 12.7 is in the mathematical description of the frequency response of the filter (stage 5).

Stage 5. Filter Frequency Response

poles_0 := 4   number of poles in filter

par := −3   dB, peak-to-valley parameter for Chebyshev filter

$$b_m := poles\_0 \cdot acosh\left(\frac{x_m}{BW}\right)$$

$$H\_FIL_m := \cfrac{1}{\sqrt{1 + \left[\left(\cfrac{1}{\left(10^{\frac{par}{20}}\right)^2}\right) - 1\right] \cdot \left(\cfrac{e^{b_m} + e^{-b_m}}{2}\right)^2}}$$

$$HF_m := 20 \cdot log\left(\left|H\_FIL_m + 10^{-10}\right|\right)$$ dB, filter frequency response

$$HFdb := HF - max(HF)$$ dB, normalized filter frequency response

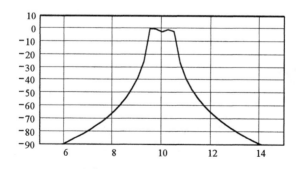

Figure 12.40 Normalized filter frequency response (dB) vs. frequency (Hz) for Chebyshev filter.

With this change, the plots of the input and output waveforms are as follows:

Signal vs. time (s)   *Signal spectrum vs. frequency (Hz)*

Input signal

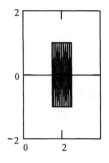   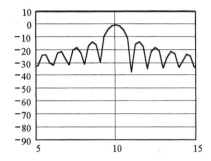

Filter frequency response

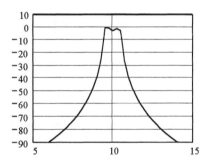

Output signal

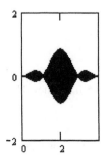   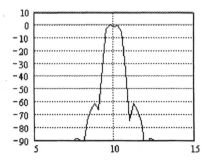

Figure 12.41 Frequency-selective filtering stages for the Chebyshev filter.

Problem 12.10 Butterworth Filter. Solve Problem 12.7 for a Butterworth 4-pole filter.

Solution. The only difference with respect to Problem 12.7 is in the mathematical description of the frequency response of the filter (stage 5).

Stage 5. Filter Frequency Response

poles_0 := 4   number of poles in filter

$$H\_FIL_m := \frac{1}{\sqrt{1 + \left(\dfrac{x_m}{BW}\right)^{2 \cdot poles\_0}}}$$

$HF_m := 20 \cdot \log\left(\left|H\_FIL_m + 10^{-10}\right|\right)$ dB, filter frequency response

$HFdb := HF - \max(HF)$ dB, normalized filter frequency response

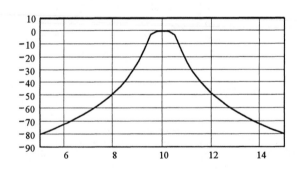

Figure 12.42 Normalized filter frequency response (dB) vs. frequency (Hz) for the Butterworth filter.

With this change the plots of the input and output waveforms are as follows:

Signal vs. time (s)   *Signal spectrum vs. frequency (Hz)*

Input signal

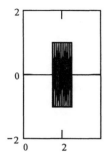 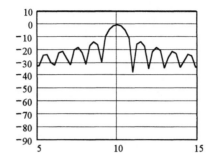

Filter frequency response

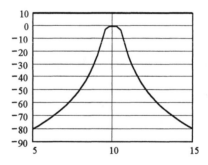

Output signal

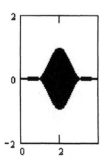 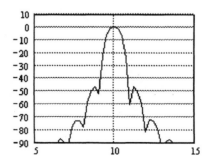

Figure 12.43 Frequency-selective filtering stages for the Butterworth filter.

Chapter 13

STOCHASTIC RADIO ENGINEERING

Problem 13.1 Probability of Events. Suppose that two radars detect a target independently. The probability of detection by the first radar is 0.8. The probability of detection by the second radar is 0.9. Find the probability of detection of the aircraft by both radars simultaneously, and the probability of detection by at least one radar.

Solution. Let us denote A is the event that aircraft is detected by the first radar, B is the event that the aircraft is detected by the second radar, $C1$ is the event that the aircraft is detected by both radars simultaneously and $C2$ is the event that the aircraft is detected by at least one radar. Since events A and B are independent ones, the probability of detecting the aircraft by both radars is (see Section 8.1.8):

$$P_A := 0.8 \qquad P_B := 0.9 \qquad P_{C1} := P_A \cdot P_B \qquad P_{C1} = 0.72$$

Using the formula from Section 8.1.9 we find that the probability of detecting the aircraft by at least one radar is:

$$P_{C2} := P_A + P_B - P_{C1} \qquad P_{C2} = 0.98$$

Another approach to solve this problem is to use complement events. Since A and B are independent events, let us denote the event complement to A as A_, and B as B_:

$$P_{C2} := 1 - P_{A\_} \cdot P_{B\_} \qquad \text{or} \qquad P_{C2} := 1 - \left(1 - P_A\right) \cdot \left(1 - P_B\right) \qquad P_{C2} = 0.98$$

Problem 13.2 Bayes' Rule. Two six-digit codes are transmitted through the communication channel: 111111 and 000000. The a priori probabilities that each of these codes was transmitted are equal: 0.5 for code # 1 (111111) and 0.5 for code # 2 (000000). Due to interference in the channel, the probability of receiving each symbol correctly (to receive 1 when 1 was sent, or to receive 0 when 0 was sent) is 0.7, and each symbol in the code can be distorted independently. The received code was 101101. Find the code that was more likely transmitted.

Solution. Let us denote A as the event to receive the code 10110. There are two hypotheses that lead to the occurrence of this event: H_1 that the code 11111 was transmitted and H_2 that code 00000 was transmitted. As given:

$$P_{H_1} := 0.5 \qquad P_{H_2} := 0.5 \qquad P_c := 0.7$$

The conditional probability to receive 101101 instead of 111111 is:

$$P_{A\_H_1} := P_c \cdot (1 - P_c) \cdot P_c \cdot P_c \cdot (1 - P_c) \cdot P_c \qquad\qquad P_{A\_H_1} = 0.022$$

The conditional probability to receive 101101 instead of 000000 is:

$$P_{A\_H_2} := (1 - P_c) \cdot P_c \cdot (1 - P_c) \cdot (1 - P_c) \cdot P_c \cdot (1 - P_c) \qquad P_{A\_H_2} = 0.004$$

Using the equation from Section 8.1.12 we can find the conditional probabilities of both hypotheses:

$$P_{H\_A_1} := \frac{P_{H_1} \cdot P_{A\_H_1}}{\displaystyle\sum_{i=1}^{2} P_{H_i} \cdot P_{A\_H_i}} \qquad\qquad P_{H\_A_1} = 0.845$$

$$P_{H\_A_2} := \frac{P_{H_2} \cdot P_{A\_H_2}}{\displaystyle\sum_{i=1}^{2} P_{H_i} \cdot P_{A\_H_i}} \qquad\qquad P_{H\_A_2} = 0.155$$

Thus, it is more likely (with probability 0.845 versus 0.155) that the code 111111 was transmitted.

Problem 13.3 Multiple Hypotheses. The probability that during a time frame for communication with a crew of the space station, the subsystems of the communication equipment will fail to be within operational margins is as follows: 0.05 for antenna, 0.1 for transmitter, and 0.15 for receiver. If the parameters of a single subsystem are out of operational margins, then communication will not be established with probability 0.2; if the parameters of two subsystems are out of operational margins, then communication will not be established with probability 0.25; if the parameters of three subsystems are out of operational margins, then communication will not be established with probability 0.3. Find the probability that communication with the crew will be established.

Solution. Let us denote event A as event in which the parameters of the antenna will be out of operational margins; B as event in which the parameters of the transmitter will be out of operational margins; C as event in which the parameters of the receiver will be out of operational margins; E is the event in which communication will not be established; F is the event in which communication will be established. As given:

$$P_A := 0.05 \qquad P_B := 0.1 \qquad P_C := 0.15$$

There are three hypotheses that lead to event E: H_1 is that parameters of a single subsystem are out of margins; H_2 is that parameters of two subsystems are out of margins; H_3 is that parameters of all three subsystems are out of margins. Probability of these hypotheses can be found as:

$$P_{H_1} := P_A \cdot (1 - P_B) \cdot (1 - P_C) + P_B \cdot (1 - P_A) \cdot (1 - P_C) + P_C \cdot (1 - P_A) \cdot (1 - P_B)$$

$$P_{H_2} := P_A \cdot P_B \cdot (1 - P_C) + P_A \cdot P_C \cdot (1 - P_B) + P_B \cdot P_C \cdot (1 - P_A)$$

$$P_{H_3} := P_A \cdot P_B \cdot P_C$$

$$P_{H_1} = 0.247 \qquad P_{H_2} = 0.025 \qquad P_{H_3} = 0.001$$

The conditional probabilities of the event E with respect to hypothesis H_1 (E_H_1), H_2 (E_H_2) and H_3 (E_H_3) are given as:

$$P_{E\_H_1} := 0.2 \qquad P_{E\_H_2} := 0.25 \qquad P_{E\_H_3} := 0.3$$

Using the equation from Section 8.1.12 we can find the probability of event E:

$$P_E := \sum_{i=1}^{3} P_{H_i} \cdot P_{E\_H_i} \qquad P_E = 0.056$$

Since F and E are complimentary events, the probability that communication with the crew will be established is:

$$P_F := 1 - P_E \qquad P_F = 0.944$$

Problem 13.4   Poisson Distribution and Random Flow. In 24 hours the company computer network received 10,000 messages contaminated with viruses. The probability of receiving a contaminated message follows the Poisson distribution law. Find the probability that in the next minute the network will be attacked by viruses five times.

Solution. The number of messages received during time t = 24 hr is M = 10,000. The average number a of messages contaminated with viruses received by the network per second is:

$$t := 24 \qquad M := 10000 \qquad a := \frac{M}{t \cdot 3600} \qquad a = 0.116$$

Intensity of the messages in t_0 = 60 seconds is likely to be:

$$t_0 := 60 \qquad \lambda := a \cdot t_0 \qquad \lambda = 6.944$$

Thus, probability P to be attacked by viruses k = 5 times during time t_0 is (see Appendix 14):

$$k := 5 \qquad P := \frac{\lambda^k}{k!} \cdot \exp(-\lambda) \qquad P = 0.13$$

Problem 13.5   Bernoulli Distribution and Independent Tests. Due to fading in a communication channel the signal might fall to the level when it is not possible to discriminate between the signal and noise (the failed transmission). The probability that this might happen in each independent transmission is 0.1. Find the probability that 0, 5, 10 transmissions out of 10 will be the failed ones.

Solution. The probability P that in n independent tests event A will occur exactly k times (provided the probability of event A in each test is constant and equal to p), follows the binomial (Bernoulli) distribution law (see Appendix 14). In case there is no single failed transmission out of 10, n = 10, k = 0, p = 0.1. Thus,

$$p := 0.1 \quad n := 10 \quad k := 0 \quad C_{n\_k} := \frac{n!}{k! \cdot (n-k)!} \qquad C_{n\_k} = 1$$

$$P := C_{n\_k} \cdot p^k \cdot (1-p)^{n-k} \qquad P = 0.349$$

Analogously, in case one-half of transmissions (5) and all transmissions (10) are failed:

$$p := 0.1 \qquad n := 10 \qquad k := 5 \quad C_{n\_k} := \frac{n!}{k! \cdot (n-k)!}$$

$$P := C_{n\_k} \cdot p^k \cdot (1-p)^{n-k} \qquad P = 1.488 \times 10^{-3}$$

$$p := 0.1 \qquad n := 10 \qquad k := 10 \quad C_{n\_k} := \frac{n!}{k! \cdot (n-k)!}$$

$$P := C_{n\_k} \cdot p^k \cdot (1-p)^{n-k} \qquad P = 1 \times 10^{-10}$$

Problem 13.6   Exponential Distribution and Reliability. In a solid-state transmitter, the time to failure is the random function with exponential probability distribution law. The intensity of failures $\lambda = 2.5 \ast 10^{-5}$ 1/hour. Find the following characteristics: probability of nonfailure operation and probability of failure for the first 5,000 hr; probability of nonfailure operation and probability of failure in the interval from 5,000 hr to 15,000 hr; mean time to failure (MTTF).

Solution. When the system failures in the interval $(0, t_0)$ are described by the probability distribution function $F_0 = F(t_0)$, the probability of nonfailure operation of the system $P(t_0) = P_0$ is given by the equation:

$$P_0 := 1 - F_0$$

The exponential PDF is given by the equation (see Appendix 14):

$$F := 1 - \exp(-\lambda \cdot t) \quad \text{Since} \quad t_0 := 5000 \quad \lambda := 2.5 \cdot 10^{-5}$$

$$P_0 := \exp(-\lambda \cdot t_0) \quad P_0 = 0.882$$

The probability of failure $Q_0 = Q(t_0)$:

$$Q_0 := 1 - P_0 \quad Q_0 = 0.118$$

The probability of nonfailure operation P_Δ and probability of failure Q_Δ in the interval $\Delta t = t_2 - t_1$ under the condition that there were no failures until the moment t_1 is:

$$t_1 := 5000 \quad t_2 := 15000 \quad \Delta t := t_2 - t_1 \quad \Delta t = 10000$$

$$P_\Delta := \exp(-\lambda \cdot \Delta t) \quad P_\Delta = 0.779 \quad Q_\Delta := 1 - P_\Delta \quad Q_\Delta = 0.221$$

Mean time to failure (MTTF) is:

$$T := \frac{1}{\lambda} \quad T = 40000 \text{ [hr]}$$

Problem 13.7 Gaussian (Normal) Distribution. Find the probability that the values of the random variable with Gaussian distribution $N(m=0, \sigma=1)$ will fall within the intervals $(-\sigma, \sigma)$, $(-2\sigma, 2\sigma)$, $(-3\sigma, 3\sigma)$.

Solution. The probability P_{ab} that the random variable X with the PDF $F(x)$ will fall into interval (a, b) is given by the equation (see Section 8.1.14):

$$P_{ab} := F(b) - F(a)$$

For the Gaussian distribution the PDF is given by the error function integral $\Phi(x)$ and thus the probability P_{ab} is (see Section 8.1.16):

$$P_{ab} := \Phi_b - \Phi_a \quad \Phi_a := \Phi\left(\frac{a - m}{\sigma}\right) \quad \Phi_b := \Phi\left(\frac{b - m}{\sigma}\right)$$

Since $m := 0 \quad \sigma := 1$ for $a := -\sigma \quad b := \sigma$ by definition (see Appendix 15):

$$\Phi_a := \frac{1}{\sqrt{2 \cdot \pi \cdot \sigma}} \cdot \left[\int_{-\infty}^{a} \exp\left[\frac{-(t-m)^2}{2 \cdot \sigma^2} \right] dt \right] \qquad \Phi_a = 0.159$$

$$\Phi_b := \frac{1}{\sqrt{2 \cdot \pi \cdot \sigma}} \cdot \left[\int_{-\infty}^{b} \exp\left[\frac{-(t-m)^2}{2 \cdot \sigma^2} \right] dt \right] \qquad \Phi_b = 0.841$$

$$P_{ab} := \Phi_b - \Phi_a \qquad P_{ab} = 0.683$$

Analogously for $a := -2\sigma \quad b := 2\sigma$

$$\Phi_a := \frac{1}{\sqrt{2 \cdot \pi \cdot \sigma}} \cdot \left[\int_{-\infty}^{a} \exp\left[\frac{-(t-m)^2}{2 \cdot \sigma^2} \right] dt \right] \qquad \Phi_a = 0.023$$

$$\Phi_b := \frac{1}{\sqrt{2 \cdot \pi \cdot \sigma}} \cdot \left[\int_{-\infty}^{b} \exp\left[\frac{-(t-m)^2}{2 \cdot \sigma^2} \right] dt \right] \qquad \Phi_b = 0.977$$

$$P_{ab} := \Phi_b - \Phi_a \qquad P_{ab} = 0.954$$

For $a := -3\sigma \quad b := 3\sigma$

$$\Phi_a := \frac{1}{\sqrt{2 \cdot \pi \cdot \sigma}} \cdot \left[\int_{-\infty}^{a} \exp\left[\frac{-(t-m)^2}{2 \cdot \sigma^2} \right] dt \right] \qquad \Phi_a = 1.35 \times 10^{-3}$$

$$\Phi_b := \frac{1}{\sqrt{2 \cdot \pi \cdot \sigma}} \cdot \left[\int_{-\infty}^{b} \exp\left[\frac{-(t-m)^2}{2 \cdot \sigma^2} \right] dt \right] \qquad \Phi_b = 0.999$$

$$P_{ab} := \Phi_b - \Phi_a \qquad P_{ab} = 0.997$$

Problem 13.8 Gaussian Distribution and Measurement Errors. The error of the measurement of the distance to a test target by a laser range finder is the random variable with Gaussian distribution, with the following parameters: mean value m = 0.25 m, and rms deviation σ = 0.33 m. Find the probability that during the test all measurement errors will stay within the interval from - 0.5 m to + 0.5 m.

Solution. Since $m := 0.25$ $\sigma := 0.33$ for $a := -0.5$ $b := 0.5$

$$\Phi_a := \frac{1}{\sqrt{2 \cdot \pi \cdot \sigma}} \cdot \left[\int_{-\infty}^{a} \exp\left[\frac{-(t-m)^2}{2 \cdot \sigma^2} \right] dt \right] \qquad \Phi_a = 0.012$$

$$\Phi_b := \frac{1}{\sqrt{2 \cdot \pi \cdot \sigma}} \cdot \left[\int_{-\infty}^{b} \exp\left[\frac{-(t-m)^2}{2 \cdot \sigma^2} \right] dt \right] \qquad \Phi_b = 0.776$$

$$P_{ab} := \Phi_b - \Phi_a \qquad P_{ab} = 0.764$$

Problem 13.9 Gaussian Distribution and Noncoherent Radar Detection. A radar performs automatic detection of the return signal at the background of white Gaussian noise. Signal-to-noise power ratio (SNR) at the input of the detector is 10 dB. Detection threshold [the ratio of the threshold-to-noise power $(V_0/\sigma_n)^2$] is set at 12 dB. Find the probability of detection of the target by a radar for the case of noncoherent single-pulse detection.

Solution. Given, power threshold in decibels, the detection threshold voltage to the rms noise voltage $r = V_0/\sigma_n$ can be found as:

$$r\_pow\_dB := 12 \qquad r := 10^{\frac{r\_pow\_dB}{20}} \qquad r = 3.981$$

Given SNR in dB, actual SNR can be found as:

$$SNR\_dB := 10 \qquad SNR := 10^{\frac{SNR\_dB}{10}} \qquad SNR = 10$$

Probability of false alarm can be found as (see Problem 13.13):

$$P_{fa} := \exp\left(\frac{-1}{2} \cdot r^2\right) \qquad P_{fa} = 3.618 \times 10^{-4}$$

Denoting:

$$\Phi(\text{arg}) := \frac{1}{\sqrt{2 \cdot \pi}} \cdot \left(\int_{-\infty}^{\text{arg}} \exp\left(\frac{-t^2}{2}\right) dt\right)$$

the formula for probability of single-pulse noncoherent detection P_d at the background of white Gaussian noise is:

$$P_d := 1 - \Phi\left(\sqrt{2 \cdot \ln\left(\frac{1}{P_{fa}}\right)} - \sqrt{1 + 2 \cdot SNR}\right) \qquad P_d = 0.726$$

Problem 13.10   Gaussian   Distribution   and   Coherent   Radar Detection. Solve the previous Problem 13.9 for the case of single-pulse coherent detection with SNR = 10 dB and detection threshold [the ratio of the threshold-to-noise power $(V_0/\sigma_n)^2$] set at 8 dB.

Solution. Given power threshold in decibels, the detection threshold voltage to the rms noise voltage $r = V_0/\sigma_n$ can be found as:

$$r\_pow\_dB := 8 \qquad r := 10^{\frac{r\_pow\_dB}{20}} \qquad r = 2.512$$

Given SNR in decibels, actual SNR can be found as:

$$SNR\_dB := 10 \qquad SNR := 10^{\frac{SNR\_dB}{10}} \qquad SNR = 10$$

The formula for probability of single-pulse coherent detection P_d at the background of white Gaussian noise is:

$$P_d := 1 - \Phi\left(\sqrt{2} \cdot r - \sqrt{2 \cdot SNR}\right) \qquad P_d = 0.821$$

Problem 13.11 Gaussian Distribution and Statistical Parameters. The number of faulty transistors in the set of 1,000 is the random quantity that has the Gaussian distribution with probability density function:

$$f_1(x) := \frac{1}{7.52} \cdot \exp\left[\frac{-(x-5)^2}{18}\right]$$

Find the mathematical expectation m and rms deviation σ of the possible quantity of faulty transistors, and the amount of faulty transistors in the set with probability 0.997.

Solution. Using the equations from Section 8.1.17 and integrating pdf with the aid of a computer we find:

$$m := \int_{-\infty}^{\infty} x \cdot f_1(x)\, dx \qquad m = 5$$

$$D := \int_{-\infty}^{\infty} (x-m)^2 \cdot f_1(x)\, dx \qquad D = 9 \qquad \sigma := \sqrt{D} \qquad \sigma = 3$$

The amount of faulty transistors k with probability 0.997 is given by the range $(-3\sigma, 3\sigma)$ (see Problem 13.7). Thus:

$$k := \text{ceil}(6 \cdot \sigma) \qquad k = 18$$

Note. For given Gaussian probability density function, parameters m and σ are explicitly defined in this function:

$$f_1(x) := \frac{1}{\sigma \cdot \sqrt{2 \cdot \pi}} \cdot \exp\left[\frac{-(x-m)^2}{2 \cdot \sigma^2}\right]$$

Thus, in the example above: $m := 5 \qquad \sigma := \sqrt{\dfrac{18}{2}} \qquad \sigma = 3$

which matches results based on numeric integration of pdf with the aid of a computer.

Problem 13.12 Rayleigh Distribution. Find the probability that the values of the random variable with Rayleigh distribution with parameter $\sigma=1$ will fall within the intervals $(0, 3\sigma)$.

Solution. Analogously, as in Problem 13.7, the probability P_{ab} that the random variable X with the PDF F(x) will fall into interval (a, b) is given by the formula:

$$P_{ab} := F_b - F_a \qquad F_b := F(b) \qquad F_a := F(a) \qquad \sigma := 1 \qquad a := 0 \qquad b := 3\sigma$$

For Rayleigh distribution (see Appendix 14):

$$F_a := 1 - \exp\left(\frac{-a^2}{2 \cdot \sigma^2}\right) \qquad F_a = 0$$

$$F_b := 1 - \exp\left(\frac{-b^2}{2 \cdot \sigma^2}\right) \qquad F_b = 0.989$$

$$P_{ab} := F_b - F_a \qquad\qquad P_{ab} = 0.989$$

Problem 13.13   Rayleigh Distribution and False Alarms. The envelope of noise voltage in a radar receiver is the random process with Rayleigh distribution. Find the probability of false alarm at the output of the radar receiver if the ratio of the detection threshold voltage to the rms noise voltage $r = V_0/\sigma_n$ is set to 1, 5, 10.

Solution. Probability of false alarm P_{fa} is the probability that noise alone (without signal) will exceed the detection threshold; thus, an erroneous decision about target detection will be made. It is equal to the probability $P_{fa} = P(V_n > V_0)$ that noise voltage V_n will exceed the threshold voltage V_0. Since noise voltage is Rayleigh distributed, then (see Appendix 14):

$$P_{fa} := \int_{V_0}^{\infty} \frac{V_n}{\sigma_n^2} \cdot \exp\left(\frac{-V_n^2}{2 \cdot \sigma_n^2}\right) dV_n$$

With the aid of a computer (see Chapter 9) we can evaluate this integral in the general form:

$$P_{fa} := \exp\left(\frac{-1}{2} \cdot \frac{V_0^2}{\sigma_n^2}\right)$$

Thus:

$$r := 1 \qquad P_{fa} := \exp\left(\frac{-1}{2} \cdot r^2\right) \qquad P_{fa} = 0.607$$

$$r := 5 \qquad P_{fa} := \exp\left(\frac{-1}{2} \cdot r^2\right) \qquad P_{fa} = 3.727 \times 10^{-6}$$

$$r := 10 \qquad P_{fa} := \exp\left(\frac{-1}{2} \cdot r^2\right) \qquad P_{fa} = 0$$

Problem 13.14 Rayleigh Distribution and Statistical Parameters.
Noise at the input of the amplitude detector is the Gaussian random process with
zero mean and rms amplitude $\sigma_n = 1$ V. Find the mean value m and rms amplitude
σ of the noise at the output of the amplitude detector.

Solution. In this case, amplitude detector transforms the input noise with Gaussian
distribution into output noise with Rayleigh distribution (see Appendix 14). Thus,
the pdf at the output of the amplitude detector:

$$\sigma_n := 1 \qquad f_1(x) := \frac{x}{\sigma_n^2} \cdot \exp\left(\frac{-x^2}{2 \cdot \sigma_n^2}\right)$$

Parameters m and σ can be found as:

$$m := \int_0^\infty x \cdot f_1(x)\, dx \qquad m = 1.253 \ [V]$$

$$D := \int_0^\infty (x - m)^2 \cdot f_1(x)\, dx \qquad D = 0.429 \qquad \sigma := \sqrt{D} \qquad \sigma = 0.655 \quad [V]$$

Note. Another way to find these parameters is to use formulas from Appendix 16:

$$m := \sigma_n \sqrt{\frac{\pi}{2}} \qquad m = 1.253 \qquad [Volt]$$

$$D := \sigma_n^2 \cdot \left(2 - \frac{\pi}{2}\right) \qquad D = 0.429 \qquad \sigma := \sqrt{D} \qquad \sigma = 0.655 \quad [Volt]$$

Problem 13.15   **Ricean Distribution and Statistical Parameters.** A nonrandom signal with amplitude A = 1 V and Gaussian random noise with zero mean and rms amplitude σ_n = 1 V passes through the amplitude detector. Find the mean value *m* and root-mean-square amplitude σ of the signal plus noise oscillation at the output of the amplitude detector.

Solution. In this case, amplitude detector transforms the input noise with Gaussian distribution into output noise with Ricean distribution (see Appendix 14). Thus, the pdf at the output of the amplitude detector:

$$\sigma_n := 1 \quad A := 1 \quad f_1(x) := \frac{x}{\sigma_n^{\,2}} \cdot \exp\left[\frac{-\left(x^2 + A^2\right)}{2 \cdot \sigma_n^{\,2}} \right] \cdot I0\left(\frac{A}{\sigma_n^{\,2}} \cdot x \right)$$

Parameters m and σ can be found as in the Problem 13.14:

$$m := \int_0^\infty x \cdot f_1(x)\, dx \qquad m = 1.549 \quad \text{[Volt]}$$

$$D := \int_0^\infty (x - m)^2 \cdot f_1(x)\, dx \quad D = 0.602 \quad \sigma := \sqrt{D} \quad \sigma = 0.776 \quad \text{[Volt]}$$

Note. When there is no signal at the input (A = 0), the Ricean distribution reduces to a Rayleigh distribution. Thus,

$$\sigma_n := 1 \quad A := 0 \quad f_1(x) := \frac{x}{\sigma_n^{\,2}} \cdot \exp\left[\frac{-\left(x^2 + A^2\right)}{2 \cdot \sigma_n^{\,2}} \right] \cdot I0\left(\frac{A}{\sigma_n^{\,2}} \cdot x \right)$$

$$m := \int_0^\infty x \cdot f_1(x)\, dx \qquad m = 1.253 \quad \text{[Volt]}$$

$$D := \int_0^\infty (x - m)^2 \cdot f_1(x)\, dx \quad D = 0.429 \quad \sigma := \sqrt{D} \quad \sigma = 0.655 \quad \text{[Volt]}$$

(compare to Problem 13.14)

Problem 13.16 Linear Transform and Statistical Parameters.
Gaussian noise X(t) with parameters $m_x = 1$ V and $\sigma_x = 3$ V passes through the circuit, which performs its linear transformation described by equation

$$Y := aX + b \qquad a := 4 \qquad b := 7$$

Find parameters m_y and σ_y of the transformed noise.

Solution. The pdf of the input noise is:

$$m_x := 1 \qquad \sigma_x := 3 \qquad f_1(x) := \frac{1}{\sigma_x \cdot \sqrt{2 \cdot \pi}} \cdot \exp\left[\frac{-(x - m_x)^2}{2 \cdot \sigma_x^2}\right]$$

According to Appendix 17, the transformed pdf will be:

$$f_1(y) := \frac{1}{|a|} \cdot f_1\left(\frac{y - b}{a}\right) \qquad \text{or} \qquad f_1(y) := \frac{1}{4\sigma_x \cdot \sqrt{2 \cdot \pi}} \cdot \exp\left[\frac{-\left(\dfrac{y - 7}{4} - m_x\right)^2}{2 \cdot \sigma_x^2}\right]$$

$$m_y := \int_{-\infty}^{\infty} y \cdot f_1(y) \, dy \qquad m_y = 11 \qquad [V]$$

$$D_y := \int_{-\infty}^{\infty} (y - m_y)^2 \cdot f_1(y) \, dy \qquad D_y = 144 \qquad \sigma_y := \sqrt{D_y} \qquad \sigma_y = 12 \qquad [V]$$

Note. The linear transform does not change the distribution of the random variables. For example, if the input noise was Gaussian, then after the linear transform the output noise also has a Gaussian distribution but the parameters of distribution change. Another way to find m_y and σ_y is to use the equations stating that for a linear transform

$$Y := aX + b \qquad m_y := a \cdot m_x + b \qquad \sigma_y := |a| \cdot \sigma_x$$

which gives the same results: $\quad m_y = 11 \qquad \sigma_y = 12$

Problem 13.17 Square Transform and Statistical Parameters.
Rayleigh noise $X(t)$ with parameter $\sigma_x = 2$ V passes through the circuit which performs its square transformation described by equation:

$$Y := X^2$$

Find parameters m_y and σ_y of the transformed noise.

Solution. According to Appendix 17, in this case the Rayleigh distribution transforms to the exponential one:

$$\sigma_x := 2 \qquad f_1(y) := \frac{1}{2\sigma_x^2} \cdot \exp\left(\frac{-y}{2 \cdot \sigma_x^2}\right)$$

Thus:

$$m_y := \int_0^\infty y \cdot f_1(y)\, dy \qquad m_y = 8 \qquad [V]$$

$$D_y := \int_0^\infty \left(y - m_y\right)^2 \cdot f_1(y)\, dy \qquad D_y = 64 \qquad \sigma_y := \sqrt{D_y} \qquad \sigma_y = 8 \qquad [V]$$

Problem 13.18 Exponential Correlation Function and Spectral Density. The correlation function of the Markovian random stationary signal is given by the correlation function and parameters:

$$\sigma := 1 \quad [V] \qquad \alpha := 5 \quad [1/s] \qquad K(\tau) := \sigma^2 \cdot \exp(-\alpha \cdot |\tau|)$$

Plot the correlation function and the spectrum of the process. Find the correlation interval and the effective spectrum width of the signal.

Solution. The normalized correlation function $K(\tau)/K(0)$ is given in Figure 13.1:

The correlation interval τ_c can be found as (see Section 8.2.7):

$$R(\tau) := \frac{K(\tau)}{K(0)} \qquad \tau_c := \frac{1}{2} \cdot \int_{-\infty}^\infty R(\tau)\, d\tau \qquad \tau_c = 0.2 \qquad [s]$$

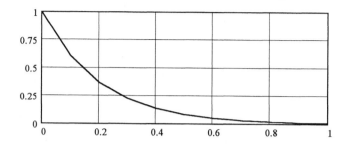

Figure 13.1 Normalized correlation function vs. time (s).

The spectrum of the random process with this correlation function is (see Appendix 19):

$$S(\omega) := \frac{\sigma^2}{\pi} \cdot \frac{\alpha}{\alpha^2 + \omega^2} \qquad [\text{W/Hz}]$$

The normalized spectrum $S(\omega)/S(0)$ is given in Figure 13.2:

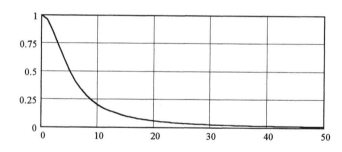

Figure 13.2 Normalized spectrum vs. circular frequency ω (Hz).

The effective spectrum width is:

$$\Delta\omega := \frac{1}{S(0)} \cdot \int_{-\infty}^{\infty} S(\omega) \, d\omega \qquad \Delta f := \frac{\Delta\omega}{2\cdot\pi} \qquad \Delta f = 2.5 \qquad [\text{Hz}]$$

Problem 13.19 Oscillating Correlation Function and Spectral Density.
Solve the previous problem for the stationary random signal with correlation function and parameters:

$$K(\tau) := \sigma^2 \cdot \exp(-\alpha \cdot |\tau|) \cdot \cos(\beta \cdot \tau) \qquad \sigma := 10 \ [\text{V}] \qquad \alpha := 5 \ [1/\text{s}] \qquad \beta := 10 \ [1/\text{s}]$$

Solution. The normalized correlation function K(τ)/K(0) is given in Figure 13.3:

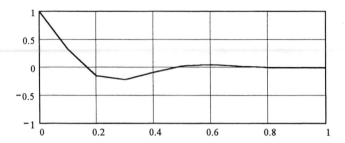

Figure 13.3 Normalized correlation function vs. time (s).

The correlation interval τ_c :

$$R(\tau) := \frac{K(\tau)}{K(0)} \qquad \tau_c := \frac{1}{2} \cdot \int_{-\infty}^{\infty} R(\tau)\, d\tau \qquad \tau_c = 0.04 \qquad [s]$$

The spectrum of the random process with this correlation function is (see Appendix 19):

$$S(\omega) := \frac{\sigma^2 \cdot \alpha}{\pi} \cdot \frac{\alpha^2 + \beta^2 + \omega^2}{\left(\omega^2 - \beta^2 - \alpha^2\right)^2 + 4 \cdot \alpha^2 \cdot \omega^2} \qquad [W/Hz]$$

The normalized spectrum S(ω)/S(0) is given in Figure 13.4:

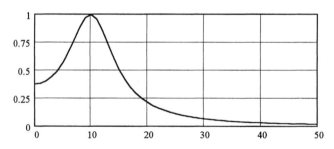

Figure 13.4 Normalized spectrum vs. circular frequency ω (Hz).

The effective spectrum width is:

$$\Delta\omega := \frac{1}{S_{max}} \cdot \int_{-\infty}^{\infty} S(\omega)\, d\omega \qquad \Delta f := \frac{\Delta\omega}{2 \cdot \pi} \qquad \Delta f = 4.722 \quad [Hz]$$

Problem 13.20 Simulation of a Random Process: Gaussian Distribution. Simulate the one second long sample of the uncorrelated Gaussian noise with correlation interval $\tau_c = 0.01$ s, mean $m = 5$ V, rms magnitude $\sigma = 1$ V.

Solution. For the uncorrelated random process, the sampling interval can be taken equal to the correlation interval $\Delta t = \tau_c$. Since the sample length $T = 1$ s, the number of points N in the noise sample can be found as:

$$\tau_c := 0.01 \quad \Delta t := \tau_c \quad \Delta t = 0.01 \quad T := 1 \quad N := \text{floor}\left(\frac{T}{\Delta t}\right) \quad N = 100$$

An algorithm to simulate the Gaussian random variable is presented in Appendix 21. First the set of pairs of the uniformly distributed random variables at the interval (0,1) in $n = 0, 1,...$ N-1 points $u_{n,0}, u_{n,1}$ has to be simulated using standard MATHCAD function rnd(1).

$$n := 0 .. N - 1 \quad k := 0 .. 1 \quad u_{n,k} := \text{rnd}(1)$$

Then the set of random variables with standard Gaussian distribution in $n = 0, 1,...$ N-1 points $z0G_n$ has to be simulated.

$$z0G_n := \sqrt{-2 \cdot \ln\left(u_{n,0} + 10^{-10}\right)} \cdot \cos\left(2 \cdot \pi \cdot u_{n,1}\right)$$

Note. Factor 10^{-10} is added to avoid undefined function ln(0) in case the random generator returns the value $u_{n,0} = 0$.

Finally, the Gaussian process with specified parameters can be simulated (Figure 13.5):

$$m := 5 \quad \sigma := 1 \quad zG_n := m + \sigma \cdot z0G_n$$

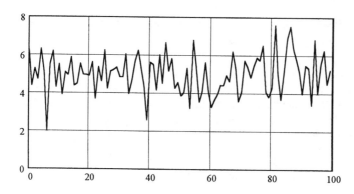

Figure 13.5 Noise magnitude (V) vs. number of samples for the uncorrelated random process with Gaussian distribution.

Problem 13.21   Simulation of a Random Process: Rayleigh Distribution. Simulate the sample of the envelope of the uncorrelated Gaussian random noise with correlation interval $\tau_c = 0.0005$ s, sample length $T = 0.1$ s. The envelope has Rayleigh pdf with parameter $\sigma = 30$ Volts.

Solution. As in the previous Problem 13.20, the number of points N in the noise sample and basic random variables z_n can be found as:

$$\tau_c := 0.0005 \quad \Delta t := \tau_c \quad \Delta t = 5 \times 10^{-4} \quad T := 0.1 \quad N := floor\left(\frac{T}{\Delta t}\right) \quad N = 200$$

$$n := 0 .. N - 1 \quad z_n := rnd(1)$$

Then the set of random variables with Rayleigh distribution in n = 0, 1,..., N-1 points zR_n can to be simulated (see Appendix 21):

$$\sigma := 30 \quad zR_n := \sigma \cdot \sqrt{-2 \cdot \ln\left(z_n + 10^{-10}\right)}$$

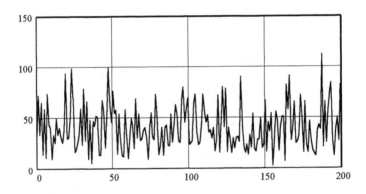

Figure 13.6 Envelope magnitude (V) vs. number of samples for the uncorrelated random process with Rayleigh distribution.

Problem 13.22   Simulation of a Random Process: Ricean Distribution. Simulate the sample of the sum of the nonrandom signal with amplitude A = 10 V and uncorrelated Gaussian random noise with correlation interval $\tau_c = 0.001$ s, sample length $T = 0.5$ s, rms magnitude $\sigma = 2$ V. The sum has Ricean pdf.

Solution. As in Problem 13.20, the number of points N in the noise sample and basic random variables $u_{n,0}, u_{n,1}$ can be found as:

$$\tau_c := 0.001 \quad \Delta t := \tau_c \quad \Delta t = 1 \times 10^{-3} \quad T := 0.5 \quad N := floor\left(\frac{T}{\Delta t}\right) \quad N = 500$$

$n := 0 .. N - 1 \quad k := 0 .. 1 \quad u_{n,k} := rnd(1)$

Then the set of random variables with Ricean distribution in n = 0, 1,..., N-1 points zRN_n can be simulated (see Appendix 21):

$\sigma := 2 \quad A := 10 \quad \alpha := A \quad \alpha = 10$

$$zRN_n := \begin{vmatrix} p \leftarrow 2 \cdot \alpha \cdot \sigma \cdot \sqrt{-2 \cdot \ln\left(u_{n,0} + 10^{-10}\right)} \cdot \cos\left(2 \cdot \pi \cdot u_{n,1}\right) \\ mag_n \leftarrow \sqrt{\alpha^2 - 2 \cdot \sigma^2 \cdot \ln\left(u_{n,0} + 10^{-10}\right)} - p \end{vmatrix}$$

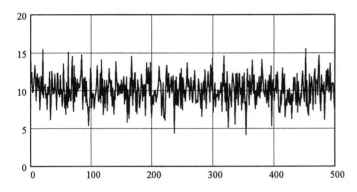

Figure 13.7 Sum of signal and noise magnitude (V) vs. number of samples for the uncorrelated random process with Ricean distribution.

Problem 13.23 Simulation of a Random Process: Exponential Distribution. The distribution of power returns from precipitation observed by a weather radar is the exponential one with mean power $1/\lambda = 10^{-3}$ W. Simulate the sample of the return power with correlation interval $\tau_c = 0.05$ s, sample length T = 5 s.

Solution. As in Problem 13.20, the number of points N in the noise sample and basic random variables z_n can be found as:

$$\tau_c := 0.05 \quad \Delta t := \tau_c \quad \Delta t = 0.05 \quad T := 5 \quad N := floor\left(\frac{T}{\Delta t}\right) \quad N = 100$$

$n := 0 .. N - 1 \quad z_n := rnd(1)$

Then the set of random variables with exponential distribution in n = 0, 1,..., N-1 points zE_n can be simulated (see Appendix 21):

$$\lambda := \frac{1}{0.001} \qquad \lambda = 1 \times 10^3 \qquad zE_n := \frac{1}{\lambda}\cdot-\ln(z_n)$$

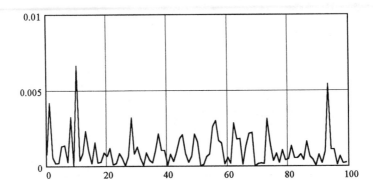

Figure 13.8 Radar return power (W) vs. number of samples for the uncorrelated random process with exponential distribution.

Problem 13.24 Simulation of a Random Process: Log-Normal Distribution. The distribution of amplitude returns from rough sea observed with a high-resolution marine radar is the log-normal one with mean amplitude $A = 10^{-6}$ Volts and mean-to-median ratio $M = 10$. Simulate the sample of the return amplitude with correlation interval $\tau_c = 0.05$ s, sample length $T = 5$ s.

Solution. Like in Problem 13.20, the number of points N in the noise sample and basic random variables $u_{n,0}$, $u_{n,1}$ can be found as:

$$\tau_c := 0.05 \qquad \Delta t := \tau_c \qquad \Delta t = 0.05 \qquad T := 5 \qquad N := \text{floor}\left(\frac{T}{\Delta t}\right) \qquad N = 100$$

$$n := 0..N-1 \qquad k := 0..1 \qquad u_{n,k} := \text{rnd}(1)$$

Then the set of random variables with log-normal distribution in $n = 0, 1,..., N-1$ points zL_n can be simulated (see Appendix 21):

$$A := 10^{-6} \qquad M := 10$$

$$zL_n := \begin{vmatrix} p \leftarrow \sqrt{-2\cdot\ln(u_{n,0})}\cdot\cos(2\cdot\pi\cdot u_{n,1}) \\[2mm] mag_n \leftarrow A + A\cdot\sqrt{M^2 - 1}\cdot\exp(p) \end{vmatrix}$$

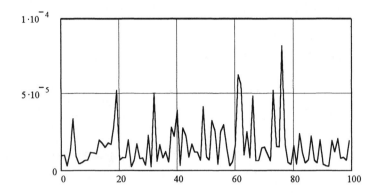

Figure 13.9 Radar return amplitude (Volts) vs. number of samples for the uncorrelated random process with log-normal distribution.

Problem 13.25 Simulation of a Correlated Random Process. The error of the range measurement by the laser range finder is the stationary Gaussian random process with zero mean and correlation function $K(\tau) = \sigma^2 \exp(-\alpha|\tau|)$ with parameters $\sigma = 0.1$ m, $\alpha = 5$ 1/s. The pulse repetition frequency (PRF) of the laser range finder is 1000 Hz. Simulate the sample of the range error with the length $T = 1s$.

Solution. First, we find the correlation interval τ_c :

$$\sigma := 0.1 \text{ [m]} \qquad \alpha := 5 \quad [1/s] \qquad K(\tau) := \sigma^2 \cdot \exp(-\alpha \cdot |\tau|)$$

$$R(\tau) := \frac{K(\tau)}{K(0)} \qquad \tau_c := \frac{1}{2} \cdot \int_{-\infty}^{\infty} R(\tau)\, d\tau \qquad \tau_c = 0.2 \quad [s]$$

Then, we define the sampling interval Δt based on PRF and the number of points within the length T:

$$\text{PRF} := 1000 \quad [\text{Hz}] \qquad \Delta t := \frac{1}{\text{PRF}} \quad [s] \qquad \Delta t = 1 \times 10^{-3} \qquad T := 1 \quad [s]$$

$$N := \text{floor}\left(\frac{T}{\Delta t}\right) \qquad N = 1 \times 10^3$$

Since the sampling interval Δt is less than the correlation interval τ_c, the process has to be simulated as the correlated one. Then we simulate the set of independent random variables with standard Gaussian distribution z0G in n = 0, 1,..., N-1 points as in Problem 13.20:

$$n := 0 .. N - 1 \quad k := 0 .. 1 \quad u_{n,k} := rnd(1)$$

$$z0G_n := \sqrt{-2 \cdot \ln\left(u_{n,0} + 10^{-10}\right)} \cdot \cos\left(2 \cdot \pi \cdot u_{n,1}\right)$$

The first sample of the simulated random process is:

$$z_0 := \sigma \cdot z0G_0$$

The rest of the samples are simulated based on the recurrent formula (Appendix 21):

$$n := 1 .. N - 1 \quad a_1 := \exp(-\alpha \cdot \Delta t) \quad b_1 := \sigma \cdot \sqrt{1 - \exp(-2 \cdot \alpha \cdot \Delta t)}$$

$$z_n := a_1 \cdot z_{n-1} + b_1 \cdot z0G_n$$

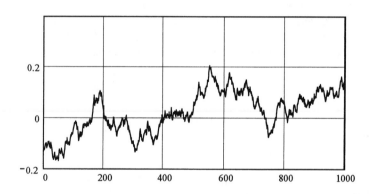

Figure 13.10 Range error (meters) vs. number of samples for the correlated random process with Gaussian distribution.

Problem 13.26 Applied Statistics: Parameters Estimation. The sample of the error of the range measurement by the laser range finder recorded during the test is given by the function in Figure 13.10. It can be considered a stationary ergodic random process. Find the mean, rms deviation, and correlation function of the range error.

Solution. The parameters of the ergodic stationary process can be found based on equations from Appendix 22:

Mean value m:

$$m := \frac{1}{N} \cdot \sum_{n=0}^{N-1} z_n$$

$m = 0.022$ [m]

Rms deviation σ:

$$\sigma := \sqrt{\frac{1}{N} \cdot \sum_{n=0}^{N-1} (z_n)^2 - \frac{1}{N^2} \cdot \left(\sum_{n=0}^{N-1} z_n \right)^2}$$

$\sigma = 0.082$ [m]

Correlation function $K(\tau) = K(r\Delta t) = K_r$:

$r := 0 .. N - 1$

$$K_r := \frac{1}{N-r} \cdot \sum_{n=0}^{N-1-r} z_n \cdot z_{n+r} - \frac{1}{(N-r)^2} \cdot \sum_{n=0}^{N-1-r} z_n \sum_{n=r}^{N-1} z_n$$

The normalized correlation function is (Figure 13.11):

$$R_r := \frac{K_r}{K_0}$$

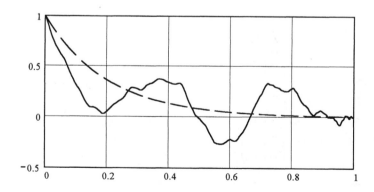

Figure 13.11 Estimate of the normalized correlation function R_r vs. time (s); dash line is the theoretical normalized correlation function $R(\tau) = K(\tau)/K(0)$, $K(\tau) = \sigma^2 \exp(-\alpha|\tau|)$, $\sigma = 0.1$ m, $\alpha = 5$ 1/s.

Note. In order to increase the accuracy of the estimate, the sample length T has to be increased (in this problem up to 10 s). In this case $m \to 0$, $\sigma \to 0.1$, $K_r \to \sigma^2 \exp(-\alpha|\tau|)$ but computation time also increases considerably.

Problem 13.27 Applied Statistics: Least Mean Square Method.
During the test of $N = 14$ groups of transistors the amount of transistors which parameters do not meet specification y_n versus the amount of transistors in *n*th group x_n were as given below.

$x_1 := 120$ $x_2 := 131$ $x_3 := 140$ $x_4 := 161$ $x_5 := 174$ $x_6 := 180$ $x_7 := 200$

$x_8 := 214$ $x_9 := 219$ $x_{10} := 241$ $x_{11} := 250$ $x_{12} := 268$ $x_{13} := 281$ $x_{14} := 300$

$y_1 := 5$ $y_2 := 6$ $y_3 := 7$ $y_4 := 8$ $y_5 := 8$ $y_6 := 9$ $y_7 := 10$

$y_8 := 11$ $y_9 := 12$ $y_{10} := 14$ $y_{11} := 15$ $y_{12} := 16$ $y_{13} := 18$

$y_{14} := 20$

Find the parabolic function $y = ax^2 + bx + c$ that the best matches the function describing the amount of the faulty transistors depending on the amount of the transistors in the test group.

Solution. The test results can be represented as the table:

$X = $

	0
0	0
1	120
2	131
3	140
4	161
5	174
6	180
7	200
8	214
9	219
10	241
11	250
12	268
13	281
14	300

$y = $

	0
0	0
1	5
2	6
3	7
4	8
5	8
6	9
7	10
8	11
9	12
10	14
11	15
12	16
13	18
14	20

$N := 14$ number of test groups

$n := 1 .. N$

Using the procedure from Section 8.4.8, we can find coefficients of the matrix A and parameters P. Coefficients of the matrix A are:

$$a_{11} := \frac{1}{N} \cdot \sum_n (x_n)^4 \qquad a_{11} = 2.593 \times 10^9$$

$$a_{12} := \frac{1}{N} \cdot \sum_n (x_n)^3 \qquad a_{12} = 1.06 \times 10^7 \quad a_{21} := a_{12} \qquad a_{21} = 1.06 \times 10^7$$

$$a_{13} := \frac{1}{N} \cdot \sum_n (x_n)^2 \qquad a_{13} = 4.535 \times 10^4 \quad a_{31} := a_{13} \qquad a_{31} = 4.535 \times 10^4$$

$$a_{23} := \frac{1}{N} \cdot \sum_n x_n \qquad a_{23} = 205.643 \qquad a_{32} := a_{23} \qquad a_{32} = 205.643$$

$$a_{22} := a_{13} \qquad\qquad a_{22} = 4.535 \times 10^4 \quad a_{33} := 1 \qquad a_{33} = 1$$

Thus,

$$A := \begin{pmatrix} a_{11} & a_{12} & a_{13} \\ a_{21} & a_{22} & a_{23} \\ a_{31} & a_{32} & a_{33} \end{pmatrix} \qquad A = \begin{pmatrix} 2.593 \times 10^9 & 1.06 \times 10^7 & 4.535 \times 10^4 \\ 1.06 \times 10^7 & 4.535 \times 10^4 & 205.643 \\ 4.535 \times 10^4 & 205.643 & 1 \end{pmatrix}$$

Parameters P can be found as:

$$P_1 := \frac{1}{N} \cdot \sum_n (x_n)^2 \cdot y_n \qquad P_1 = 6.181 \times 10^5$$

$$P_2 := \frac{1}{N} \cdot \sum_n x_n \cdot y_n \qquad P_2 = 2.58 \times 10^3$$

$$P_3 := \frac{1}{N} \cdot \sum_n y_n \qquad P_3 = 11.357$$

The equation to find the X = (a, b, c) is:

$$A \cdot x := \begin{pmatrix} P_1 \\ P_2 \\ P_3 \end{pmatrix}$$

With the aid of a computer (see Chapter 9), we can find the solution of this equation:

$$X := A^{-1} \cdot \begin{pmatrix} P_1 \\ P_2 \\ P_3 \end{pmatrix} \qquad X = \begin{pmatrix} 2.016 \times 10^{-4} \\ -3.886 \times 10^{-3} \\ 3.016 \end{pmatrix}$$

Thus, the approximation function is:

$$Y_n := 2.016 \cdot 10^{-4} \cdot \left(x_n\right)^2 - 3.886 \cdot 10^{-3} \cdot x_n + 3.016$$

Actual dependence of the amount of the faulty transistors upon the amount of transistors in the test group y_n (solid line), and approximation function Y_n (dash line) are given in Figure 13.12.

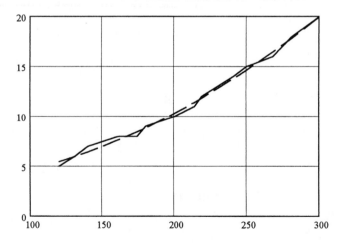

Figure 13.12 Dependence of the amount of the faulty transistors y_n (solid line) and approximation function Y_n (dash line) vs. amount of transistors in the test group.

PART III

REFERENCE DATA

Appendix 1 Common Constants, Units, and Conversion Factors

Constants

- $\pi = 3.141592654$ $2\pi = 6.283185307$
- $e = 2.718281828$
- Factor to convert degrees to radians $z = \dfrac{\pi}{180} = 0.01745329252$
- Speed of light $c = 2.997925 \cdot 10^8$ m/s
- Boltzmann's constant $k = 1.38054 \cdot 10^{-23} \dfrac{W*s}{{}^0K}$
- Mean radius of Earth $= 6{,}371{,}030$ m
- Mean radius of Earth with 4/3 approximation $= 8{,}493{,}333$ m

Metric Units

• Exa	E	10^{18}		• Deci	d	10^{-1}
• Peta	P	10^{15}		• Centi	c	10^{-2}
• Tera	T	10^{12}		• Milli	m	10^{-3}
• Giga	G	10^{9}		• Micro	μ	10^{-6}
• Mega	M	10^{6}		• Nano	n	10^{-9}
• Kilo	k	10^{3}		• Pico	p	10^{-12}
• Hecto	h	10^{2}		• Femto	f	10^{-15}
• Deca	da	10		• Atto	a	10^{-18}

Angular Measure

- 1 degree (deg or °) $= 0.01745329$ rad
- 1 radian (rad) $= 57.29578° = 3437'.747 = 206264''.8$
- 1 miliradian (mrad) $= 3'26''.26$
- 1 steradian (sr) $= 3282.8$ deg^2
- 1 arc. minute in longitude/latitude = 1 nautical mile = 1852 m
- 1 degree in longitude/latitude = 60 nautical miles = 111.12 km

Length

- 1 micron (μm) = 1 micrometer $= 10^{-6}$ m $= 0.0001$ cm $= 10{,}000 \,\overset{\circ}{A}$
- 1 angstrom $(\overset{\circ}{A}) = 10^{-10}$ m $= 0.0001 \mu$m
- 1 mil $= 0.001$ inch (in)
- 1 inch (in) $= 2.54$ cm
- 1 foot (ft) $= 30.48$ cm

- 1 yard (yd) = 0.9144 m
- 1 metre (m) = 39.37008 in = 3.280840 ft = 1.093613 yd
- 1 mile (mi) = 1.609344 km = 1760 yd = 5280 ft = 0.8689762 international nautical mile (nm)
- 1 kilometer (km) = 0.6213712 mi
- 1 international nautical mile (nm) = 1852 m

Velocity and Angular Velocity

- 1 mile per hour (m.p.h. or mi/hr) = 1.6093 km/hr = 88 ft/min = 1.4667 ft/s = 44.7040 cm/s
- 1 ft/s = 30.48 cm/s = 0.6818 mi/h = 1.097 km/hr
- 1 (international) knot = 1 nm/hr
- 1 U.K. knot = 1 U.K. nm/hr = 1.151515 mile/hr = 1.688889 ft/s = 1.85318 km/hr = 51.48 cm/s = 1.00064 knots (international)
- 1 km/hr = 0.6214 m.p.h. = 0.9113 ft/s = 27.78 cm/s
- 1 velo = 1 cm/s = 0.0328 ft/s
- 1 Mach = velocity of sound at a given height
- 1 Mach at ground level (15°C and 14.7 lb/in^2) = 1116.1 ft/s = 1224.7 km/hr = 760.97 m.p.h. = 660.85 U.K. knot = 661.27 knots
- Unit angular velocity = 1 rad/s = 9.549 rev/min = 0.1592 rev/s = 57.3 deg/s = 2.06265 x 10^5 deg/hr
- 1°/hr = $4.848 \cdot 10^{-6}$ rad/s

Acceleration

- Standard g = 9.80665 m/s^2 = 32.1740 ft/s^2
- 1 Gal (or 1 celo) = 1 cm/s^2 = 0.0328 ft/s^2

Area

- 1 in^2 = 6.4516 cm^2
- 1 cm^2 = 0.1550 in^2
- 1 ft^2 = 0.0929 m^2
- 1 yd^2 = 9 ft^2 = 0.8361 m^2
- 1 m^2 = 10.764 ft^2 = 1.196 yd^2
- 1 acre = 43,560 ft^2 = 4,840 yd^2
- 1 mi^2 = 640 acres = 258.999 hectares = 2.58999 km^2
- 1 hectare = 10,000 m^2 = 2.471 acres
- 1 barn (b) = 10^{-24} cm^2
- 1 sq. $\overset{\circ}{A}$ = 10^{-16} cm^2 = 100 Mb

Volume

- $1 \text{ in}^3 = 16.387 \text{ cm}^3$
- $1 \text{ m}^3 = 35.31 \text{ ft}^3 = 1.308 \text{ yd}^3$
- $1 \text{ yd}^3 = 0.7646 \text{ m}^3$
- $1 \text{ litre (L)} = 1 \text{ dm}^3$
- $1 \text{ UK gallon (U.K. gal)} = 4.54609 \text{ L} = 277.42 \text{ in}^3$
- $1 \text{ US gallon (U.S. gal)} = 3.785411 \text{ L} = 231 \text{ in}^3$
- $1 \text{ ft}^3 = 6.229 \text{ U.K. gal} = 7.481 \text{ U.S. gallons}$

Energy

- $1 \text{ ft lb wt (= loosely, 1 ft lb)} = 32.16 \text{ ft pdl} = 192 \text{ in oz wt}$
  $= 13,825 \text{ cm g wt} = 1.356 \cdot 10^7 \text{ dyn cm (= erg.)} = 1.356 \text{ Nm (or mN)}$
- $1 \text{ oz wt in} = 72.01 \text{ g wt cm}$
- $1 \text{ Joule (J)} = 1 \text{ Nm} = 10^7 \text{ erg} = 0.7376 \text{ lb wt ft}$
- $1 \text{ calorie } 15°C \text{ (c)} = 4.1855 \text{ J}$
- $1 \text{ calorie (international)} = 4.1868 \text{ J}$
- $1 \text{ large calorie (C)} = 1,000 \text{ c}$
- $1 \text{ British Thermal Unit (Btu)} = 1,055 \text{ J} = 252 \text{ c} = 778 \text{ lb wt ft}$
- $1 \text{ Board of Trade Unit} = 1 \text{ kW hr} = 3,600 \times 10^3 \text{ J} = 2.655 \times 10^6 \text{ lb wt ft}$
  $= 3,412 \text{ Btu}$
- $1 \text{ thm} = 10^5 \text{ Btu}$

Power

- $1 \text{ watt (W)} = 1 \text{ J/s} = 10^7 \text{ erg/s}$
- $1 \text{ hp} = 745.7 \text{ W} = 2,544 \text{ Btu/hr} = 42.4 \text{ Btu/min} = 550 \text{ ft lb/s} = 178.2 \text{ c/s}$
- $1 \text{ metric hp} = 735.499 \text{ W}$

Pressure

- $1 \text{ Standard Atmosphere } (0°C) = 760 \text{ mm Hg} = 29.921 \text{ in Hg} = 14.696 \text{ lb}$
  $\text{wt./in}^2 = 1.01325 \cdot 10^6 \text{ dyn/cm}^2 = 1.01325 \text{ Bar} = 1,033.23 \text{ g wt/cm}^2$
  $= 101,325 \text{ N/m}^2$
- $1 \text{ Torr} = 1 \text{ mm Hg} = 1.359 \text{ g wt/cm}^2 = 1,333 \text{ dyn/cm}^2 = 1.333 \text{ mbar}$
- $1 \text{ g wt/cm}^2 = 980.7 \text{ dyn/cm}^2 = 98.07 \text{ N/m}^2$
- $1 \text{ lb wt/in}^2 = 70.307 \text{ g wt/cm}^2 = 68.95 \text{ mbar} = 51.72 \text{ Torr}$
- $1 \text{ millibar (mbr)} = 1,000 \text{ dyn/cm}^2 = 100 \text{ N/m}^2 = 0.0145 \text{ lb wt/m}^2$
  $= 0.7502 \text{ Torr}$

Mass

- $1 \text{ lb} = 0.45359237 \text{ kg} = 16 \text{ oz} = 7000 \text{ grain} = 1.2153 \text{ lb Troy or Apothecary}$

- 1 grain (gr) = 64.7989 mg
- 1 kg = 2.20462 lb
- 1 metric tone = 1,000 kg = 10^6 g
- 1 metric carat = 0.2 g
- 1 oz = 28.3495 g
- 1 slug = 32.174 lb = 14.5939 kg
- 1 U.S. lb = 0.45359243 kg

Force

- 1 lb wt = $4.4482 \cdot 10^5$ dyn
- 1 g wt = 980.665 dyn
- 1 poundal (pdl) = $1.3825 \cdot 10^4$ dyn
- 10^6 dyne (dyn) = 1.0197 kg wt = 2.2481 lb wt
- 1 Newton (N) = 10^5 dyn

Moment of Inertia

- 1 kg m^2 = 10^7 g cm^2 = 23.73 lb ft^2 = 0.7376 slug ft^2
- 1 slug ft^2 = 32.174 lb ft^2 = $13.558 \cdot 10^6$ g cm^2 = 1.3558 kg m^2

Density

- 1 lb/in^3 = 27.680 g/cm^3
- 1 g/cm^3 = 0.03613 lb/in^3
- 1 lb/ft^3 = 16.0185 kg/m^3
- 1 kg/m^3 = 0.06243 lb/ft^3
- Water density (4°C) = 0.999972 g/cm^3 = 62.43 lb/ft^3
- Mercury density (0°C) = 13.5951 g/cm^3

Temperature

- °C = 5(°F − 32)/9
- °F = (9°C/5) + 32
- °K = °C + 273.15
- °R = °F + 459.67

Note. °F = degrees Fahrenheit
      °C = degrees Celsius (formerly Centigrade)
      °K = degrees Kelvin
      °R = degrees Rankine

Appendix 2 Hyperbolic and Inverse Hyperbolic Functions: Basic Identities and Conversion Formulas

♦ **Basic Identities**

$$sh(-x) = -sh\,x \qquad ch(-x) = ch\,x \qquad th(-x) = -th\,x$$

$$ch^2x - sh^2x = 1 \qquad (ch\,x \pm sh\,x)^n = ch\,nx \pm sh\,nx$$

$$sh\,x = \sqrt{ch^2x - 1} = \frac{th\,x}{\sqrt{1 - th^2x}} \qquad ch\,x = \sqrt{1 + sh^2x} = \frac{1}{\sqrt{1 - th^2x}}$$

$$th\,x = \frac{sh\,x}{\sqrt{1 + sh^2x}} = \frac{\sqrt{ch^2x - 1}}{ch\,x} \qquad \frac{ch\,x + sh\,x}{ch\,x - sh\,x} = \frac{1 + th\,x}{1 - th\,x}$$

$$Arsh\,x = Arch\sqrt{x^2 + 1} = Arth\frac{x}{\sqrt{x^2 + 1}}$$

$$Arch\,x = Arsh\sqrt{x^2 - 1} = Arth\frac{\sqrt{x^2 - 1}}{x} = 2Arch\sqrt{\frac{x+1}{2}} = 2Arsh\sqrt{\frac{x-1}{2}}$$

$$Arth\,x = Arsh\frac{x}{\sqrt{1 - x^2}} = Arch\frac{1}{\sqrt{1 - x^2}} = \frac{1}{2}Arsh\frac{2x}{1 - x^2} = \frac{1}{2}Arch\frac{1 + x^2}{1 - x^2} = \frac{1}{2}Arth\frac{2x}{1 + x^2}$$

$$sh\,x = \frac{e^x - e^{-x}}{2}, \quad ch\,x = \frac{e^x + e^{-x}}{2}, \quad th\,x = \frac{e^x - e^{-x}}{e^x + e^{-x}} = \frac{1 - e^{-2x}}{1 + e^{-2x}}$$

$$th\,x = 1 - \frac{e^{-x}}{ch\,x} \qquad cth\,x = 1 + \frac{e^{-x}}{sh\,x}$$

$$e^x = ch\,x + sh\,x = \frac{ch\dfrac{x}{2} + sh\dfrac{x}{2}}{ch\dfrac{x}{2} - sh\dfrac{x}{2}} = \frac{1 + th\dfrac{x}{2}}{1 - th\dfrac{x}{2}}$$

$$e^{-x} = chx - shx = \frac{ch\dfrac{x}{2} - sh\dfrac{x}{2}}{ch\dfrac{x}{2} + sh\dfrac{x}{2}} = \frac{1 - th\dfrac{x}{2}}{1 + th\dfrac{x}{2}}$$

$$Arsh\,x = \ln\left(x + \sqrt{x^2 + 1}\right) = -\ln\left(\sqrt{x^2 + 1} - x\right)$$

$$Arch\,x = \ln\left(x + \sqrt{x^2 - 1}\right) = -\ln\left(x - \sqrt{x^2 - 1}\right)$$

$$Arth\,x = \frac{1}{2}\ln\frac{1 + x}{1 - x} \qquad Arth\frac{1}{x} = \frac{1}{2}\ln\frac{x + 1}{x - 1}$$

$$\ln x = Arsh\frac{x^2 - 1}{2x} = Arch\frac{x^2 + 1}{2x} = Arth\frac{x^2 - 1}{x^2 + 1}$$

◆ Functions of the Multiple Arguments

$$sh\,2x = 2\,sh\,x\,chx = \frac{2\,th\,x}{1 - th^2 x}$$

$$sh\,3x = 4\,sh^3 x + 3\,sh\,x = sh\,x\left(4\,ch^2 x - 1\right)$$

$$sh(n + 1)x = 2\,chx\,ch\,nx - sh(n - 1)x$$

$$sh\,nx = n\,sh\,x\,ch^{n-1}x + \binom{n}{3}sh^3 x\,ch^{n-3}x + \binom{n}{5}sh^5 x\,ch^{n-5}x + \ldots$$

$$ch\,2x = ch^2 x + sh^2 x = \frac{1 + th^2 x}{1 - th^2 x}$$

$$ch\,3x = 4ch^3 x - 3\,chx = chx\left(4\,sh^2 x + 1\right)$$

$$ch(n + 1)x = 2\,chx\,ch\,nx - ch(n - 1)x$$

$$ch\,nx = ch^n x + \binom{n}{2}sh^2 x\,ch^{n-2}x + \binom{n}{4}sh^4 x\,ch^{n-4}x + \ldots$$

$$th\,2x = \frac{2\,th\,x}{1+th^2 x} \qquad th\,3x = \frac{th^3 x + 3\,th\,x}{3\,th^2 x + 1} \qquad 2\,cth\,2x = th\,x + cth\,x$$

$$th\frac{x}{2} = \sqrt{\frac{ch\,x - 1}{ch\,x + 1}} = \frac{sh\,x}{ch\,x + 1} = \frac{ch\,x - 1}{sh\,x}$$

◆ Powers

$$2\,sh^2 x = ch\,2x - 1$$

$$4\,sh^3 x = sh\,3x - 3\,sh\,x$$

$$8\,sh^4 x = ch\,4x - 4\,ch\,2x + 3$$

$$16\,sh^5 x = sh\,5x - 5\,sh\,3x + 10\,sh\,x$$

$$32sh^6 x = ch\,6x - 6\,ch\,4x + 15\,ch\,2x - 10$$

$$64\,sh^7 x = sh\,7x - 7\,sh\,5x + 21\,sh\,3x - 35\,sh\,x$$

$$128\,sh^8 x = ch\,8x - 8\,ch\,6x + 28\,ch\,4x - 56\,ch\,2x + 35$$

$$2\,ch^2 x = ch\,2x + 1$$

$$4\,ch^3 x = ch\,3x + 3\,ch\,x$$

$$8\,ch^4 x = ch\,4x + 4\,ch\,2x + 3$$

$$16\,ch^5 x = ch\,5x + 5\,ch\,3x + 10\,ch\,x$$

$$32\,ch^6 x = ch\,6x + 6\,ch\,4x + 15\,ch\,2x + 10$$

$$64\,ch^7 x = ch\,7x + 7\,ch\,5x + 21\,ch\,3x + 35\,ch\,x$$

$$128\,ch^8 x = ch\,8x + 8\,ch\,6x + 28\,ch\,4x + 56\,ch\,2x + 35$$

◆ Formulas of Summation

$$sh(x \pm y) = sh\,x\,ch\,y \pm ch\,x\,sh\,y$$

$$ch(x \pm y) = ch\,x\,ch\,y \pm sh\,x\,sh\,y$$

$$th(x \pm y) = \frac{th\,x \pm th\,y}{1 \pm th\,x\,th\,x}$$

$$sh\,x + sh\,y = 2\,sh\frac{x+y}{2}\,ch\frac{x-y}{2}$$

$$ch\,x + ch\,y = 2\,ch\frac{x+y}{2}\,ch\frac{x-y}{2}$$

$$ch\,x - ch\,y = 2\,sh\frac{x+y}{2}\,sh\frac{x-y}{2}$$

$$th\,x \pm th\,y = \frac{sh(x \pm y)}{ch\,x\,ch\,y}$$

$$2\,sh\,x\,sh\,y = ch(x+y) - ch(x-y)$$

$$2\,ch\,x\,ch\,y = ch(x+y) + ch(x-y)$$

$$2\,sh\,x\,ch\,y = sh(x+y) + sh(x-y)$$

$$2\,ch\,x\,sh\,y = sh(x+y) - sh(x-y)$$

$$Arsh\,x \pm Arsh\,y = Arsh\left(x\sqrt{1+y^2} \pm y\sqrt{1+x^2}\right)$$

$$Arch\,x \pm Arch\,y = Arch\left(xy \pm \sqrt{(x^2-1)(y^2-1)}\right)$$

$$Arth\,x \pm Arth\,y = Arth\frac{x \pm y}{1 \pm xy}$$

$$A\,ch\,x + B\,sh\,x = \sqrt{A^2 - B^2}\,ch\left(x + Arth\frac{B}{A}\right) = \sqrt{B^2 - A^2}\,sh\left(x + Arth\frac{A}{B}\right)$$

Appendix 3 **Delta Function**

The *delta function* (δ-function or Dirac function) is the impulse of infinitesimally short duration that has infinitely large amplitude at the moment $t = t_0$ and an area equal to unity. By Dirac's definition:

$$\delta(t-t_0) = \begin{cases} \infty, & t = t_0 \\ 0, & t \neq t_0 \end{cases}$$

$$\int_{-\infty}^{\infty} \delta(t-t_0)\,dt = 1$$

In electrical engineering these two definitions are typically complemented by the other two definitions:

$$\delta(t) = \delta(-t)$$

which means it is an even function of the argument, and

$$\lim_{T \to \infty} \frac{1}{T} \int_{-T/2}^{T/2} \delta^2(t) = 1$$

which means that it has a unity average power at the infinite interval T. With these definitions the following equations are valid:

$$\int_{-\infty}^{0} \delta(t)\,dt = \int_{0}^{\infty} \delta(t)\,dt = \frac{1}{2}$$

$$\int_{-\infty}^{t} \delta(t-t_0)\,dt = \begin{cases} 0, & t < t_0 \\ 1/2, & t = t_0 \\ 1, & t > t_0 \end{cases}$$

Delta function has the following properties:

• It is an *even function* of its argument:

$$\delta(t-t_0) = \delta(t_0 - t)$$

- *Selectivity* feature: delta function selects the value of the function $x(t)$ only in the moment $t = t_0$:

$$x(t) \cdot \delta(t - t_0) = x(t_0) \cdot \delta(t - t_0)$$

- The variation of timescale:

$$\delta(at) = \frac{1}{|a|} \cdot \delta(t) \quad (\text{e.g.,} \quad \delta(\omega) = \frac{1}{2\pi} \delta(f))$$

- Power and energy of delta function is finite at any infinite interval T:

$$P = \frac{1}{T} \int_{-\frac{T}{2}}^{T/2} \delta^2(t) dt = \lim_{\Delta t \to 0} \frac{1}{T} \int_{-T/2}^{T/2} \left(\frac{1}{\Delta t}\right)^2 dt = \lim_{\Delta t \to 0} \frac{1}{T \cdot \Delta t}$$

- The second power of delta function is equal to delta function:

$$\delta^2(t) = \delta(t)$$

- Delta function is always *orthogonal* at noncoincident moments of time:

$$\lim_{T \to \infty} \int_{-T/2}^{T/2} \delta(t - t_1) \cdot \delta(t - t_2) dt = \begin{cases} 0, & t_1 \neq t_2 \\ 1, & t_1 = t_2 \end{cases}$$

- *Filtration* feature: the following equation is valid for any function $x(t)$ that is continuous and limited in point $t = t_0$:

$$\int_{t_a}^{t_b} x(t) \cdot \delta(t - t_0) dt = \begin{cases} x(t_0), & t_a < t_0 < t_b \\ \frac{1}{2} x(t_0), & t_0 = t_a, \quad t_0 = t_b \\ 0, & t_0 < t_a, \quad t_0 > t_b \end{cases}$$

Appendix 4   Trigonometric and Inverse Trigonometric Functions: Basic Identities and Conversion Formulas

◆ Trigonometric Functions

$$\cos^2 x + \sin^2 x = 1 \qquad 1 + \tan^2 x = \sec^2 x \qquad 1 + \cot^2 x = \csc^2 x$$

$$\sin(x + y) = \sin x \cos y + \cos x \sin y$$

$$\sin(x - y) = \sin x \cos y - \cos x \sin y$$

$$\cos(x + y) = \cos x \cos y - \sin x \sin y$$

$$\cos(x - y) = \cos x \cos y + \sin x \sin y$$

$$\tan(x + y) = \frac{\tan x + \tan y}{1 - \tan x \tan y}$$

$$\tan(x - y) = \frac{\tan x - \tan y}{1 + \tan x \tan y}$$

$$\sin 2x = 2 \sin x \cos x$$

$$\cos 2x = \cos^2 x - \sin^2 x = 1 - 2 \sin^2 x = 2 \cos^2 x - 1$$

$$\sin x + \sin y = 2 \sin \frac{1}{2}(x + y) \cos \frac{1}{2}(x - y)$$

$$\sin x - \sin y = 2 \cos \frac{1}{2}(x + y) \sin \frac{1}{2}(x - y)$$

$$\cos x + \cos y = 2 \cos \frac{1}{2}(x + y) \cos \frac{1}{2}(x - y)$$

$$\cos x - \cos y = -2 \sin \frac{1}{2}(x + y) \sin \frac{1}{2}(x - y)$$

$$\sin x \cos y = \frac{1}{2}\left[\sin(x + y) + \sin(x - y)\right]$$

$$\cos x \sin y = \frac{1}{2}\left[\sin(x+y) - \sin(x-y)\right]$$

$$\cos x \cos y = \frac{1}{2}\left[\cos(x+y) + \cos(x-y)\right]$$

$$\sin x \sin y = \frac{1}{2}\left[\cos(x-y) - \cos(x+y)\right]$$

$$\sin 3x = 3\sin x - 4\sin^3 x$$

$$\cos 3x = 4\cos^3 x - 3\cos x$$

♦ Inverse Trigonometric Functions

For $0 < x < 1$

$$a \sin x = a \cos \sqrt{1-x^2} = a \tan \frac{x}{\sqrt{1-x^2}} = a \cot \frac{\sqrt{1-x^2}}{x}$$

$$a \cos x = a \sin \sqrt{1-x^2} = a \tan \frac{\sqrt{1-x^2}}{x} = a \cot \frac{x}{\sqrt{1-x^2}}$$

For $0 < x < \infty$

$$a \tan x = a \sin \frac{x}{\sqrt{1+x^2}} = a \cos \frac{1}{\sqrt{1+x^2}} = a \cot \frac{1}{x}$$

$$a \cot x = a \sin \frac{1}{\sqrt{1+x^2}} = a \cos \frac{x}{\sqrt{1+x^2}} = a \tan \frac{1}{x}$$

For $-1 \le x \le 1$

$$a \sin x + a \cos x = \frac{\pi}{2}$$

For $-\infty < x < \infty$

$$\text{a}\tan x + \text{a}\cot x = \frac{\pi}{2}$$

$$a\sin x + a\sin y =$$

$$= \begin{cases} a\sin\left(x\sqrt{1-y^2} + y\sqrt{1-x^2} \right), & \text{if } xy \leq 0 \text{ and } x^2 + y^2 \leq 1 \\ \pi - a\sin\left(x\sqrt{1-y^2} + y\sqrt{1-x^2} \right), & \text{if } x > 0, y > 0 \text{ and } x^2 + y^2 > 1 \\ -\pi - a\sin\left(x\sqrt{1-y^2} + y\sqrt{1-x^2} \right), & \text{if } x < 0, y < 0 \text{ and } x^2 + y^2 > 1 \end{cases}$$

$$a\sin x - a\sin y =$$

$$= \begin{cases} a\sin\left(x\sqrt{1-y^2} - y\sqrt{1-x^2} \right), & \text{if } xy \geq 0 \text{ or } x^2 + y^2 \leq 1 \\ \pi - a\sin\left(x\sqrt{1-y^2} - y\sqrt{1-x^2} \right), & \text{if } x > 0, y < 0 \text{ and } x^2 + y^2 > 1 \\ -\pi - a\sin\left(x\sqrt{1-y^2} - y\sqrt{1-x^2} \right), & \text{if } x < 0, y > 0 \text{ and } x^2 + y^2 > 1 \end{cases}$$

$$a\cos x + a\cos y \equiv \begin{cases} a\cos\left(xy - \sqrt{(1-x^2)(1-y^2)} \right), & \text{if } x + y \geq 0 \\ 2\pi - a\cos\left(xy - \sqrt{(1-x^2)(1-y^2)} \right), & \text{if } x + y < 0 \end{cases}$$

$$a\cos x - a\cos y \equiv \begin{cases} -a\cos\left(xy + \sqrt{(1-x^2)(1-y^2)} \right), & \text{if } x \geq y \\ a\cos\left(xy + \sqrt{(1-x^2)(1-y^2)} \right), & \text{if } x < y \end{cases}$$

$$a\tan x + a\tan y \equiv \begin{cases} a\tan \dfrac{x+y}{1-xy}, & \text{if } xy < 1 \\ \pi + a\tan \dfrac{x+y}{1-xy}, & \text{if } x > 0 \text{ and } xy > 1 \\ -\pi + a\tan \dfrac{x+y}{1-xy}, & \text{if } x < 0 \text{ and } xy > 1 \end{cases}$$

$$a \tan x - a \tan y \equiv \begin{cases} a \tan \dfrac{x-y}{1+xy}, & \text{if } xy > -1 \\[3mm] \pi + a \tan \dfrac{x-y}{1+xy}, & \text{if } x > 0 \text{ and } xy < -1 \\[3mm] -\pi + a \tan \dfrac{x-y}{1+xy}, & \text{if } x < 0 \text{ and } xy < -1 \end{cases}$$

$$2 a \cos x = \begin{cases} a \cos\!\left(2x^2 - 1\right), & \text{if } 0 \le x \le 1 \\[2mm] 2\pi - a \cos\!\left(2x^2 - 1\right), & \text{if } -1 \le x < 0 \end{cases}$$

$$2 a \sin x = \begin{cases} a \sin\!\left(2x\sqrt{1-x^2}\right), & \text{if } |x| \le \dfrac{\sqrt{2}}{2} \\[3mm] \pi - a \sin\!\left(2x\sqrt{1-x^2}\right), & \text{if } \dfrac{\sqrt{2}}{2} < x \le 1 \\[3mm] -\pi - a \sin\!\left(2x\sqrt{1-x^2}\right), & \text{if } -1 \le x \le -\dfrac{\sqrt{2}}{2} \end{cases}$$

$$2 a \tan x = \begin{cases} a \tan \dfrac{2x}{1-x^2}, & \text{if } |x| < 1 \\[3mm] \pi + a \tan \dfrac{2x}{1-x^2}, & \text{if } x > 1 \\[3mm] -\pi + a \tan \dfrac{2x}{1-x^2}, & \text{if } x < -1 \end{cases}$$

$$\frac{1}{2} a \sin x \equiv \begin{cases} a \sin \sqrt{\dfrac{1-\sqrt{1-x^2}}{2}}, & \text{if } 0 \le x \le 1 \\[4mm] -a \sin \sqrt{\dfrac{1-\sqrt{1-x^2}}{2}}, & \text{if } -1 \le x < 0 \end{cases}$$

$$\frac{1}{2} a \cos x \equiv a \cos \sqrt{\frac{1+x}{2}}, \quad \text{if } -1 \le x \le 1$$

$$\frac{1}{2} \arctan x \equiv \begin{cases} \arctan \dfrac{\sqrt{1+x^2}-1}{x}, & \text{if } x \ne 0 \\[3mm] 0, & \text{if } x = 0 \end{cases}$$

Appendix 5 Spherical Triangles: Basic Identities and Conversion Formulas

Note. For notations see Section 1.8.

$$\frac{\sin a}{\sin \alpha} = \frac{\sin b}{\sin \beta} = \frac{\sin c}{\sin \gamma}$$

$$\cos c = \cos a \cdot \cos b + \sin \alpha \sin b \cos \gamma$$

$$\cos \gamma = -\cos \alpha \cos \beta + \sin \alpha \sin \beta \cos c$$

$$\tan \frac{\gamma}{2} = \sqrt{\frac{\sin(p-a)\sin(p-b)}{\sin p \sin(p-c)}}$$

$$\sin \frac{\gamma}{2} = \sqrt{\frac{\sin(p-a)\sin(p-b)}{\sin a \cdot \sin b}}$$

$$\cos \frac{\gamma}{2} = \sqrt{\frac{\sin p \sin(p-c)}{\sin a \cdot \sin b}}$$

where $p = (a+b+c)/2$

$$\tan \frac{c}{2} = \sqrt{\frac{-\sin P \sin(P-\gamma)}{\sin(P-\alpha)\sin(P-\beta)}}$$

$$\sin \frac{c}{2} = \sqrt{-\frac{\sin P \sin(P-\gamma)}{\sin \alpha \sin \beta}}$$

$$\cos \frac{c}{2} = \sqrt{\frac{\sin(P-\alpha)\sin(P-\beta)}{\sin \alpha \sin \beta}}$$

where $P = (\alpha + \beta + \gamma - \pi)/2 = \varepsilon/2$

$$\tan \frac{c}{2} \cos \frac{\alpha - \beta}{2} = \tan \frac{a+b}{2} \cos \frac{\alpha + \beta}{2}$$

$$\tan\frac{c}{2}\sin\frac{\alpha-\beta}{2} = \tan\frac{a-b}{2}\sin\frac{\alpha+\beta}{2}$$

$$\cot\frac{\gamma}{2}\cos\frac{a-b}{2} = \tan\frac{\alpha+\beta}{2}\cos\frac{a+b}{2}$$

$$\cot\frac{\gamma}{2}\sin\frac{a-b}{2} = \tan\frac{\alpha-\beta}{2}\sin\frac{a+b}{2}$$

$$\sin\frac{\gamma}{2}\sin\frac{a+b}{2} = \sin\frac{c}{2}\cos\frac{\alpha-\beta}{2}$$

$$\sin\frac{\gamma}{2}\cos\frac{a+b}{2} = \cos\frac{c}{2}\cos\frac{\alpha+\beta}{2}$$

$$\cos\frac{\gamma}{2}\sin\frac{a-b}{2} = \sin\frac{c}{2}\sin\frac{\alpha-\beta}{2}$$

$$\cos\frac{\gamma}{2}\cos\frac{a-b}{2} = \cos\frac{c}{2}\sin\frac{\alpha+\beta}{2}$$

$$\cot R = \sqrt{-\frac{\sin(P-\alpha)\sin(P-\beta)\sin(P-\gamma)}{\sin P}}$$

$\cot R = \cot\frac{a}{2}\sin(\alpha-P)$, where R is a circumradius,

$$P = (\alpha+\beta+\gamma-\pi)/2 = \varepsilon/2$$

$$\tan r = \sqrt{\frac{\sin(p-a)\sin(p-b)\sin(p-c)}{\sin p}}$$

$\tan r = \tan\frac{\alpha}{2}\sin(p-a)$, where r is a radius of incircle, $p = (a+b+c)/2$

$$\tan\frac{P}{2} = \tan\frac{\varepsilon}{4} = \sqrt{\tan\frac{p}{2}\tan\frac{p-a}{2}\tan\frac{p-b}{2}\tan\frac{p-c}{2}}, \quad 2P = \varepsilon$$

Appendix 6 Common Formulas of Coordinate Conversion

♦ Two-Dimensional (2D) Coordinates

- Conversion from the Cartesian coordinates (x, y) to the polar ones (ρ, α) (Figure A6.1)

$$x = \rho \cos \alpha; \quad y = \rho \sin \alpha$$

$$\rho = \sqrt{x^2 + y^2} \quad \alpha = \operatorname{atan}(y/x)$$

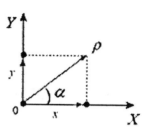

Figure A6.1

- Parallel shift of the Cartesian coordinates (Figure A6.2)

$$x_1 = x - a; \quad y_1 = y - b;$$
$$x = x_1 + a; \quad y = y_1 + b$$

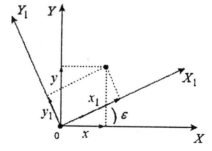

Figure A6.2

- Rotation of the Cartesian coordinates (Figure A6.3)

$$x_1 = x \cdot \cos \varepsilon + y \cdot \sin \varepsilon$$
$$y_1 = -x \cdot \sin \varepsilon + y \cdot \cos \varepsilon$$

(see Section 6.1.14)

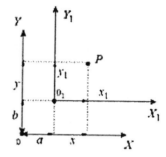

Figure A6.3

◆ Three-Dimensional (3D) Coordinates

• Conversion from Cartesian coordinates (x, y, z) to spherical ones (ρ, φ, θ) (Figure A6.4)

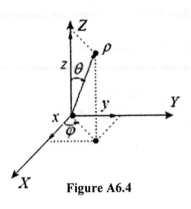

$$x = \rho \sin \theta \cos \varphi$$
$$y = \rho \sin \theta \sin \varphi$$
$$z = \rho \cos \theta$$
$$\rho = \sqrt{x^2 + y^2 + z^2}$$
$$\varphi = a \tan(y/x)$$

Figure A6.4

$$\theta = a \tan\left(\sqrt{x^2 + y^2}\Big/z\right)$$

• Conversion from Cartesian coordinates (x, y, z) to cylindrical ones (ρ, φ, z) (Figure A6.5)

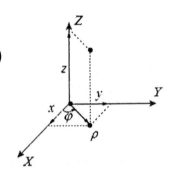

$$x = \rho \cos \varphi$$
$$y = \rho \sin \varphi$$
$$z = z$$
$$\rho = \sqrt{x^2 + y^2}$$
$$\varphi = a \tan(y/x)$$

Figure A6.5

• Parallel shift of the Cartesian coordinates (Figure A6.6). The new coordinates (x_1, y_1, z_1) are linked to the previous ones (x, y, z):

$$x_1 = x - a; \quad y_1 = y - b; \quad z_1 = z - c$$
$$x = x_1 + a; \quad y = y_1 + b; \quad z = z_1 + c$$

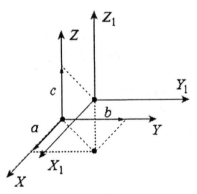

Figure A6.6

• Rotation of Cartesian coordinates around its axes.

If coordinates are rotated with respect to all three axes by angles as shown in Figure A6.7 (by angle φ around axes Z, by angle ψ around axes X, and by angle θ around axes Y), then the new coordinates (x_1, y_1, z_1) and (x, y, z) are linked as:

$$\begin{vmatrix} x_1 \\ y_1 \\ z_1 \end{vmatrix} = C_r \cdot \begin{vmatrix} x \\ y \\ z \end{vmatrix}, \quad C_r = C_\varphi \cdot C_\theta \cdot C_\psi = \begin{vmatrix} c_{11} & c_{12} & c_{13} \\ c_{21} & c_{22} & c_{23} \\ c_{31} & c_{32} & c_{33} \end{vmatrix},$$

where

$$c_{11} = \cos \varphi \cos \theta$$
$$c_{12} = \sin \varphi \cos \psi - \cos \varphi \sin \theta \sin \psi$$
$$c_{13} = \sin \varphi \sin \psi + \cos \varphi \cos \psi \sin \theta$$
$$c_{21} = -\sin \varphi \cos \theta$$
$$c_{22} = \cos \varphi \cos \psi + \sin \varphi \sin \theta \sin \psi$$
$$c_{23} = \cos \varphi \sin \psi - \sin \varphi \cos \psi \sin \theta$$
$$c_{31} = -\sin \theta$$
$$c_{32} = -\sin \psi \cos \theta$$
$$c_{33} = \cos \psi \cos \theta$$

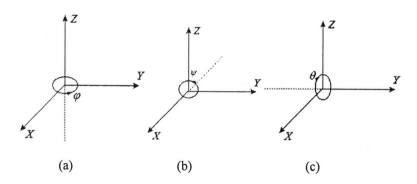

(a)   (b)   (c)

Figure A6.7 Rotation of coordinates around (a) axes Z, (b) axes X, and (c) axes Y.

Appendix 7 Common Antenna Coordinates

♦ Linear Antenna or Array

The typical coordinates for linear antenna or array of length l is given in Figure A7.1.

Here x_m is the coordinate of the mth element of antenna, and θ is the angle between normal to the antenna and direction of radiation. Sometimes angle $\varphi = \pi/2 - \theta$ is used in the formulas for antenna patterns rather than angle θ.

In these coordinates, the normalized radiation pattern of the linear array (array factor) has the form:

$$f(\theta) = \frac{\left| \sum_{m=0}^{M-1} I_m \cdot \exp(jkx_m \sin\theta) \right|}{\sum_{m=0}^{M-1} A_m}$$

where A_m, Φ_m is the amplitude and phase distribution of the illuminating field of the mth element of the array;

$I_m = A_m \cdot e^{j\Phi_m}$ is the amplitude-phase distribution of the illuminating field of the mth element;

$k = 2\pi/\lambda$ is the wave number.

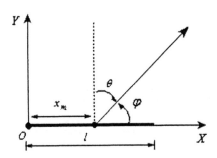

Figure A7.1 Linear array coordinates.

♦ **Planar Antenna or Array**

The typical coordinates for a planar antenna or array are given in Figure A7.2.
Here $u = \cos\varphi\sin\theta$, $v = \sin\varphi\sin\theta$ are *guide cosines* of the direction of radiation.

In these coordinates the normalized radiation pattern of the planar array takes the form:

$$f(\varphi,\theta) = \frac{\left|\displaystyle\sum_{n=0}^{N-1}\sum_{m=0}^{M-1} I_{nm}\cdot\exp[jk(x_n u + y_m v)]\right|}{\displaystyle\sum_{n=0}^{N-1}\sum_{m=0}^{M-1} A_{nm}}$$

where $A_{nm}, \Phi_{nm}, I_{nm} = A_{nm}\cdot e^{j\Phi_{nm}}$ are amplitude, phase, and amplitude-phase distribution of the illuminating field of the *nm*th element of the array; N, M are the number of elements in the dimensions X and Y correspondingly.

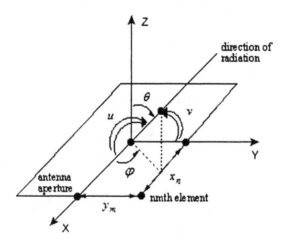

Figure A7.2 Planar array coordinates.

Appendix 8 Common Series

♦ Finite Series

$$1+2+3+\ldots+n = \frac{n(n+1)}{2}$$

$$p+(p+1)+(p+2)+\ldots+(p+n) = \frac{(n+1)(2p+n)}{2}$$

$$1+3+5+\ldots+(2n-1) = n^2$$

$$2+4+6+\ldots+2n = n(n+1)$$

$$1^2+2^2+3^2+\ldots+n^2 = \frac{n(n+1)(2n+1)}{6}$$

$$1^3+2^3+3^3+\ldots+n^3 = \frac{n^2(n+1)^2}{4}$$

$$1^2+3^2+5^2+\ldots+(2n-1)^2 = \frac{n(4n^2-1)}{3}$$

$$1^3+3^3+5^3+\ldots+(2n-1)^3 = n^2(2n^2-1)$$

$$1^4+2^4+3^4+\ldots+n^4 = \frac{n(n+1)(2n+1)(3n^2+3n-1)}{30}$$

♦ Infinite Series

$$\frac{1}{1+x} = 1-x+x^2-x^3+\ldots+(-1)^n x^n+\ldots \quad (-1<x<1)$$

$$\frac{1}{1-x} = 1+x+x^2+x^3+\ldots+x^n+\ldots \quad (-1<x<1)$$

$$(1+x)^k = 1+\binom{k}{1}x+\binom{k}{2}x^2+\binom{k}{3}x^3+\ldots+\binom{k}{n}x^n+\ldots \quad (-1<x<1)$$

$$\ln(1+x) = x - \frac{x^2}{2} + \frac{x^3}{3} - \frac{x^4}{4} + \ldots + (-1)^n \frac{x^{n+1}}{n+1} + \ldots \quad (-1 < x \le 1)$$

$$-\ln(1-x) = x + \frac{x^2}{2} + \frac{x^3}{3} + \frac{x^4}{4} + \ldots + \frac{x^{n+1}}{n+1} + \ldots \quad (-1 \le x < 1)$$

$$\ln\frac{1+x}{1-x} = 2\left(x + \frac{x^3}{3} + \frac{x^5}{5} + \ldots + \frac{x^{2n+1}}{2n+1} + \ldots\right) \quad (-1 < x < 1)$$

$$e^x = 1 + \frac{x}{1!} + \frac{x^2}{2!} + \frac{x^3}{3!} + \ldots + \frac{x^n}{n!} + \ldots \quad (\text{all } x)$$

$$e^{-x} = 1 - \frac{x}{1!} + \frac{x^2}{2!} - \frac{x^3}{3!} + \frac{x^4}{4!} + \ldots + (-1)^n \frac{x^n}{n!} + \ldots \quad (\text{all } x)$$

$$\cosh x = 1 + \frac{x^2}{2!} + \frac{x^4}{4!} + \frac{x^6}{6!} + \ldots + \frac{x^{2n}}{(2n)!} + \ldots \quad (\text{all } x)$$

$$\sinh x = x + \frac{x^3}{3!} + \frac{x^5}{5!} + \frac{x^7}{7!} + \ldots + \frac{x^{2n+1}}{(2n+1)!} + \ldots \quad (\text{all } x)$$

$$\cos x = 1 - \frac{x^2}{2!} + \frac{x^4}{4!} - \frac{x^6}{6!} + \ldots + (-1)^n \frac{x^{2n}}{(2n)!} + \ldots \quad (\text{all } x)$$

$$\sin x = x - \frac{x^3}{3!} + \frac{x^5}{5!} - \frac{x^7}{7!} + \ldots + (-1)^n \frac{x^{2n+1}}{(2n+1)!} + \ldots \quad (\text{all } x)$$

Note. In the last two series, x is an angle measured in radians.

Appendix 9 Common Derivatives

♦ **Common Derivatives of the First Order**

Function	Derivative	Function	Derivative
C (constant)	0	$a \sin x$	$\dfrac{1}{\sqrt{1-x^2}}$
x	1	$a \cos x$	$-\dfrac{1}{\sqrt{1-x^2}}$
x^n	nx^{n-1}	$a \tan x$	$\dfrac{1}{1+x^2}$
$\dfrac{1}{x}$	$-\dfrac{1}{x^2}$	$a \cot x$	$-\dfrac{1}{1+x^2}$
$\dfrac{1}{x^n}$	$-\dfrac{n}{x^{n+1}}$	$a \sec x$	$\dfrac{1}{x\sqrt{x^2-1}}$
$\sqrt{x}$	$\dfrac{1}{2\sqrt{x}}$	$a \csc x$	$-\dfrac{1}{x\sqrt{x^2-1}}$
$\sqrt[n]{x}$	$\dfrac{1}{n\sqrt[n]{x^{n-1}}}$	$\operatorname{sh} x$	$\operatorname{ch} x$
e^x	e^x	$\operatorname{ch} x$	$\operatorname{sh} x$
a^x	$a^x \ln a$	$\operatorname{th} x$	$\dfrac{1}{\operatorname{ch}^2 x}$
$\ln x$	$\dfrac{1}{x}$	$\operatorname{cth} x$	$-\dfrac{1}{\operatorname{sh}^2 x}$
$\log_a x$	$\dfrac{1}{x}\log_a e = \dfrac{1}{x \ln a}$	$\operatorname{Arsh} x$	$\dfrac{1}{\sqrt{1+x^2}}$
$\log_{10} x = \lg x$	$\dfrac{1}{x}\lg e \approx \dfrac{0.4343}{x}$	$\operatorname{Arch} x$	$\dfrac{1}{\sqrt{x^2-1}}$
$\sin x$	$\cos x$	$\operatorname{Arth} x$	$\dfrac{1}{1-x^2}$
$\cos x$	$-\sin x$	$\operatorname{Arcth} x$	$-\dfrac{1}{1-x^2}$
$\tan x$	$\dfrac{1}{\cos^2 x} = \sec^2 x$	$\sec x$	$\dfrac{\sin x}{\cos^2 x} = \tan x \cdot \sec x$
$\cot x$	$-\dfrac{1}{\sin^2 x} = -\csc^2 x$	$\csc x$	$-\dfrac{\cos x}{\sin^2 x} = -\cot x \cdot \csc x$

♦ Common Derivatives of the nth Order

Function	Derivative
x^m	$m(m-1)(m-2)\ldots(m-n+1)x^{m-n}$
	(if m is integer and $n > m$, nth derivative is equal to 0)
$\ln x$	$(-1)^{n-1}(n-1)!\dfrac{1}{x^n}$
$\log_a x$	$(-1)^{n-1}x\dfrac{(n-1)!}{\ln a}\dfrac{1}{x^n}$
e^{kx}	$k^n e^{kx}$
a^x	$(\ln a)^n a^x$
a^{kx}	$(k\ln a)^n a^{kx}$
$\sin x$	$\sin\left(x+\dfrac{n\pi}{2}\right)$
$\cos x$	$\cos\left(x+\dfrac{n\pi}{2}\right)$
$\sin kx$	$k^n \sin\left(kx+\dfrac{n\pi}{2}\right)$
$\cos kx$	$k^n \cos\left(kx+\dfrac{n\pi}{2}\right)$
sh x	sh x when n is even, ch x when n is odd
ch x	ch x when n is even, sh x when n is odd

Appendix 10 **Common Integrals**

$$\int x^n \, dx = \frac{x^{n+1}}{n+1} \quad (n \neq -1)$$

$$\int \frac{dx}{x} = \ln|x| \quad (x \neq 0)$$

$$\int e^x \, dx = e^x$$

$$\int a^x \, dx = \frac{a^x}{\ln a} \quad (a \neq 1)$$

$$\int \sin x \, dx = -\cos x$$

$$\int \cos x \, dx = \sin x$$

$$\int \frac{dx}{\sin^2 x} = -\cot x \quad (x \neq k\pi)$$

$$\int \frac{dx}{\cos^2 x} = \tan x \quad [x \neq (2k+1)(\pi/2)]$$

$$\int \frac{dx}{\sqrt{1-x^2}} = a\sin x$$

$$\int \frac{dx}{\sqrt{a^2-x^2}} = a\sin \frac{x}{a}$$

$$\int \frac{dx}{1+x^2} = a\cot x$$

$$\int \frac{dx}{a^2+x^2} = \frac{1}{a} a\cot \frac{x}{a}$$

$$\int \tan x \, dx = -\ln|\cos x|$$
$$[x \neq (2k+1)(\pi/2)]$$

$$\int \cot x \, dx = \ln|\sin x| \quad (x \neq 2k\pi)$$

$$\int \frac{dx}{\cos^2 x} = \tan x [x \neq (2k+1)\frac{\pi}{2}]$$

$$\int \frac{dx}{\sin^2 x} = -\cot x \quad (x \neq k\pi)$$

$$\int \text{sh } x \, dx = \text{ch } x$$

$$\int \text{ch } x \, dx = \text{sh } x$$

$$\int \text{th } x \, dx = \ln \text{ch } x$$

$$\int \text{cth } x \, dx = \ln|\text{sh } x| \quad (x \neq 0)$$

$$\int \frac{dx}{\text{ch}^2 x} = \text{th } x$$

$$\int \frac{dx}{\text{sh}^2 x} = -\text{cth } x \, (x \neq 0)$$

$$\int \frac{dx}{\sqrt{a^2-x^2}} = \text{asin } \frac{x}{a} \quad (|x| < a)$$

$$\int \frac{dx}{\sqrt{a^2+x^2}} = \begin{cases} \text{arsh}(x/a), \\ \ln\left(x + \sqrt{a^2+x^2}\right) \end{cases}$$

Note. Constant of integration is omitted.

Appendix 11   Common Formulas for Gradient, Divergence, and Curl

$$grad(\vec{a} \cdot \vec{b}) = \vec{a} \times rot\,\vec{b} + \vec{b} \times rot\,\vec{a} + \frac{d\vec{a}}{d\vec{b}} + \frac{d\vec{b}}{d\vec{a}}$$

$$div(f\vec{a}) = \vec{a} \cdot grad\,f + f\,div\,\vec{a}$$

$$div(\vec{a} \times \vec{b}) = \vec{b} \cdot rot\,\vec{a} - \vec{a} \cdot rot\,\vec{b}$$

$$div(grad\,f) = \Delta f$$

$$div(rot\,\vec{a}) = 0$$

$$div(\Delta\vec{a}) = \Delta(div\,\vec{a})$$

$$rot(grad\,f) = 0$$

$$rot(f\vec{a}) = grad\,f \times \vec{a} + f\,rot\,\vec{a}$$

$$rot(\vec{a} \times \vec{b}) = \vec{a}\,div\,\vec{b} - \vec{b}\,div\,\vec{a} + \frac{d\vec{a}}{d\vec{b}} - \frac{d\vec{b}}{d\vec{a}}$$

$$rot\,rot\,\vec{a} = grad\,div\,\vec{a} - \Delta\vec{a}$$

Appendix 12 Common Transforms

♦ Fourier Transform

For two functions $x(t)$ and $X(\omega)$, the Fourier transform pair is:

$$X(\omega) = \int_{-\infty}^{\infty} x(t)\, e^{-j\omega t} dt$$

$$x(t) = \frac{1}{2\pi} \int_{-\infty}^{\infty} X(\omega)\, e^{j\omega t} d\omega$$

In electrical engineering, typically $x(t) = s(t)$ is the signal and $X(\omega) = S(\omega)$ is its spectrum. Further, we will amplify the properties of the Fourier transform in notations of the pair of functions $s(t)$ and $S(\omega)$ linked as:

$$S(\omega) = \int_{-\infty}^{\infty} s(t)\, e^{-j\omega t} dt$$

$$s(t) = \frac{1}{2\pi} \int_{-\infty}^{\infty} S(\omega)\, e^{j\omega t} d\omega$$

For discrete signals $x(n\,\Delta t) = s(t) \cdot y_\Delta(t)$, the sampling function $y_\Delta(t)$ can be represented via a Fourier series:

$$y_\Delta(t) = \frac{1}{\Delta t} \sum_{m=-\infty}^{\infty} \exp(jm\omega_1 t), \qquad \omega_1 = \frac{2\pi}{\Delta t}$$

that gives the spectrum of the discrete signal as:

$$X(e^{j\omega\Delta t}) = \frac{1}{\Delta t} \sum_{m=-\infty}^{\infty} S\left[j(\omega + m\frac{2\pi}{\Delta t}) \right]$$

Thus the spectrum of a discrete signal is a sequence of spectra $S(\omega)$ of the initial signal $s(t)$ shifted with respect to each other by $\Delta\omega = 2\pi/\Delta t$.

In practical applications another representation of the discrete signal and its spectrum is often used that is termed a *discrete Fourier transform* (DFT):

direct DFT: $\quad X(k) = \displaystyle\sum_{n=0}^{N-1} x(n\,\Delta t) \cdot W_N^{kn} \qquad k = 0,...,N-1;$

inverse DFT: $\quad x(n\Delta t) = \dfrac{1}{N}\displaystyle\sum_{k=0}^{N-1} X(k) \cdot W_N^{-kn} \qquad n = 0,...,N-1$

where $x(n\Delta t)$ is a sequence of N time-domain counts, $X(k)$ is a sequence of N frequency-domain counts,

$$W_N = e^{-j\cdot\frac{2\pi}{N}}, \quad j = \sqrt{-1}$$

The basic features of DFT are as follows:

- *Linearity*: If $X(k) = DFT[x(n\Delta t)]$ and $Y(k) = DFT[y(n\Delta t)]$,   then $DFT[ax(n\Delta t) + by(n\Delta t)] = aX(n) + bY(k)$, where a, b are arbitrary constants.

- *Shift*: If $X(k) = DFT[x(n\Delta t)]$, and $y(n\Delta t)$ is obtained from $x(n\Delta t)$ by shifting it by n_0 counts, then $Y(k) = DFT[y(n\Delta t)] = X(k) \cdot W_N^{n_0 k}$.

- *Symmetry*: $\mathrm{Re}[X(k)] = \mathrm{Re}[X(N-k)]$, $\mathrm{Im}[X(k)] = -\mathrm{Im}[X(N-k)]$, $|X(k)| = |X(N-k)|$, $\arg X(k) = -\arg X(N-k)$.

- *Convolution*: If $X(k) = DFT[x(n\Delta t)]$ and $Y(k) = DFT[y(n\Delta t)]$,   if $u(n\Delta t) = \displaystyle\sum_{l=0}^{N-1} x(l\Delta t) \cdot y[(n-l)\Delta t]$, then $DFT[u(n\Delta t)] = X(k) \cdot Y(k) = U(k)$

If $u(n\Delta t) = x(n\Delta t) \cdot y(n\Delta t)$, then:

$$DFT[u(n\Delta t)] = U(k) = \dfrac{1}{N}\sum_{l=0}^{N-1} X(l)Y(k-l), \quad k = 0,...N-1.$$

- *Reciprocity*: The inverse DFT can be calculated via the direct DFT as

$$x(n\Delta t) = P^*(n\Delta t), \quad \text{where } P(n\Delta t) = \frac{1}{N}\left(\sum_{k=0}^{N-1} X^*(k)W_N^{nk}\right), \quad n = 0,\ldots,N-1.$$

In a matrix form DFT can be described by a compact model:

$$\vec{X} = \vec{W}_N \cdot \vec{x}, \quad \vec{x} = \vec{W}_N^{-1} \cdot \vec{X}$$

where $\vec{X}$ and $\vec{x}$ are N-dimensional vectors

$$\vec{X} = [X(0), X(1),\ldots, X(N-1)]^T,$$

$$\vec{x} = [x(0), x(1),\ldots, x(N-1)]^T,$$

$\vec{W}_N$ is a matrix of dimension $N \times N$ with the elements

$$d(n,k) = W_N^{nk} = W_N^{nk(\text{mod } N)}, \quad n, k = 0,\ldots, N-1.$$

$\vec{W}_N^{-1}$ is a matrix reciprocal to $\vec{W}_N$ with the elements

$$d^{-1}(n,k) = \frac{1}{N}W_N^{-nk} = \frac{1}{N}W_N^{-[nk(\text{mod } N)]}$$

The computation of DFT requires a significant number of operations to be performed. The special algorithms are used to reduce the number of operations required that are referred to as *fast Fourier transform* (FFT).

The basic FFT algorithms are as follows:

- *Time-decimation FFT (base 2)*. The sequence $x(n\Delta t)$ with length $N = 2^\gamma$ is divided into two $N/2$-points sequences, and then correspondingly into $N/4$-points ones, $N/8$-points ones, and so forth. The steps of transformation $\gamma = \log_2 N$.

$$X(k) = \sum_{n=0}^{\frac{N}{2}-1} x(2n\Delta t)W_{N/2}^{nk} + W_N^k \sum_{n=0}^{\frac{N}{2}-1} x[(2n+1)\Delta t]W_{N/2}^{nk}$$

- *Frequency-decimation FFT (base 2)*. The sequence $x(n\Delta t)$ is divided into $x_1(n\Delta t) = x(n\Delta t)$, and $x_2(n\Delta t) = x\left[\left(n + \frac{N}{2}\right)\Delta t\right]$, then even and odd DFT are performed separately:

$$X(2k) = \sum_{n=0}^{\frac{N}{2}-1} [x_1(n\Delta t) + x_2(n\Delta t)] W_{N/2}^{nk}$$

$$X(2k+1) = \sum_{n=0}^{\frac{N}{2}-1} [x_1(n\Delta t) - x_2(n\Delta t)] W_N^n W_{N/2}^{nk}$$

- *Algorithm with rotation.* If $= N = N_1 \cdot N_2$ where N_1, N_2 are any positive integers, the N-point DFT is reduced to $N_1\ N_2$ – point and $N_2\ N_1$ – point DFT and N multiplications by rotation factors W_N^l:

$$k = k_1 + k_2 N_2 \quad n = n_1 + n_2 N_1 \quad n_1, k_2 = 0,...,N_1 - 1 \quad n_2, k_1 = 0,...,N_2 - 1$$

$$X(k_1 + k_2 N_2) = \sum_{n_1=0}^{N_1-1} \left[\left\{ \sum_{n_2=0}^{N_2-1} x[(n_1 + n_2 N_1)\Delta t] W_{N_2}^{k_1 n_2} \right\} W_N^{k_1 n_1} \right] W_{N_1}^{k_2 n_1}$$

- *Algorithm with arbitrary base.* FFT model with any base different from 2 can be obtained by using the algorithm with rotation. The most common bases are 4, 8, and 16. The formulas to compare the required number of operations for FFT with bases 2, 4, 8, and 16 are given in Table A12.1.

The multidimensional (l-dimensional) DFT of l-dimensional sequences is the following pair (direct DFT and inverse DFT correspondingly):

$$X(k_1, k_2,..., k_l) = \sum_{n_1=0}^{N_1-1} ... \sum_{n_l=0}^{N_l-1} x(n_1 \Delta t,..., n_l \Delta t) W_{N_1}^{k_1 n_1} ... W_{N_l}^{k_l n_l}$$

$$x(n_1 \Delta t,..., n_l \Delta t) = \frac{1}{N_1 ... N_l} \sum_{k_1=0}^{N_1-1} ... \sum_{k_l=0}^{N_l-1} X(k_1,..., k_l) W_{N_1}^{-k_1 n_1} ... W_{N_l}^{-k_l n_l}$$

$$k_1 = 0,..., N_1 - 1,..., k_l = 0,..., N_l - 1$$

$$n_1 = 0,..., N_1 - 1,..., n_l = 0,..., N_l - 1$$

Table A12.1

The Comparison of Number of Operations Required for FFT Algorithms with Different Bases

Algorithm	Brief description	Number of real multiplications	Number of real summations
Base 2.	1. Computation of $\left(\dfrac{N}{2}\right)\gamma$	0	$2N\gamma$
$N = 2^{\gamma}$,	2-point DFT		
$\gamma \geq 1$ is an	2. Rotation	$(2\gamma - 4)N + 4$	$(\gamma - 2)N + 2$
integer	3. Complete transform $(2\gamma - 4)N + 4$		$(3\gamma - 2)N + 2$
Base 4.	1. Computation of $\left(\dfrac{N}{4}\right)\cdot\dfrac{\gamma}{2}$	0	$2N\gamma$
$N = \left(2^{2}\right)^{\gamma/2}$	4-point DFT		
$\gamma/2 \geq 1$ is	2. Rotation	$(3\gamma/2 - 4)N + 4$	$(3\gamma/4 - 2)N + 2$
an integer	3. Complete transform $(3\gamma/2 - 4)N + 4$		$(2.75\gamma/4 - 2)N + 2$
Base 8.	1. Computation of $\left(\dfrac{N}{8}\right)\cdot\dfrac{\gamma}{3}$	$N\gamma/6$	$13N\gamma/6$
$N = \left(2^{3}\right)^{\gamma/3}$	8-point DFT		
$\gamma/3 \geq 1$ is	2. Rotation	$(7\gamma/6 - 4)N + 4$	$(7\gamma/12 - 2)N + 2$
an integer	3. Complete transform $(4\gamma/3 - 4)N + 4$		$(2.75\gamma - 4)N + 4$
Base 16.	1. Computation of $\left(\dfrac{N}{16}\right)\cdot\dfrac{\gamma}{4}$	$5N\gamma/8$	$37N\gamma/16$
$N = \left(2^{4}\right)^{\gamma/4}$	16-point DFT		
$\gamma/4 \geq 1$ is	2. Rotation	$(15\gamma/16 - 4)N + 4$	$(15\gamma/32 - 2)N + 2$
an integer	3. Complete transform $(25\gamma/16 - 4)N + 4$		$(89\gamma/32 - 2)N + 2$

Note. It is assumed that a basic operation N-point DFT is performed by the algorithm with a minimum number of operations.

♦ The Winograd-Fourier Transform

The Winograd-Fourier transform (WFT) is a special algorithm of DFT that minimizes the number of multiplications required for a given n-point FFT. It is based on representation of the matrix $N = N_1 \times N_2 \times ... N_Y$ -point DFT (where all N_l are mutually prime numbers) as a direct product of the matrices of an N-point DFT:

$$\vec{W}_N \Rightarrow \vec{W}_{N_1} con \vec{W}_{N_2} con...\vec{W}_{N_l}$$

where *con* is a convolution sign. Typically, the Winograd-Fourier transform is more efficient than classical FFT algorithms, and, while the number of additions is comparable to that of conventional FFT, the number of multiplications is approximately 80% less.

In WFT the computation of an N_l-point DFT is done by calculation of circular convolutions based on the following approach. The sequence $y(n\Delta t)$ that is a circular convolution of sequences $x(n\Delta t)$ and $h(n\Delta t)$, $n = 0, ..., N - 1$, is a sequence of polynomial coefficients:

$$Y(z) = X(z) H(z)[\text{mod } (z^N - 1)]$$

where $X(z) = \sum_{l=0}^{N-1} x(l \Delta t) z^l$; $H(z) = \sum_{l=0}^{N-1} h(l \Delta t) \cdot z^l$; $Y(z) = \sum_{l=0}^{N-1} y(l \Delta t) z^l$

The computation of $Y(z)$ is based on the Chinese remainder theorem. The polynomial $z^N - 1$ can be broken down into k mutually prime polynomials:

$$z^N - 1 = \prod_{l=1}^{k} P_l(z), \quad [P_1(z),...,P_k(z)] = 1$$

Then $Y(z) = \left(\sum_{l=1}^{k} Y_l(z) Q_l(z) S_l(z) \right) \left(\text{mod}(z^N - 1) \right)$

where $Y_l(z) = X(z) H(z)[\text{mod } P_l(z)]$; $S_l(z) = (z^N - 1)/P_l(z)$,

and the polynomials $Q_l(z)$ must meet the equality:

$$Q_l(z)S_l(z) = 1[\bmod P_l(z)], l = 1, ..., k$$

♦ Digital Convolution

The concept of linear (aperiodic) and circular (periodic) discrete convolutions is widely used in fast Fourier transforms and Winograd-Fourier transforms. The *linear convolution* of two finite sequences $x(n\Delta t)$ and $h(n\Delta t)$ each having N_1 and N_2 counts correspondingly is a sequence $y(n\Delta t)$:

$$y(n\Delta t) = \sum_{l=0}^{n} h(l\Delta t)x[(n-l)\Delta t] = \sum_{l=0}^{n} x(l\Delta t)h[(n-l)\Delta t]$$

$$n = 0, ..., N_1 + N_2 - 2$$

The sequence $y(n\Delta t)$ is also a finite one and has a length of $N_1 + N_2 - 1$ counts.

The circular convolution of the periodic sequences $x(n\Delta t)$ and $h(n\Delta t)$ having N-count periods each is a sequence $y(n\Delta t)$:

$$y(n\Delta t) = \sum_{l=0}^{N-1} h(l\Delta t)x[(n-l)\Delta t] = \sum_{l=0}^{N-1} x(l\Delta t)h[(n-l)\Delta t]$$

The sequence $y(n\Delta t)$ is also a periodic one with a period equal to N counts. So it can be modeled only for a single period, e.g., for $n = 0, ..., N-1$. In a matrix representation the circular convolution can be modeled as:

$$\vec{y} = \vec{H} \cdot \vec{x} = \vec{X} \cdot \vec{h}$$

where

$$\vec{y} = \{y(0), y(\Delta t), ..., y[(N-1)\Delta t]\}^T$$
$$\vec{h} = \{h(0), h(\Delta t), ..., h[(N-1)\Delta t]\}^T$$
$$\vec{x} = \{x(0), x(\Delta t), ..., x[(N-1)\Delta t]\}^T$$

are finite vectors, and $\vec{H}, \vec{x}$ are periodic matrices $N \times N$ of the following type $(\vec{A} = \vec{H} \text{ or } \vec{A} = \vec{X})$:

$$\vec{A} = \begin{bmatrix} a(0), & a(\Delta t) & ...a[(N-1)\Delta t] \\ a(\Delta t), & a(2\Delta t) & ...a(0) \\ a(2\Delta t), & a(3\Delta t) & ...a(\Delta t) \\ .... & .... & .... \\ a[(N-1)\Delta t], & a[(N-2)\Delta t] & ... \end{bmatrix}$$

If $x_1(n\Delta t)$ and $h_1(n\Delta t)$ have length $N_1 + N_2 - 1$ counts and

$$x_1(n\Delta t) = \begin{cases} x(n\Delta t), & n = 0,..., N_1 - 1 \\ 0, & n = N_1,..., N_1 + N_2 - 2 \end{cases}$$

$$y_1(n\Delta t) = \begin{cases} y(n\Delta t), & n = 0,..., N_2 - 1 \\ 0, & n = N_2,..., N_1 + N_2 - 2 \end{cases}$$

then a linear convolution of $x(n\Delta t)$ and $y(n\Delta t)$ is equal to an $(N_1 + N_2 - 1)$-point circular convolution of the sequences $x_1(n\Delta t)$ and $h_1(n\Delta t)$:

$$y(n\Delta t) = \sum_{l=0}^{N_1+N_2-2} x_1(l\Delta t) h_1[(n-l)\Delta t], n = 0,..., N_1 + N_2 - 2$$

and can be calculated by using the $(N_1 + N_2 - 1)$-point DFT. The following operations are done to perform a circular convolution of sequences $x_1(n\Delta t)$ and $h_1(n\Delta t)$ by DFT:

- $X_1(k) = \sum_{n=0}^{N-1} x_1(n\Delta t) W_N^{nk}$, $k = 0, ..., N-1$

- $H_1(k) = \sum_{n=0}^{N-1} h_1(n\Delta t) W_N^{nk}$, $k = 0, ..., N-1$

- $Y(k) = X_1(k) \cdot H_1(k)$, $k = 0,..., N-1$

- $y(n\Delta t) = \dfrac{1}{N} \sum_{k=0}^{N-1} Y(k) W_N^{-nk}$, $n = 0,..., N-1$

If one sequence is much longer than another one ($N_1 \gg N_2$, or $N_2 \gg N_1$), the long sequence is broken into short sections and partial convolutions are calculated. Then these partial convolutions are used to obtain the final one.

◆ The Laplace and Z-Transforms

The Laplace transform of the continuous function $s(t)$ is defined as:

$$F(p) = \int_0^\infty s(t)\, e^{-pt}\, dt$$

where $p = \sigma + j\omega$ is a complex variable. The basic advantage of going from a real variable ω to a complex p is that all restrictions imposed by the requirement that the function $s(t)$ has to be completely integrable are eliminated. The discrete Laplace transform is:

$$X(p) = \sum_{n=0}^\infty x(n\Delta t) \cdot e^{-pn\Delta t}$$

The change of variables $z = e^{p\Delta t}$, $p = \dfrac{1}{\Delta t}\ln z$ changes the Laplace transform to a *Z-transform*. The direct Z-transform is defined as:

$$X(z) = Z\{x(n\Delta t)\} = \sum_{n=0}^\infty x(n\Delta t) \cdot z^{-n}$$

The main features of a direct Z-transform are:

- Linearity. If $x_3(n\Delta t) = c_1 x_1(n\Delta t) + c_2 x_2(n\Delta t)$, then

$$X_3(z) = c_1 X_1(z) + c_2 X_2(z)$$

- Shift theorem. If $x_1(-m\Delta t) = x_1((-m+1)\Delta t) = \ldots x_1(-\Delta t) = 0$, then

$$X_2(z) = Z^{-m} \cdot X_1(z)$$

- If $x_2(n\Delta t) = x_1[(n-m)\Delta t]$, then

$$X_2(z) = x_1(-m\Delta t) + x_1[(-m+1)\Delta t] \cdot z^{-1} + \ldots + z^{-m} \cdot X_1(z)$$

- If $x_3(n\Delta t) = x_1(n\Delta t) \cdot x_2(n\Delta t)$, then

$$X_3(z) = \frac{1}{2\pi j} \int_C X_1(\vartheta) X_2(z/\vartheta) \vartheta^{-1} d\vartheta$$

An *inverse* Z-transform is defined as:

$$x(n\Delta t) = Z^{-1}\{X(z)\} = \frac{1}{2\pi j} \int_C X(z) z^{n-1} dz$$

where C is a closed-loop contour in the Z plane.
There are two basic formulas to evaluate the inverse Z-transform:

$$x(n\Delta t) = \sum_{k=1}^{P} \frac{1}{(l_k-1)!} \lim_{z \to z_k^{(l_k)}} \frac{d^{l_k-1}\left[(z-z^{(l_k)})^{(l_k)} \cdot F(z)\right]}{dz^{l_k-1}} \qquad (A12.1)$$

$$x(n\Delta t) = \frac{1}{n!}\left[\frac{d^n X(z^{-1})}{dz^n}\right]_{z=0} \qquad (A12.2)$$

where $F(z) = X(z) \cdot z^{n-1}$, $z_1^{(l_1)}$, $z_2^{(l_2)}$, ... $z_P^{(l_p)}$ are $F(z)$ poles not equal to each other; l_k is a multiple of pole $Z_k^{(l_k)}$, and $0! = 1$, $d^o \varphi(z)/dz^o = \varphi(z)$. Equation (A12.1) makes it possible to derive analytical dependence of $x(n\Delta t)$ upon n, and calculate it for an arbitrary n while the (A12.2) makes it possible to calculate $x(n\Delta t)$ without finding the poles of the function $F(z)$. Z-transform images for some common functions $f(t)$ are given below in Table A12.2.

♦ **The Hilbert Transform**

The Hilbert transform relates the function $u(t)$ and its image $\hat{u}(t)$ as:

$$\hat{u}(t) = \frac{1}{\pi} \int_{-\infty}^{\infty} \frac{u(z)}{t-z} dz$$

Table A12.2

Z-Transform of Some Common Functions

$x(n\Delta t)$	$Z\{x(n\Delta t)\}$
1	$z/(z-1)$
$(-1)^n$	$1/(1+z^{-1})$
n	$z^{-1}(1-z^{-1})^2$
n^2	$(z^{-1}+z^{-2})(1-z^{-1})^3$
a^n	$1/(1-az^{-1})$
na^{n-1}	$z^{-1}/(1-az^{-1})^2$
$e^{n\tau}$	$z/(z-e^{\tau})$
$a^n \sin(n\tau)$	$(az^{-1}\sin\tau)/(1-2az^{-1}\cos\tau+a^2z^{-2})$
$a^n \cos(n\tau)$	$(1-az^{-1}\cos\tau)/(1-2az^{-1}\cos\tau+a^2z^{-2})$
$\cosh(n\tau)$	$[z(z-ch\tau)]/(z^2-2zch\tau+1)$
$\sinh(n\tau)$	$(zsh\tau)/(z^2-2zch\tau+1)$

$$u(t) = \frac{1}{\pi} \int_{-\infty}^{\infty} \frac{\hat{u}(z)}{z-t}\, dz$$

In electrical engineering, it is typically used to model complex signal $\overline{U}(t)$ as:

$$\overline{U}(t) = u(t) + j\,\hat{u}(t)$$

and real signal:

$$u(t) = \mathrm{Re}\left\{\overline{U}(t)\right\} = \frac{1}{2}\left[\overline{U}(t) + \overline{U}^{*}(t)\right]$$

♦ The Hartley Transform

The Hartley transform is typically applied to real discrete signals. The convenience of this transform results from the fact that direct and inverse transforms match with accuracy up to the constant multiplier $1/N$:

$$X(k) = \sum_{n=0}^{N-1} x(n\Delta t)\left[\cos\left(\frac{2\pi nk}{N}\right) + \sin\left(\frac{2\pi nk}{N}\right)\right]$$

$$x(n\Delta t) = \frac{1}{N}\sum_{k=0}^{N-1} X(k)\left[\cos\left(\frac{2\pi nk}{N}\right) + \sin\left(\frac{2\pi nk}{N}\right)\right]$$

♦ The Wavelet Transform

The wavelet transform relates two functions $x(t)$ and $X(a,b)$ as:

$$X(a,b) = \int x(t)\cdot\overline{\Psi}_{a,b}(t)\,dt$$

$$x(t) = C_{\psi}^{-1}\int_{0}^{\infty}\frac{da}{a^{2}}\int_{-\infty}^{\infty} X(a,b)\Psi_{a,b}(t)\,dt$$

where $\Psi_{a,b}(t) = \dfrac{1}{\sqrt{a}}\Psi\left(\dfrac{t-b}{a}\right),\quad a > 0$

$\Psi\left(\dfrac{t-b}{a}\right)$ is the fixed function called "mother wavelet," and

$$C_{\psi} = \int_{-\infty}^{\infty}\frac{\left|\hat{\Psi}(\omega)\right|^{2}}{|\omega|} < \infty$$

$$\hat{\Psi}(\omega) = \int_{-\infty}^{\infty} \Psi(t) e^{-j\omega t} dt$$

is the Fourier transform of the function $\Psi(\omega)$, and $\overline{\Psi}_{a,b}(t)$ is the complex conjugate of the function $\Psi_{a,b}(t)$.

Function $\Psi_{a,b}(t)$ plays the same role as $\exp(j\omega t)$ in the definition of the Fourier transform. In electrical engineering, it is mainly used in analysis of nonstationary signals (as the Fourier transform for stationary signals).

Appendix 13 Common Special Functions

♦ Trigonometric and Hyperbolic Functions of the Complex Variable

For complex variable $z = x + jy = \rho \cdot e^{j\varphi}$:

- $\sin z = r \cdot e^{j\psi} = u_1 + j\vartheta_1 \quad \cos z = \mu \cdot e^{j\theta} = u_2 + j\vartheta_2 = \sin\left(\dfrac{\pi}{2} \pm z\right)$

$u_1 = \sin x \cdot \text{ch } y; \quad \vartheta_1 = \cos x \cdot \text{sh } y; \quad u_2 = \cos x \cdot \text{ch } y; \quad \vartheta_2 = -\sin x \cdot \text{sh } y;$

$r^2 = \sin^2 x + \text{sh}^2 y; \quad \mu^2 = \cos^2 x + \text{sh}^2 y \quad 2r^2 = \text{ch } 2y - \cos 2x;$

$2\mu^2 = \text{ch } 2y + \cos 2x$

- $\tan z = \eta \cdot e^{j\alpha} = U + jV$

$U = \sin 2x / (\cos 2x + \text{ch } 2y); \quad V = \text{sh } 2x / (\cos 2x + \text{ch } 2y);$

$\eta^2 = \dfrac{\sin^2 x + \text{sh}^2 y}{\cos^2 x + \text{sh}^2 y} = \dfrac{\text{ch } 2y - \cos 2x}{\text{ch } 2y + \cos 2x} \; ; \quad \tan \alpha = \dfrac{\text{sh } 2y}{\sin 2x}$

♦ Gamma Function

Gamma function $\Gamma(z)$ is defined as the solution of the equation:

- $\Gamma z + 1 = z\Gamma(z),$ where $\Gamma(1) = 1$

<u>Integrals with</u> $\Gamma(z)$:

For $\text{Re}(z) > 0$

$$\Gamma(z) = \int_0^\infty e^{-t} t^{z-1} dt = \int_0^1 \left(\ln\frac{1}{t}\right)^{z-1} dt \qquad \text{(Euler's integral of the second kind)}$$

- For $z \neq 0, \pm 1, \pm 2$:

$$\Gamma(z) = \frac{1}{e^{j2\pi z} - 1} \int_{\infty}^{0} e^{-t} t^{z-1} dt \quad (0 < \arg t \le 2\pi)$$

- For any z:

$$\frac{1}{\Gamma(z)} = \frac{1}{j2\pi} \int_{-\infty}^{0} e^{t} t^{z-1} dt \quad (-\pi \le \arg t \le \pi)$$

- For $\operatorname{Re} z > 0$, $\operatorname{Re} \omega > 0$:

$$\int_{0}^{\infty} \frac{t^{z-1}}{(1+z)^{\omega}} dt = \frac{\Gamma(z)\Gamma(\omega - z)}{\Gamma(\omega)} \quad (\arg t = 0)$$

$$\int_{0}^{\infty} e^{-\omega t^{n}} t^{z-1} dt = \frac{\Gamma\left(\dfrac{z}{n}\right)}{n\omega^{2/n}} \quad (\arg t = 0,\ n = 1, 2, \ldots)$$

$$\int_{0}^{1} \frac{dt}{\sqrt[n]{1 - t^{m}}} = \frac{\Gamma\left(1 + \dfrac{1}{m}\right)\Gamma\left(1 - \dfrac{1}{n}\right)}{\Gamma\left(1 + \dfrac{1}{m} - \dfrac{1}{n}\right)}, \quad m = 1, 2, \ldots; n = 2, 3, \ldots$$

$$\int_{0}^{1} \frac{dt}{\sqrt{1 - t^{m}}} = \sqrt{\pi}\, \frac{\Gamma\left(1 + \dfrac{1}{m}\right)}{\Gamma\left(\dfrac{1}{2} + \dfrac{1}{m}\right)}, \quad m = 1, 2, \ldots$$

- For $\operatorname{Re} \alpha > -1$, $\operatorname{Re} \beta > -1$:

$$\int_{0}^{\pi/2} \sin^{\alpha}\varphi \cdot \cos^{\beta}\varphi\, d\varphi = \frac{\Gamma\left(\dfrac{\alpha+1}{2}\right)\Gamma\left(\dfrac{\beta+1}{2}\right)}{2\Gamma\left(\dfrac{\alpha+\beta}{2} + 1\right)}$$

Products and series:

- For $z \neq 0, -1, -2$

$$\Gamma(z) = \lim_{n \to \infty} \frac{n! z^n}{z(z+1)\dots(z+n)} \qquad \Gamma(z) = \frac{1}{2}\prod_{n=1}^{\infty}\left(1+\frac{1}{n}\right)^2\left(1+\frac{z}{n}\right)^{-1}$$

$$\Gamma(z)\cdot\Gamma\left(z+\frac{1}{n}\right)\dots\Gamma\left(z+\frac{n-1}{n}\right) = \sqrt{(2\pi)^{n-1}}\,n\,\frac{\Gamma(nz)}{n^{nz}}$$

if $n = 2$

$$\Gamma(2z) = \frac{1}{\sqrt{\pi}}2^{2z-1}\Gamma(z)\Gamma\left(z+\frac{1}{2}\right)$$

Recurrent and complement formulas:

$$\Gamma(z+1) = z\Gamma(z); \qquad\qquad \Gamma(z-1) = \frac{1}{z-1}\Gamma(z), \quad n = 1, 2, \dots$$

$$\Gamma(z+n) = z(z+1)\dots(z+n-1)\Gamma(z)$$

$$\Gamma(z-n) = \frac{\Gamma(z)}{(z-1)(z-2)\dots(z-n)};$$

$$\Gamma(z)\cdot\Gamma(-z) = \frac{-\pi}{z\cdot\sin \pi z} \qquad\qquad \Gamma(z)\cdot\Gamma(1-z) = \frac{\pi}{\sin \pi z}$$

$$\Gamma\left(\frac{1}{2}+z\right)\cdot\Gamma\left(\frac{1}{2}-z\right) = \frac{\pi}{\cos \pi z} \qquad \Gamma(1+z)\cdot\Gamma(1-z) = \frac{\pi z}{\sin \pi z}$$

Partial cases:

$$\Gamma(1) = 1; \quad \Gamma(2) = 1; \quad \Gamma(n) = (n-1)!; \quad \Gamma\left(\frac{1}{2}\right) = \sqrt{\pi}; \quad \Gamma\left(-\frac{1}{2}\right) = -2\sqrt{\pi}$$

$$\Gamma\left(n+\frac{1}{2}\right) = \sqrt{\pi}\,\frac{1\cdot3\cdot5\dots(2n-1)}{2^n}; \quad \Gamma\left(-n+\frac{1}{2}\right) = \sqrt{\pi}\,\frac{(-2)^n}{1\cdot3\cdot5\dots(2n-1)}$$

Incomplete gamma function:

$$\Gamma(\alpha,z) = \int_z^\infty e^{-t} t^{\alpha-1} dt; \quad \gamma(\alpha,z) = \Gamma(\alpha) - \Gamma(\alpha,z)$$

◆ Beta Function

Beta function (Euler's integral of the first kind):

$$B(z,\omega) = \frac{\Gamma(z)\Gamma(\omega)}{\Gamma(z+\omega)} = \int_0^1 t^{z-1}(1-t)^{\omega-1} dt$$

◆ Integral Exponential Function

$$Ei(z) = -\Gamma\left(0, ze^{-j\pi}\right) = -\int_z^\infty \frac{e}{t} dt$$

where the path of integration from z to ∞ is such that

$$\lim_{t\to\infty} \arg t = \beta, \quad \frac{\pi}{2} \le \beta \le \frac{3\pi}{2}$$

and $\mathrm{Re}(t)$ is right-limited.

◆ Integral Logarithm

$$li(z) = \int_0^z \frac{dt}{\ln t}$$

The following formulas are valid:

$$li(z) = Ei(\ln z), \quad Ei(z) = li(e^z)$$

◆ Integral Sine and Cosine

$$si(z) = \frac{1}{2j}[Ei(jz) - Ei(-jz)] = \int_\infty^z \frac{\sin t}{t} dt$$

$$ci(z) = \frac{1}{2}[Ei(jz) + Ei(-jz)] = \int_{\infty}^{z} \frac{\cos t}{t} dt$$

The following definitions are also used:

$$Si(z) = si(z) + \frac{\pi}{2} = \int_{0}^{z} \frac{\sin t}{t} dt \qquad Ci(z) = ci(z)$$

◆ **Integral Hyperbolic Sine and Cosine**

$$shi(z) = \int_{0}^{z} \frac{\sinh t}{t} dt = -j\, Si(jz)$$

$$chi(z) = C + \ln z + \int_{0}^{z} \frac{\cosh t - 1}{t} dt = Ci(jz) - \frac{j\pi}{2}$$

◆ **Error Function**

$$\Phi(z) = \frac{2}{\sqrt{\pi}} \int_{0}^{z} e^{-t^2} dt; \quad \Phi(-z) = -\Phi(z) \text{ For other definitions see Appendix 15}$$

◆ **Fresnel Integrals**

$$C(z) = \frac{1}{\sqrt{2\pi}} \int_{0}^{z} \frac{\cos t}{\sqrt{t}} dt = \sqrt{\frac{2}{\pi}} \int_{0}^{\sqrt{z}} \cos t^2 dt$$

$$S(z) = \frac{1}{\sqrt{2\pi}} \int_{0}^{z} \frac{\sin t}{\sqrt{t}} dt = \sqrt{\frac{2}{\pi}} \int_{0}^{\sqrt{z}} \sin t^2 dt$$

Fresnel integrals are related to the error function as:

$$C(z) + jS(z) = \frac{1}{\sqrt{2}} e^{j\frac{\pi}{4}} \Phi\left(e^{-j\frac{\pi}{4}} \sqrt{z}\right) \quad C(z) - jS(z) = \frac{1}{\sqrt{2}} e^{-j\frac{\pi}{4}} \Phi\left(e^{j\frac{\pi}{4}} \sqrt{z}\right)$$

Another definition of Fresnel integrals:

$$C\left(\frac{\pi}{2}z^2\right) = \int_0^z \cos\frac{\pi}{2}t^2\,dt \qquad S\left(\frac{\pi}{2}z^2\right) = \int_0^z \sin\frac{\pi}{2}t^2\,dt$$

♦ **Bessel Function**

The Bessel function $Z_\gamma(z)$ is the solution of Bessel's differential equation

$$z^2\frac{\partial^2\omega}{\partial z^2} + z\frac{\partial\omega}{\partial z} + \left(z^2 - \gamma^2\right)\omega = 0$$

where z is the complex number, and γ is the index, which also can be a complex number; if γ is real, then typical notation is $\gamma = n$.

For $n = 0, 1, 2$, the Bessel functions are defined as:

$$J_n(z) = \frac{1}{\pi}\int_0^\pi \cos(z\sin t - nt)\,dt = \frac{1}{2\pi}\int_{-\pi}^\pi \exp\left[j(z\sin t - nt)\,dt\right]$$

The modified Bessel functions are defined as:

$$I_\gamma(z) = e^{-\gamma j\frac{\pi}{2}}J_\gamma\left(z\cdot e^{j\frac{\pi}{2}}\right) = \left(\frac{z}{2}\right)^\gamma \sum_{k=0}^\infty \frac{\left(\frac{z}{2}\right)^{2k}}{k!\,\Gamma(\gamma + k + 1)}$$

For $x = \mathrm{Re}\{z\}$ if $0 < x \le 1$

$$I_\gamma(z) \approx \frac{1}{\Gamma(\gamma + 1)}\left(\frac{x}{2}\right)^\gamma$$

For $n = 1, 2, \ldots$

$$I_n(-z) = (-1)^n I_n(z)$$

♦ Function of a Parabolic Cylinder

Function of a parabolic cylinder $D_p(z)$ is the solution of the differential equation:

$$\frac{d^2\omega}{dz^2} + \left(p + \frac{1}{2} - \frac{1}{4}z^2\right)\omega = 0$$

where p is the parameter.

For integer $p = n = 0, 1, 2, \ldots$

$$D_n(z) = 2^{-n/2}e^{-z^2/4}H_n\left(\frac{z}{\sqrt{2}}\right)$$

where $H_n(u) = (-1)^n e^{z^2}\frac{d^n}{dz^n}\left(e^{-z^2}\right)$

♦ Hypergeometric Function

$$rF_s(\alpha_1, \ldots, \alpha_r, \gamma_1, \ldots, \gamma_s, x) = \frac{\Gamma(\gamma_1) \ldots \Gamma(\gamma_s)}{\Gamma(\alpha_1) \ldots \Gamma(\alpha_r)} \times \sum_{n=1}^{\infty} \frac{\Gamma(\alpha + n) \ldots \Gamma(\alpha_r + n)}{\Gamma(\gamma_1 + n) \ldots \Gamma(\gamma_s + n)} \cdot \frac{x^n}{n!}$$

The confluent hypergeometric function $(r = s = 1)$ is:

$$_1F_1(\alpha, \gamma, x) = 1 + \frac{\alpha}{\gamma}x + \frac{\alpha(\alpha+1)}{\gamma(\gamma+1)} \cdot \frac{x^2}{2!} + \frac{\alpha(\alpha+1)(\alpha+2)}{\gamma(\gamma+1)(\gamma+2)} \cdot \frac{x^3}{3!} \ldots$$

The following formulas are valid:

$$_1F_1\left(\frac{1}{2}, 1, -x\right) = e^{x/2} I_0(x/2)$$

$$\int_0^{\infty} t^{\mu-1}J_\gamma(\alpha t)e^{-\beta^2 t^2}\,dt = \frac{\Gamma\left(\dfrac{\mu+\nu}{2}\right)\left(\dfrac{\alpha}{2\beta}\right)^\gamma}{2\beta^\mu\Gamma(\gamma+1)}{}_1F_1\left(\frac{\mu+\gamma}{2}, \gamma+1, -\frac{\alpha^2}{4\beta^2}\right),$$

$\alpha > 0, \beta > 0, \mu + \gamma > 0$

The nth derivative of the function

$$\varphi(x) = \frac{1}{\sqrt{2\pi}} e^{-x^2/2}$$

is:

$$\varphi^{(n)}(x) = 2^{\frac{n-1}{2}} \left\{ \frac{1}{\Gamma[(1-n)/2]} \, {}_1F_1\left(\frac{1+n}{2}, \frac{1}{2}, -\frac{x^2}{2}\right) + \frac{x\sqrt{2}}{\Gamma(-n/2)} \, {}_1F_1\left(\frac{2+n}{2}, \frac{3}{2}, -\frac{x^2}{2}\right) \right\}$$

◆ Chebyshev Polynomials

Chebyshev polynomials are the solutions of the differential equation

$$\left(1 - z^2\right)\frac{\partial^2 \omega}{\partial z^2} - z\frac{d\omega}{dz} + n^2 z = 0$$

The Chebyshev polynomials of the first kind:

$$T_n(z) = \cos(n\, a\cos z) = \frac{1}{2}\left[\left(z + j\sqrt{1-z^2}\right)^n + \left(z - j\sqrt{1-z^2}\right)^n\right]$$

The Chebyshev polynomials of the second kind:

$$U_n(z) = \sin(n\, a\cos z) = \frac{1}{2j}\left[\left(z + j\sqrt{1-z^2}\right)^n - \left(z - j\sqrt{1-z^2}\right)^n\right]$$

The orthogonality rule at the interval $-1 \le x \le 1$ is:

$$\int_{-1}^{1} \frac{T_m(x)T_n(x)}{\sqrt{1-x^2}}\,dx = \begin{cases} 0 \\ \frac{\pi}{2} \\ \pi \end{cases} \qquad \int_{-1}^{1}\frac{U_m(x)U_n(x)}{\sqrt{1-x^2}}\,dx = \begin{cases} 0, & m \ne n \\ \pi/2, & m = n \ne 0 \\ 0, & m = n = 0 \end{cases}$$

◆ Laquerre Polynomials

Laquerre polynomials are the solution of the differential equations:

$$z\frac{\partial^2 \omega}{\partial z^2} + (\alpha + 1 - z)\frac{d\omega}{dz} + n\omega = 0$$

when $n = 0, 1, 2, \ldots$, α is a complex number. The Laquerre polynomials are defined as:

$$L_n^{(\alpha)}(z) = \frac{e^z z^{-\alpha}}{n!} \frac{d^n}{dz^n} \left(e^{-z} z^{n+\alpha} \right)$$

Sometimes the function $l_n(x)$ is defined as:

$$l_n(x) = e^{-x/2} L_n(z)$$

If α is real, and $\alpha > -1$ the orthogonality rule is:

$$\int_0^\infty e^{-x} x^\alpha L_m^{(\alpha)}(x) L_n^{(\alpha)}(x) dx = \begin{cases} 0, & m \neq n \\ \Gamma(1+\alpha) \binom{n+\alpha}{n}, & m = n \end{cases}$$

$$\int_0^\infty l_m(x) l_n(x) dx = \begin{cases} 0, & m \neq n \\ 1, & m = n \end{cases}$$

♦ **Hermittian Polynomials**

Hermittian polynomials are the solutions of the differential equation:

$$\frac{\partial^2 \omega}{\partial z^2} - 2z \frac{d\omega}{dz} + 2n\omega = 0$$

where $n = 0, 1, 2, \ldots$.

The Hermittian polynomials are defined as:

$$H_n(z) = (-1)^n e^{z^2} \frac{d^n}{dz^n} \left(e^{-z^2} \right)$$

They are linked to Laquerre polynomials as:

$$H_{2m}(z) = (-1)^m 2^{2m} m! L_m^{(-1/2)}(z^2) \qquad H_{2m+1}(z) = (-1)^m 2^{2m+1} m! z L_m^{(1/2)}(z^2)$$

For $x = \mathrm{Re}\{z\}$ the orthogonality rule is:

$$\int_{-\alpha}^{\infty} e^{-x^2} H_m(x) H_n(x) dx = \begin{cases} 0, & m \neq n \\ 2^n \sqrt{\pi} n! & m = n \end{cases}$$

Thus, the functions

$$\varphi_n(z) = \frac{e^{-z^2/2} H_n(z)}{\sqrt{n! 2^n} \sqrt{\pi}}$$

are orthogonal and normalized:

$$\int_{-\alpha}^{\alpha} \varphi_m(x) \varphi_n(x) dx = \begin{cases} 0, & m \neq n \\ 1, & m = n \end{cases}$$

◆ Legendre Polynomials

Legendre polynomials (sometimes called *Legendre functions* or *spherical functions*) are the solutions of the differential equations

$$\left(1 - z^2\right) \frac{\partial^2 \omega}{\partial z^2} - 2z \frac{d\omega}{dz} + \left[\gamma(\gamma+1) - \frac{\mu}{1 - z^2}\right] \omega = 0$$

z is the complex variable, and μ, γ are indices that can be complex numbers. The common assumption is that $\mu = m$, $\gamma = n$ are real positive integers.

The Legendre polynomial of order n is defined as:

$$P_n(z) = \frac{1}{2^n \cdot n!} \frac{d^n}{dz^n} \left(z^2 - 1\right)^n$$

The orthogonality rule is:

$$\int_{-1}^{1} P_n(x) P_m(x) dx = \begin{cases} 0, & m \neq n \\ \dfrac{2}{2n+1}, & m = n \end{cases}$$

Appendix 14 Common Distribution Laws for Random Variables

♦ The Distribution Laws for the Discrete Random Variables

1 *Binomial (Bernoulli) Distribution*

Variable variation range: $k = 0, 1, ..., n$

Distribution function: $P_n(k) = C_n^k p^k (1-p)^{n-k}$

Parameters: n, p

Characteristic function: $\left[1 + p\left(e^{j\vartheta} - 1\right)\right]^n$

2 *Binomial (Negative) Distribution*

Variable variation range: $k = n, n+1, ...$ or $k = 0, 1, 2, ...$

Distribution function: $P(k) = C_{k-1}^{n-1} \cdot p^n \cdot (1-p)^{k-n}$ or $P(k) = C_{k+n}^{n-1} \cdot p^n \cdot (1-p)^k$

Parameters: n, p

Characteristic function: $p^n \cdot \left[1 - (1-p)e^{j\vartheta}\right]^{-n}$

3 *Geometric (Farry) Distribution* (a partial case of Pascal distribution function when $n = 1$)

Variable variation range: $k = 0, 1, 2, ...$

Distribution function: $P_n(k) = p(1-p)^k$

Parameters: p

Characteristic function: $p\left[1 - (1-p)e^{j\vartheta}\right]^{-1}$

4 *Hypergeometric Distribution*

Variable variation range: $k = 0, 1, 2, ... \min(M, n)$

Distribution function: $P_n(k) = \dfrac{C_M^k C_{N-M}^{n-k}}{C_N^n}$

Parameters: N, M, n

Characteristic function: $\displaystyle\sum_{k=0}^{\min(M,n)} \dfrac{C_M^k C_{N-M}^{n-k}}{C_N^n} \cdot e^{jk\vartheta}$

5 *Pascal Distribution*

Variable variation range: $k = 0, 1, 2, \ldots$

Distribution function: $P_n(k) = C_{k+n-1}^{n-1} \cdot p^n \cdot (1-p)^k$

Parameters: p, n, k

6 *Poisson Distribution*

Variable variation range: $k = 0, 1, 2, \ldots$

Distribution function: $P_n(k) = \dfrac{\lambda^k}{k!} \cdot e^{-\lambda}$

Parameter: λ

Characteristic function: $e^{\lambda(e^{j\vartheta}-1)}$

7 *Polya Distribution*

Variable variation range: $k = 0, 1, 2, \ldots$ if $\alpha = 0$; $k = 1, 2, \ldots$ if $\alpha > 0$

Distribution function: $P(k) = P_0 \cdot \left(\dfrac{\lambda}{1+\alpha\cdot\lambda}\right)^k \times \dfrac{1\cdot(1+\alpha)\ldots[1+(k-1)\cdot\alpha]}{k!}$

$$\alpha \geq 0,\ \lambda > 0; \quad P_0 = P(0) = (1+\alpha\cdot\lambda)^{-1/\alpha}$$

Parameters: α, λ

Characteristic function: $\left(1 + \alpha\lambda\left(1 - e^{j\vartheta}\right)\right)^{-1/\alpha}$

8 *Polynomial Distribution*

Variable variation range:

$$k_1 = 0, 1, ..., n \quad k_2 = 0, 1, ..., n \quad ... \quad k_m = 0, 1, ..., n; \quad \sum_{i=1}^{m} k_i = n$$

Distribution function: $P_n(k_1, k_2, ... k_m) = \dfrac{n!}{k_1! k_2! ... k_m!} p_1^{k_1} ... p_m^{k_m}$

$$p_1 + p_2 + ... + p_m = 1$$

Parameters: n and any $n-1$ values from $p_1, p_2, ..., p_m$

9 *Uniform Distribution*

Variable variation range: $k = 1, 2, ..., n$

Distribution function: $P(k) = \dfrac{1}{n}$

Parameter: n

Characteristic function: $\dfrac{e^{j\vartheta}\left(1 - e^{j\vartheta n}\right)}{n\left(1 - e^{j\vartheta}\right)}$

♦ The Distribution Laws for the Continuous Random Variables

1 *Asin Distribution*

Variable variation range: $-a < x < a$

Distribution function: $F_1(x) = \begin{cases} 0 & x \le -a \\ \dfrac{1}{2} + \dfrac{1}{\pi} \cdot \arcsin \dfrac{x}{a} & -a < x < a \\ 1 & x \ge a \end{cases}$

Density function: $f_1(x) = \dfrac{1}{\pi\sqrt{a^2 - x^2}}$

Characteristic function: $\dfrac{1}{\pi}\displaystyle\int_{-a}^{a} \dfrac{e^{j\vartheta x}}{\sqrt{a^2 - x^2}}\, dx$

Parameters: a

2 *Beta Distribution*

Variable variation range: $0 < x < 1$

Distribution function: $F_1(x) = \begin{cases} 0 & x \le 0 \\ \dfrac{Bx(a,b)}{B(a,b)} & 0 < x < 1 \\ 1 & x \ge 1 \end{cases}$

Density function: $f_1(x) = \dfrac{1}{B(a,b)} \cdot x^{a-1} \cdot (1-x)^{b-1}$

$B(a,b) = \Gamma(a)\cdot\Gamma(b)/\Gamma(a+b)$; $B_x(a,b) =$ incomplete beta function

Characteristic function: $\dfrac{\Gamma(a+b)}{\Gamma(a)} \times \displaystyle\sum_{m=0}^{\infty} \dfrac{(j\vartheta)^m}{m!} \dfrac{\Gamma(a+m)}{\Gamma(a+b+m)}$

Parameters: $a > 0, b > 0$

3 *Chi Distribution*

Variable variation range: $0 < x < \infty$

Distribution function: $F_1(x) = \begin{cases} 0 & x \le 0 \\ \dfrac{\Gamma\left(\dfrac{n}{2}, \dfrac{x^2}{2}\right)}{\Gamma\left(\dfrac{n}{2}\right)} & x > 0 \end{cases}$

Density function: $f_1(x) = \dfrac{1}{2^{\frac{n}{2}-1}\,\Gamma\!\left(\dfrac{n}{2}\right)} \cdot x^{n-1} \cdot e^{-x^2/2}$

Characteristic function: $\dfrac{\Gamma(n)}{2^{\frac{n}{2}-1}\cdot\Gamma\!\left(\dfrac{n}{2}\right)} \times e^{-\vartheta^2/4} \cdot D_{-n}(-j\vartheta)$

$D_p(z)$—function of parabolic cylinder (see Appendix 13).

Parameters: n

4 *Chi-squared Distribution*

Variable variation range: $0 < x < \infty$

Distribution function: $F_1(x) = \begin{cases} 0 & x \le 0 \\[2mm] \dfrac{\Gamma\!\left(\dfrac{n}{2},\dfrac{x}{2}\right)}{\Gamma\!\left(\dfrac{n}{2}\right)} & x > 0 \end{cases}$

Density function: $f_1(x) = \dfrac{1}{2^{\frac{n}{2}}\,\Gamma\!\left(\dfrac{n}{2}\right)} \cdot x^{\frac{n}{2}-1} \cdot e^{-x/2}$

Characteristic function: $(1 - 2j\vartheta)^{-n/2}$

Parameters: n

5 *Erlang Distribution*

Variable variation range: $0 < x < \infty$

Distribution function: $F_1(x) = \begin{cases} 0 & x \le 0 \\[2mm] \dfrac{\Gamma[(k+1),\lambda x]}{\Gamma(k+1)} & x > 0 \end{cases}$

Density function: $f_1(x) = \dfrac{\lambda^{k+1}}{\Gamma(k+1)} \cdot x^k \cdot e^{-\lambda x}$

Characteristic function: $\left(1 - \dfrac{j\vartheta}{\lambda}\right)^{-(k+1)}$

Parameters: λ, k (k is integer)

6 *Exponential Distribution*

Variable variation range: $0 < x < \infty$

Distribution function: $F_1(x) = \begin{cases} 0 & x \le 0 \\ 1 - e^{-\lambda x} & x > 0 \end{cases}$

Density function: $f_1(x) = \lambda \cdot e^{-\lambda x}$

Characteristic function: $\dfrac{\lambda}{\lambda - j\vartheta}$

Parameters: λ

7 *Exponential Distribution, Double*

Variable variation range: $-\infty < x < \infty$

Distribution function: $F_1(x) = \exp\left(-c \cdot e^{-\alpha x}\right)$

Density function: $f_1(x) = c \cdot \alpha \cdot \exp\left(-\alpha x - c \cdot e^{-\alpha x}\right)$

Parameters: $c > 0, \alpha > 0$

8 *Exponential Distribution, Power*

Variable variation range: $0 < x < \infty$

Distribution function: $F_1(x) = \begin{cases} 0 & x \le 0 \\ \dfrac{\Gamma[(m+1), x]}{\Gamma(m+1)} & x > 0 \end{cases}$

Density function: $f_1(x) = \dfrac{x^m}{m!} e^{-x}$

Characteristic function: $(1 - j\vartheta)^{-(m+1)}$

Parameters: m

9 *Fisher Distribution (F-Distribution)*

Variable variation range: $0 < x < \infty$

Density function:

$$f_1(x) = \frac{\Gamma\left(\dfrac{n_1 + n_2}{2}\right)}{\Gamma\left(\dfrac{n_1}{2}\right) \cdot \Gamma\left(\dfrac{n_2}{2}\right)} \times \left(\dfrac{n_1}{n_2}\right)^{n_1/2} \times x^{(n_1/2)-1} \times \left(1 + \dfrac{n_1}{n_2} \cdot x\right)^{-\frac{n_1 + n_2}{2}}$$

Parameters: n_1, n_2

10 *Gamma Distribution*

Variable variation range: $0 < x < \infty$

Distribution function: $F_1(x) = \begin{cases} 0 & x \leq 0 \\ \dfrac{\Gamma\left[(\alpha + 1), \dfrac{x}{\beta}\right]}{\Gamma(\alpha + 1)} & x > 0 \end{cases}$

Density function: $f_1(x) = \dfrac{1}{\beta^{\alpha+1} \cdot \Gamma(\alpha + 1)} \cdot x^\alpha \cdot e^{-x/\beta}$

Characteristic function: $(1 - j\vartheta\beta)^{-(\alpha+1)}$

Parameters: $\alpha > -1, \beta > 0$

11 *Gaussian (Normal) Distribution*

Variable variation range: $-\infty < x < \infty$

Distribution function: $F_1(x) = \dfrac{1}{\sigma\sqrt{2\pi}} \displaystyle\int_{-\infty}^{x} e^{-\frac{(t-m)^2}{2\sigma^2}}\, dt = \Phi\!\left(\dfrac{x-m}{\sigma}\right)$

Density function: $f_1(x) = \dfrac{1}{\sigma\sqrt{2\pi}}\exp\!\left[-\dfrac{(x-m)^2}{2\sigma^2}\right]$

Characteristic function: $e^{jm\vartheta - \frac{\sigma^2\vartheta^2}{2}}$

Parameters: m, σ

12 *Gaussian (Standard) Distribution*

Variable variation range: $-\infty < x < \infty$

Distribution function: $F_1(x) = \dfrac{1}{\sqrt{2\pi}} \displaystyle\int_{-\infty}^{x} e^{-t^2/2}\, dt = \Phi(x)$

Density function: $f_1(x) = \dfrac{1}{\sqrt{2\pi}} \cdot e^{-x^2/2}$

Characteristic function: $e^{-\vartheta^2/2}$

Parameters: $m = 0,\ \sigma = 1$

13 *Gaussian (Log-Normal) Distribution*

Variable variation range: $0 < x < \infty$

Distribution function: $F_1(x) = \begin{cases} 0 & x \le 0 \\ \Phi\!\left(\dfrac{\log x - m}{\sigma}\right) & x > 0 \end{cases}$

Density function: $f_1(x) = \dfrac{\log e}{x\sigma\sqrt{2\pi}} \cdot \exp\!\left(-\dfrac{(\log x - m)^2}{2\sigma^2}\right)$

Parameters: m, σ

14 *Gaussian (Truncated) Distribution*

Variable variation range: $a < x < b$

Distribution function: $F_1(x) = \begin{cases} 0 & x \le a \\ c\left[\Phi\left(\dfrac{x-m}{\sigma}\right) - \Phi\left(\dfrac{a-m}{\sigma}\right)\right], & a < x < b \\ 1 & x > b \end{cases}$

Density function: $f_1(x) = \dfrac{c}{\sigma\sqrt{2}}\exp\left[-\dfrac{(x-m)^2}{2\sigma^2}\right]$

$c = \left[\Phi\left(\dfrac{b-m}{\sigma}\right) - \Phi\left(\dfrac{a-m}{\sigma}\right)\right]^{-1}$

Characteristic function: $c \times \exp(z) \times \Phi\left[\left(\dfrac{b-m-u}{\sigma}\right) - \left(\dfrac{a-m-u}{\sigma}\right)\right]$

$z = jm\vartheta - \dfrac{\vartheta^2\sigma^2}{2} \quad u = j\vartheta\sigma^2$

Parameters: m, σ, a, b

15 *Cauchy Distribution*

Variable variation range: $-\infty < x < \infty$

Distribution function: $F_1(x) = \dfrac{1}{2} + \dfrac{1}{\pi}a\tan\left(\dfrac{x-x_0}{h}\right)$

Density function: $f_1(x) = \dfrac{1}{\pi} \cdot \dfrac{h}{h^2 + (x-x_0)^2}$

Characteristic function: $e^{jx_0\vartheta - h|\vartheta|}$

Parameters: h, x_0

16 *Laplace Distribution*

Variable variation range: $-\infty < x < \infty$

Distribution function: $F_1(x) = \begin{cases} \dfrac{1}{2} e^{-\lambda(x-\mu)}, & -\infty < x < \mu \\ 1 - \dfrac{1}{2} e^{-\lambda(x-\mu)}, & \mu < x < \infty \end{cases}$

Density function: $f_1(x) = \dfrac{\lambda}{2} e^{-\lambda|x-\mu|}$

Characteristic function: $\dfrac{\lambda^2 \cdot e^{j\theta\mu}}{\lambda^2 + \vartheta^2}$

Parameters: λ, μ

17 *M-Distribution (Nakagami)*

Variable variation range: $0 < x < \infty$

Distribution function: $F_1(x) = \begin{cases} 0 & x \leq 0 \\ \dfrac{\Gamma\!\left(m, \dfrac{mx^2}{\sigma^2}\right)}{\Gamma(m)} & x > 0 \end{cases}$

Density function: $f_1(x) = \Gamma\!\left(\dfrac{2}{m}\right) \cdot \left(\dfrac{m}{\sigma^2}\right)^m \cdot x^{2m-1} \times \exp\!\left(-\dfrac{m}{\sigma^2} \cdot x^2\right)$ $m \geq 1/2$

Characteristic function: $\dfrac{\Gamma(2m)}{2^{m-1}\Gamma(m)} \times e^{\frac{-\vartheta^2\sigma^2}{8m}} \times D_{-2m}\!\left(-\dfrac{j\vartheta\sigma}{\sqrt{2m}}\right)$

$D_p(z)$—function of a parabolic cylinder (see Appendix 13)

Parameters: m, σ

18 *Maxwell Distribution*

Variable variation range: $0 < x < \infty$

Distribution function: $F_1(x) = \begin{cases} 0 & x \le 0 \\ \dfrac{2}{\sqrt{\pi}}\Gamma\left(\dfrac{3}{2}, \dfrac{x^2}{2\sigma^2}\right) & x > 0 \end{cases}$

Density function: $f_1(x) = \sqrt{\dfrac{2}{\pi}} \cdot \dfrac{1}{\sigma^3} x^2 e^{-x^2/2\sigma^2}$

Characteristic function: $\left(1 - \sigma^2 \vartheta^2\right) \cdot W\left(\dfrac{\sigma\vartheta}{\sqrt{2}}\right) + j\sigma\vartheta\sqrt{\dfrac{2}{\pi}}$

$W(z) = e^{-z^2}\left(1 + \dfrac{2j}{\pi}\displaystyle\int_0^z e^{t^2}\,dt\right)$ is a probability integral.

Parameters: σ

19 *The Modulus of a Gaussian Variable*

Variable variation range: $0 < x < \infty$

Distribution function: $F_1(x) = \begin{cases} 0 & x \le 0 \\ \Phi\left(\dfrac{x - m}{\sigma}\right) + \Phi\left(\dfrac{x + m}{\sigma}\right) - 1 & x > 0 \end{cases}$

Density function: $f_1(x) = \dfrac{1}{\sigma\sqrt{2\pi}} \times \{\exp(A) + \exp(B)\}$

$A = -\dfrac{(x - m)^2}{2\sigma^2} \quad B = -\dfrac{(x + m)^2}{2\sigma^2}$

Characteristic function: $e^{-\frac{\sigma^2\vartheta^2}{2}} \times \left[e^{j\vartheta m}\Phi\left(\dfrac{m}{\sigma} + j\vartheta\sigma\right) + e^{-j\vartheta m}\Phi\left(-\dfrac{m}{\sigma} + j\vartheta\sigma\right)\right]$

Parameters: m, σ

20 *The Modulus of a Multidimensional Vector*

Variable variation range: $0 < x < \infty$

Distribution function: $F_1(x) = \begin{cases} 0 & x \le 0 \\ \dfrac{\Gamma\left(\dfrac{n}{2}, \dfrac{x^2}{2\sigma^2}\right)}{\Gamma\left(\dfrac{n}{2}\right)} & x > 0 \end{cases}$

Density function: $f_1(x) = \dfrac{2x^{n-1} \cdot e^{-x^2/2\sigma^2}}{\left(2\sigma^2\right)^{n/2} \cdot \Gamma\left(\dfrac{n}{2}\right)}$

Characteristic function: $\dfrac{\Gamma(n)}{2^{\frac{n}{2}-1}\Gamma\left(\dfrac{n}{2}\right)} \exp\left(-\dfrac{\vartheta^2\sigma^2}{4}\right) D_{-n}\left(j\vartheta\sigma\right)$

Parameters: n, σ

21 Rayleigh Distribution

Variable variation range: $0 < x < \infty$

Distribution function: $F_1(x) = \begin{cases} 0 & x \le 0 \\ 1 - e^{-x^2/2\sigma^2} & x > 0 \end{cases}$

Density function: $f_1(x) = \dfrac{x}{\sigma^2} e^{-\frac{x^2}{2\sigma^2}}$

Characteristic function: $1 + ja\vartheta\sqrt{\dfrac{\pi}{2}} W\left(\dfrac{a\vartheta}{\sqrt{2}}\right)$

Parameters: σ

22 Ricean (Generalized Rayleigh) Distribution

Variable variation range: $0 < x < \infty$

Distribution function: $F_1(x) = \begin{cases} 0 & x \le 0 \\ e^{-\frac{a^2}{2\sigma^2}} \sum_{k=0}^{\infty} \frac{x}{(k!)^2} \left(\frac{a^2}{2\sigma^2}\right)^k \Gamma\left(k+1, \frac{x^2}{2\sigma^2}\right) & x > 0 \end{cases}$

Density function: $f_1(x) = \frac{x}{\sigma^2} \exp\left(-\frac{x^2 + a^2}{2\sigma^2}\right) I_0\left(\frac{ax}{\sigma^2}\right)$

Parameters: a, σ

23 *Ch-squared Distribution*

Variable variation range: $-\infty < x < \infty$

Distribution function: $F_1(x) = \frac{1}{2} + \frac{1}{2} th(ax)$

Density function: $f_1(x) = \frac{a}{2 ch^2(ax)}$

Characteristic function: $\dfrac{\vartheta\pi}{2a \cdot sh\left(\dfrac{\vartheta\pi}{2a}\right)}$

Parameters: a

24 *Simpson (Triangular) Distribution*

Variable variation range: $a < x < b$

Distribution function: $F_1(x) = \begin{cases} 0 & -\infty < x < a \\ \dfrac{2(x-a)^2}{(b-a)^2} & a < x < \dfrac{a+b}{2} \\ 1 - \dfrac{2(b-x)^2}{(b-a)^2} & \dfrac{a+b}{2} < x < b \\ 1 & b < x < \infty \end{cases}$

Density function: $f_1(x) = \begin{cases} 0 & -\infty < x < a \\ \dfrac{4(x-a)}{(b-a)^2} & a < x < \dfrac{b+a}{2} \\ \dfrac{4(b-x)}{(b-a)^2} & \dfrac{a+b}{2} < x < b \\ 0 & b < x < \infty \end{cases}$

Characteristic function: $-\dfrac{4}{\vartheta^2(b-a)^2} \times \left(e^{j\vartheta\frac{b}{2}} - e^{j\vartheta\frac{a}{2}} \right)^2$

Parameters: a, b

25 Student Distribution (t-Distribution)

Variable variation range: $-\infty < x < \infty$

Distribution function: $F_1(x) = \dfrac{\Gamma\left(\dfrac{k+1}{2}\right)}{\sqrt{k\pi}\,\Gamma\left(\dfrac{k}{2}\right)} \times \int_{\infty}^{x} \left(1 + \dfrac{t^2}{k}\right)^{-\frac{k+1}{2}} dt$

Density function: $f_1(x) = \dfrac{\Gamma\left(\dfrac{k+1}{2}\right)}{\sqrt{k\pi}\,\Gamma\left(\dfrac{k}{2}\right)} \cdot \left(1 + \dfrac{x^2}{k}\right)^{-\frac{k+1}{2}}$

Characteristic function: $\Gamma\left(\dfrac{k+1}{2}\right)\sqrt{k\pi}\,\Gamma\left(\dfrac{k}{2}\right)\int_{-\infty}^{\infty} \left(1 + \dfrac{t^2}{k}\right)^{-\frac{k+1}{2}} \cdot e^{j\vartheta t} dt$

Parameters: k

26 Uniform Distribution

Variable variation range: $a < x < b$

Distribution function: $F_1(x) = \begin{cases} 0 & x < a \\ \dfrac{x-a}{b-a} & a \le x \le b \\ 1 & x > b \end{cases}$

Density function: $f_1(x) = \dfrac{1}{b-a}$

Characteristic function: $\dfrac{e^{jb\vartheta} - e^{-ja\vartheta}}{j\vartheta(b-a)}$

Parameters: a, b

27 *Weibull Distribution*

Variable variation range: $0 < x < \infty$

Distribution function: $F_1(x) = \begin{cases} 0 & x \le 0 \\ 1 - \exp(-c \cdot x^\alpha) & x > 0 \end{cases}$

Density function: $f_1(x) = c \cdot \alpha \cdot x^{\alpha-1} \cdot \exp(-c \cdot x^a)$

Characteristic function: $1 + j\vartheta \displaystyle\int_0^\infty e^{j\vartheta x - cx^\alpha} \, dx$

Parameters: $c > 0,\ \alpha > 0$

Appendix 15 Integral Forms for Gaussian Distribution

Typically, the following integral forms of the Gaussian distributions are used:

$$1 \ \Phi(x) = \frac{1}{\sqrt{2\pi}} \int_{-\infty}^{x} e^{-t^2/2} dt \qquad 2 \ \hat{\Phi}(x) = \frac{\rho}{\sqrt{\pi}} \int_{-\infty}^{x} e^{-\rho^2 t^2} dt$$

$$3 \ \Phi_0(x) = \frac{1}{\sqrt{2\pi}} \int_{0}^{x} e^{-t^2/2} dt \qquad 4 \ \hat{\Phi}_0(x) = \frac{\rho}{\sqrt{\pi}} \int_{0}^{x} e^{-\rho^2 t^2} dt$$

$$5 \ F(x) = \frac{2}{\sqrt{2\pi}} \int_{0}^{x} e^{-t^2/2} dt \qquad 6 \ \hat{F}(x) = \frac{2\rho}{\sqrt{\pi}} \int_{0}^{x} e^{-\rho^2 t^2} dt$$

$$7 \ erf(x) = \frac{2}{\sqrt{\pi}} \int_{0}^{x} e^{-t^2} dt$$

The correspondence between different forms is as follows:

$$1 \ \Phi(x) = \hat{\Phi}\left(\frac{x}{\rho\sqrt{2}}\right) = \Phi_0(x) + 0.5 = \hat{\Phi}_0\left(\frac{x}{\rho\sqrt{2}}\right) + 0.5 = \frac{1}{2}[1 + F(x)]$$

$$= \frac{1}{2}\left[1 + \hat{F}\left(\frac{x}{\rho\sqrt{2}}\right)\right] = \frac{1}{2}\left[1 + erf\left(\frac{x}{\sqrt{2}}\right)\right]$$

$$2 \ \hat{\Phi}(x) = \Phi\left(\rho\sqrt{2}x\right) = \Phi_0\left(\rho\sqrt{2}x\right) + 0.5 = \hat{\Phi}_0(x) + 0.5 = \frac{1}{2}\left[1 + F\left(\rho\sqrt{2}x\right)\right]$$

$$= \frac{1}{2}\left[1 + \hat{F}(x)\right] = \frac{1}{2}[1 + erf(\rho x)]$$

$$3 \ \Phi_0(x) = \Phi(x) - 0.5 = \hat{\Phi}\left(\frac{x}{\rho\sqrt{2}}\right) - 0.5 = \hat{\Phi}_0\left(\frac{x}{\rho\sqrt{2}}\right) = \frac{1}{2}F(x)$$

$$= \frac{1}{2}\hat{F}\left(\frac{x}{\rho\sqrt{2}}\right) = \frac{1}{2}erf\left(\frac{x}{\sqrt{2}}\right)$$

$$4 \ \hat{\Phi}_0(x) = \Phi\left(\rho\sqrt{2}x\right) - 0.5 = \hat{\Phi}(x) - 0.5 = \Phi_0\left(\rho\sqrt{2}x\right) = \frac{1}{2}F\left(\rho\sqrt{2}x\right) = \frac{1}{2}\hat{F}(x)$$

$$= \frac{1}{2}erf(\rho x)$$

5 $F(x) = 2\Phi(x) - 1 = 2\hat{\Phi}\left(\dfrac{x}{\rho\sqrt{2}}\right) - 1 = 2\Phi_0(x) = 2\hat{\Phi}_0\left(\dfrac{x}{\rho\sqrt{2}}\right)$

$= \hat{F}\left(\dfrac{x}{\rho\sqrt{2}}\right) = erf\left(\dfrac{x}{\sqrt{2}}\right)$

6 $\hat{F}(x) = 2\Phi\left(\rho\sqrt{2}x\right) - 1 = 2\hat{\Phi}(x) - 1 = 2\Phi_0\left(\rho\sqrt{2}x\right) = 2\hat{\Phi}_0(x) = F\left(\rho\sqrt{2}x\right)$

$= erf(\rho x)$

7 $erf(x) = 2\Phi\left(\sqrt{2}x\right) - 1 = 2\hat{\Phi}\left(\dfrac{x}{\rho}\right) - 1 = 2\Phi_0\left(\sqrt{2}x\right) = 2\hat{\Phi}_0\left(\dfrac{x}{\rho}\right) = F\left(\sqrt{2}x\right)$

$= \hat{F}\left(\dfrac{x}{\rho}\right)$

Appendix 16 **Moments of Random Variables**

◆ General Definitions

• *Initial Moment of Order k*:

Discrete variable: $m_k = M(X^k) = \sum_{n=1}^{N} x_n^k p_n$

Continuous variable: $m_k = \int_{-\infty}^{\infty} x^k f_1(x) dx$

• *Absolute Initial Moment of Order k*:

Discrete variable: $\beta_k = M(|X|^k) = \sum_{n=1}^{N} |x_n|^k p_n$

Continuous variable: $\beta_k = \int_{-\infty}^{\infty} |x|^k f_1(x) dx$

• *Factorial Initial Moment of Order k* *:

$m_{[k]} = M(X^{[k]}) = \sum_{n=1}^{N} x_n^{[k]} \cdot p_n$

$m_{[k]} = \int_{-\infty}^{\infty} x^{[k]} f_1(x) dx$

• *Central Moment of Order k:*

Discrete variable: $M_k = M(X_0^k) = \sum_{n=1}^{N} (x_n - m_x)^k \cdot p_n$

Continuous variable: $M_k = \int_{-\infty}^{\infty} (x - m_x)^k f_1(x) dx$

• *Absolute Central Moment of Order k*:

Discrete variable: $B_k = M(|X_0|^k) = \sum_{n=1}^{N} |x_n - m_x|^k \cdot p_n$

Continuous variable: $B_k = \int\limits_{-\infty}^{\infty} |x - m_x|^k f_1(x)dx$

- *Factorial Central Moment of Order k:*

Discrete variable: $M_{[k]} = M\left(X_0^{[k]}\right) = \sum\limits_{n=1}^{N} (x_n - m_x)^{[k]} \cdot p_n$

Continuous variable: $M_{[k]} = \int\limits_{-\infty}^{\infty} (x - m_x)^{[k]} f_1(x)dx$

Note. $z^{[k]} = z(z-1)(z-2)...(z-k+1)$

♦ **Moments of Discrete Random Variables**

1 *Binomial Distribution*

$$m = np; \quad \sigma^2 = npq; \quad \gamma_1 = \frac{q-p}{\sqrt{npq}}; \quad \gamma_2 = \frac{1-6pq}{npq}$$

$$M_{k+1} = pq \cdot \left(nkM_{k-1} + \frac{dM_k}{dp}\right); \quad q = 1-p$$

2 *Hypergeometric Distribution*

$$m = n\frac{M}{N}; \quad \sigma^2 = \frac{M(N-M)n(N-n)}{N^2(N-1)}; \quad \gamma_1 = \frac{(N-2M)(N-2n)\sqrt{N-1}}{(N-2)\sqrt{M(N-M)n(N-n)}}$$

$$M_4 = \frac{M(N-M)n(N-n)}{N^4(N-1)\cdot(N-2)\cdot(N-3)}$$
$$\times \left\{N^3(N+1)-6N^2 n(N-n)+3M(N-M)\left[N^2(n-2)-Nn^2+6n(N-n)\right]\right\}$$

3 *Poisson Distribution*

$$m = \lambda; \quad \sigma^2 = \lambda; \quad \gamma_1 = \frac{1}{\sqrt{\lambda}}; \quad \gamma_2 = \frac{1}{\lambda}; \quad M_{k+1} = \lambda_k M_{k-1} + \lambda \cdot \frac{dM_k}{d\lambda}$$

$$m_{k+1} = m_k + \lambda \cdot \sum_{n=0}^{k-1} C_n^{k-1} \cdot m_{k-n}$$

4 *Polya Distribution*

$$m = \lambda \; ; \; \sigma^2 = \lambda(1 + \alpha\lambda); \; \gamma_1 = \frac{1 + 2\alpha\lambda}{\sqrt{\lambda(1 + \alpha\lambda)}}; \; \gamma_2 = 6\alpha + \frac{1}{\lambda(1 + \alpha\lambda)}$$

5 *Uniform Distribution*

$$m = \frac{n+1}{2}; \; \sigma^2 = \frac{n^2 - 1}{12}; \; \gamma_1 = 0; \; \gamma_2 = -1.2 + \frac{4}{n^2 - 1}$$

♦ **Moments of Continuous Random Variables**

1 *Asin Distribution*

$$m = 0; \; \sigma^2 = \frac{1}{2}a^2$$

2 *Beta Distribution*

$$m = \frac{a}{a+b}; \; \sigma^2 = \frac{ab}{(a+b)^2(a+b+1)}; \; m_k = \frac{\Gamma(a+b)\Gamma(a+k)}{\Gamma(a)\Gamma(a+b+k)}$$

3 *Chi-Squared Distribution*

$$m = n; \; \sigma^2 = 2n$$

4 *Exponential Distribution*

$$m = \lambda \; ; \; \sigma^2 = \frac{1}{\lambda^2}; \; \gamma_1 = 2; \; \gamma_2 = 6; \; m_k = k! \lambda^{-k}; \; m_{k+1} = \frac{k+1}{\lambda} m_k$$

5 *Gamma Distribution*

$$m = (\alpha+1)\beta \; ; \; \sigma^2 = (\alpha+1)\beta^2 \; ; \; \gamma_1 = \frac{2}{\sqrt{\alpha+1}}; \; \gamma_2 = \frac{6}{\alpha+1}$$

$$m_{k+1} = (\alpha + k + 1) \cdot \beta \cdot m_k$$

6 *Gaussian Distribution*

$$m = m;\ \sigma^2 = \sigma^2;\ \gamma_1 = 0;\ \gamma_2 = 0;\ m_k = k! \sum_{i=0}^{[k/2]} A_i;\ M_{2k} = \frac{(2k)!}{k!} \left(\frac{\sigma^2}{2} \right)^k$$

$$M_{2k+1} = 0;\ A_i = \frac{m^{k-2i}}{(k-2i)!} \frac{\left(\dfrac{\sigma^2}{2} \right)^i}{i!};\ [k/2] = \text{floor}\left(\frac{k}{2} \right)$$

7 *Gaussian Distribution, Standard*

$$m = 0;\ \sigma^2 = 1;\ \gamma_1 = 0;\ \gamma_2 = 0;\ m_{2k} = M_{2k} = \frac{(2k)}{k!} \left(\frac{1}{2} \right)^k;\ m_{2k+1} = M_{2k+1} = 0$$

8 *Laplace Distribution*

$$m = \mu;\ \sigma^2 = \frac{2}{\lambda^2};\ \gamma_1 = \frac{3\sqrt{2}}{2};\ \gamma_2 = 3;\ m_k = k! \sum_{i=0}^{[k/2]} A_i;\ M_{2k} = (2k)!\,\lambda^{-2k}$$

$$M_{2k+1} = 0;\ A_i = \frac{M^{k-2i}}{(k-2i)!} \lambda^{-2i}$$

9 *Maxwell Distribution*

$$m = 2\sqrt{\frac{2}{\pi}}\,\sigma;\ \sigma^2 = \left(3 - \frac{8}{\pi} \right) \cdot \sigma^2$$

10 *Modulus of a Gaussian Variable*

$$m = 2 \left[m\Phi\left(\frac{m}{\sigma} \right) + \sigma\Phi\left(\frac{m}{\sigma} \right) \right] = m_1;\ \sigma^2 = \sigma^2 + m^2 - m_1^2$$

11 *Rayleigh Distribution*

$$m = \sigma\sqrt{\frac{\pi}{2}};\ \sigma^2 = \sigma^2\left(2 - \frac{\pi}{2} \right);\ \gamma_1 = (\pi - 3)\sqrt{\frac{\pi}{2}};\ \gamma_2 = 5 - \frac{3}{4}\pi^2$$

$$m_k = \left(\sigma\sqrt{2}\right)^k \Gamma\left(\frac{k+2}{2}\right); \quad M_k = \sum_{i=0}^{k} (-1)^i C_k^i \cdot m^i m_{k-i}$$

12 Simpson Distribution

$$m = \frac{a+b}{2}; \quad \sigma^2 = \frac{(b-a)^2}{6}$$

13 Student Distribution

$$m = 0 \ (k>1); \quad \sigma^2 = \frac{k}{k-2} \quad (k>2)$$

14 Uniform Distribution

$$m = (a+b)/2; \quad \sigma^2 = \frac{(b-a)^2}{12}; \quad \gamma_1 = 0; \quad \gamma_2 = -1.2; \quad m_k = \frac{b^{k+1}-a^{k+1}}{(b-a)(k+1)}$$

$$m_{k+1} = m_k \cdot \frac{\left(b^{k+2}-a^{k+2}\right)(k+1)}{\left(b^{k-1}-a^{k-1}\right)(k+2)}$$

15 Weibull Distribution

$$m = c^{-1/\alpha}\Gamma\left(1+\frac{1}{\alpha}\right); \quad \sigma^2 = c^{-2/\alpha} \times \left[\Gamma\left(1+\frac{2}{\alpha}\right)-\Gamma^2\left(1+\frac{1}{\alpha}\right)\right]; \quad \gamma_1 = \frac{M_3}{\sigma^3}$$

$$\gamma_2 = \frac{M_4}{\sigma^4} - 3; \quad m_k = c^{-k/\alpha}\Gamma\left(1+\frac{k}{\alpha}\right); \quad M_k = \sum_{i=0}^{k}(-1)^i \times C_k^i m^i m_{k-i}$$

Appendix 17 **The Functional Transform of Random Variables**

1 *Linear Transform* $(Y = aX + b)$

Initial function: $f_1(x)$

Transformed function: $f_1(y) = \dfrac{1}{|a|} f_1\left(\dfrac{y-b}{a}\right)$

2 *Square Transform* $(Y = X^2)$

Initial function: $f_1(x)$

Transformed function: $f_1(y) = \begin{cases} \dfrac{1}{2\sqrt{y}}\left[f_1\left(\sqrt{y}\right) + f_1\left(-\sqrt{y}\right)\right], & y > 0 \\ 0 & , \; y < 0 \end{cases}$

3 *Square Transform of Gaussian Variable*

Initial function: $f_1(x) = \dfrac{1}{\sigma\sqrt{2\pi}} \cdot e^{-\frac{(x-a)^2}{2\sigma^2}}$

Transformed function: $f_1(y) = \dfrac{1}{\sigma\sqrt{2\pi y}} e^{-\frac{y^2+a^2}{2\sigma^2}} ch\left(\dfrac{a\sqrt{y}}{\sigma^2}\right), \quad y > 0$

4 *Square Transform of Gaussian Variable with Zero Mean*

Initial function: $f_1(x) = \dfrac{1}{\sigma\sqrt{2\pi}} \cdot e^{-\frac{x^2}{2\sigma^2}}$

Transformed function: $f_1(y) = \dfrac{1}{\sigma\sqrt{2\pi y}} \cdot e^{-\frac{y}{2\sigma^2}}, \quad y > 0$

5 *Square Transform of Rayleigh Variable*

Initial variable: $f_1(x) = \dfrac{x}{\sigma^2} \cdot e^{-\frac{x^2}{2\sigma^2}}$ (Rayleigh pdf)

Transformed variable: $f_1(y) = \dfrac{1}{2\sigma^2} \cdot \exp\left(-\dfrac{y}{2\sigma^2}\right)$, $y > 0$ (exponential pdf)

6 *Exponential Transform* ($Y = e^X$)

Initial function: $f_1(x) = \dfrac{1}{\sigma\sqrt{2\pi}} \cdot e^{-\frac{(x-a)^2}{2\sigma^2}}$

Transformed function: $f_1(y) = \dfrac{1}{y\sqrt{2\pi\sigma^2}} \exp\left[-\dfrac{(\ln y - a)^2}{2\sigma^2}\right]$, $y \geq 0$

7 *Detector of Power* $(a > 0)$ $Y = \begin{cases} (X - x_0)^a, & X \geq x_0 \\ 0 & , & X < x_0 \end{cases}$

Initial function: $f_1(x)$

Transformed function:

$$f_1(y) = \delta(y) \int_{-\infty}^{x_0} f_1(x)dx + \left[a \cdot y^{\frac{a-1}{a}}\right]^{-1} \cdot f_1\left(x_0 + y^{1/a}\right) \quad y \geq 0$$

8 *Linear Detector* $(a = 0)$ $Y = \begin{cases} X - x_0, & X \geq x_0 \\ 0 & , & X < x_0 \end{cases}$

Initial function: $f_1(x)$

Transformed function: $f_1(y) = \delta(y) \cdot \Phi\left(\dfrac{x_0}{\delta}\right) + \dfrac{1}{\sigma\sqrt{2\pi}} e^{-\frac{(y+x_0)^2}{2\sigma^2}}$ $y \geq 0$

9 *Harmonic Oscillation with Random Phase* ($Y = a \cdot \sin X$)

Initial function: $f_1(x)$

Transformed function:

$$f_1(y) = \frac{1}{a\sqrt{1 - \left(\dfrac{y}{a}\right)^2}} \sum_{k=-\infty}^{\infty} f_1\left[\pi k + (-1)^k \arcsin \frac{y}{a}\right] \quad |y| < a$$

10 *Harmonic Oscillation with Random Phase Distributed Uniformly at* $(0, 2\pi)$

Initial function: $f_1(x) = \begin{cases} \dfrac{1}{2\pi}, & |x| \le \pi \\ 0, & |x| > \pi \end{cases}$

Transformed function: $f_1(y) = \dfrac{1}{\pi a \sqrt{1 - \left(\dfrac{y}{a}\right)^2}}, \quad |y| < a$

♦ **Probability Density Function of Sum, Difference, Product, and Quotient of Two Random Variables**

• *Sum* ($Y = X_1 + X_2$)

Dependent variables: $f_1(y) = \displaystyle\int_{-\infty}^{\infty} f_2(y - x_2, x_2)\,dx_2 = \int_{-\infty}^{\infty} f_2(y - x_1, x_1)\,dx_1$

Independent variables:

$$f_1(y) = \int_{-\infty}^{\infty} f_1(y - x_2) \cdot f_1(x_2)\,dx_2 = \int_{-\infty}^{\infty} f_1(y - x_1) \cdot f(x_1)\,dx_1$$

• *Difference* ($Y = X_1 - X_2$)

Dependent variables: $f_1(y) = \displaystyle\int_{-\infty}^{\infty} f_2(y + x_2, x_2)\,dx_2 = \int_{-\infty}^{\infty} f_2(x_1 - y, x_1)\,dx_1$

Independent variables:

$$f_1(y) = \int\limits_{-\infty}^{\infty} f_1(y + x_2) \cdot f(x_2) dx_2 = \int f_1(x_1 - y) \cdot f_1(x_1) dx_1$$

- *Product* $(Y = X_1 \cdot X_2)$

Dependent variables: $f_1(y) = \int\limits_{-\infty}^{\infty} f_2\left(\dfrac{y}{x_2}, x_2\right) \dfrac{1}{|x_2|} dx_2 = \int\limits_{-\infty}^{\infty} f_2\left(\dfrac{y}{x_1}, x_1\right) \cdot \dfrac{1}{|x_1|} dx_1$

Independent variables:

$$f_1(y) = \int\limits_{-\infty}^{\infty} f_1\left(\dfrac{y}{x_2}\right) \cdot f_1(x_2) \dfrac{1}{|x_2|} dx_2 = \int\limits_{-\infty}^{\infty} f_1\left(\dfrac{y}{x_1}\right) \cdot f_1(x_1) \cdot \dfrac{1}{|x_1|} \cdot dx_1$$

- *Quotient* $(Y = \dfrac{X_1}{X_2})$

Dependent variables: $f_1(y) = \int\limits_{-\infty}^{\infty} f_2(y \cdot x_2, x_2) \cdot |x_2| dx_2 = \int\limits_{-\infty}^{\infty} f_2\left(\dfrac{x_1}{y}, x_1\right) \left|\dfrac{x_1}{y^2}\right| dx_1$

Independent variables:

$$f_1(y) = \int\limits_{-\infty}^{\infty} f_1(y \cdot x_2) f_1(x_2) |x_2| dx = \int\limits_{-\infty}^{\infty} f_1\left(\dfrac{x_1}{y}\right) \cdot f_1(x_1) \left|\dfrac{x_1}{y^2}\right| dx_1$$

Appendix 18 **A Gaussian Random Process**

1 *An n-Dimensional Generic Process*

Density function:

$$f_n(x_1, x_2, \ldots, x_n; t_1, \ldots, t_n) = \frac{1}{\sqrt{(2\pi)^n \Delta}}$$

$$\times \exp\left\{-\frac{1}{2\Delta} \cdot \sum_{\mu=1}^{n} \sum_{\gamma=1}^{n} \Delta_{\mu\gamma} [x_\mu - m_\xi(t_\mu)] \cdot [x_\gamma - m_\xi(t_\gamma)]\right\}$$

Characteristic function:

$$\theta_n(j\vartheta_1, j\vartheta_2, \ldots, j\vartheta_n; t_1, \ldots, t_n)$$

$$= \exp\left[j \sum_{\mu=1}^{n} m_\xi(t_\mu) \cdot \vartheta_\mu + \frac{1}{2} j^2 \sum_{\mu=1}^{n} \sum_{\gamma=1}^{n} K_\xi(t_\mu, t_\gamma) \vartheta_\mu \vartheta_\gamma\right]$$

2 *A n-Dimensional Stationary Process*

Density function:

$$f_n(x_1, \ldots, x_n) = \frac{1}{\sigma^n \sqrt{(2\pi)^n D}}$$

$$\times \exp\left\{-\frac{1}{2\sigma^2 D} \cdot \sum_{\mu=1}^{n} \sum_{\gamma=1}^{n} D_{\mu\gamma} (x_\mu - m) \cdot (x_\gamma - m)\right\}$$

Characteristic function:

$$\theta_n(j\vartheta_1, \ldots, j\vartheta_n)$$

$$= \exp\left[j \cdot m \cdot \sum_{\mu=1}^{n} \vartheta_\mu + \frac{1}{2} j^2 \sigma^2 \sum_{\mu=1}^{n} \sum_{\gamma=1}^{n} R(\tau_{\mu\gamma}) \vartheta_\mu \vartheta_\gamma\right]$$

3 *A Two-Dimensional Stationary Process*

Density function:

$$f_2(x_1, x_2) = \frac{1}{2\pi\sigma^2 \sqrt{1 - R^2(\tau)}}$$

$$\times \exp\left\{ -\frac{1}{2\sigma^2 \left[1 - R^2(\tau)\right]} \times \left[(x_1 - m)^2 - 2R(\tau)(x_1 - m)(x_2 - m) + (x_2 - m)^2\right] \right\}$$

Characteristic function:

$$\theta_2(j\vartheta_1, \ldots, j\vartheta_2)$$

$$= \exp\left\{ jm(\vartheta_1 + \vartheta_2) - \frac{1}{2}\sigma^2\left[\vartheta_1^2 + 2R(\tau)\vartheta_1\vartheta_2 + \vartheta_2^2\right] \right\}$$

4 *A One-Dimensional Stationary Process*

Density function: $f_1(x) = \frac{1}{\sigma\sqrt{2\pi}} \exp\left[-\frac{(x - m)^2}{2\sigma^2} \right]$

Characteristic function: $e^{jm\vartheta - \frac{\sigma^2\vartheta^2}{2}}$

Appendix 19   **Common Correlation Functions and Spectra**

Correlation Function	*Spectral Density*

$$K(\tau) = \sigma^2 \cdot e^{-\alpha|\tau|}$$

$$S(\omega) = \frac{\sigma^2}{\pi} \cdot \frac{a}{\omega^2 + \alpha^2}$$

$$K(\tau) = \sigma^2 \cdot e^{-\alpha|\tau|} \cdot \cos\beta\tau$$

$$S(\omega) = \frac{\sigma^2\alpha}{\pi} \cdot \frac{\omega^2 + \alpha^2 + \beta^2}{\left(\omega^2 - \beta^2 - \alpha^2\right)^2 + 4\alpha^2\omega^2}$$

$$K(\tau) = \sigma^2 \cdot e^{-\alpha|\tau|} \cdot \cos\beta\tau$$
$$\times\left(\cos\beta\tau + \frac{a}{\beta}\sin\beta|\tau|\right)$$

$$S(\omega) = \frac{2\sigma^2\alpha}{\pi} \cdot \frac{\alpha^2 + \beta^2}{\left(\omega^2 - \beta^2 - \alpha^2\right)^2 + 4\alpha^2\omega^2}$$

$$K(\tau) = \sigma^2 \cdot e^{-\alpha^2\tau^2} \cdot \cos\beta\tau$$

$$S(\omega) = \frac{\sigma^2}{4\alpha\sqrt{\pi}} \cdot \left[e^{-\frac{(\omega+\beta)^2}{4\alpha^2}} + e^{-\frac{(\omega-\beta)^2}{4\alpha^2}}\right]$$

$$K(\tau) = a_0 \cdot \delta(\tau) + a_1 \cdot \delta^{(2)}(\tau)$$
$$+ \ldots a_n \cdot \delta^{(2n)}(\tau)$$

$$S(\omega) = \frac{1}{2\pi}\left[\begin{array}{c} a_0 - \omega^2 a_1 + \omega^4 a_2 + \ldots \\ + (-1)^n \omega^{2n} a_n \end{array}\right]$$

Appendix 20   The Derivatives of the Random Process

1 *A Generic Random Process*

Covariation : $B_{\xi^{(n)}}(t_1,t_2) = \dfrac{\partial^{2n} B_\xi(t_1,t_2)}{\partial t_1^n \partial t_2^n}$

Mutual covariation : $B_{\xi^{(k)}\xi^{(l)}}(t_1,t_2) = \dfrac{\partial^{k+l} K_\xi(t_1,t_2)}{\partial t_1^k \partial t_2^l}$

Correlation: $K_{\xi^{(n)}}(t_1,t_2) = \dfrac{\partial^{2n} K_\xi(t_1,t_2)}{\partial t_1^n \partial t_2^n}$

Mutual correlation: $K_{\xi^{(k)}\xi^{(l)}}(t_1,t_2) = \dfrac{\partial^{k+l} K_\xi(t_1,t_2)}{\partial t_1^k \partial t_2^l}$

2 *A Stationary Random Process*

Covariation: $B_{\xi^{(n)}}(\tau) = (-1)^n \dfrac{d^{2n} B_\xi(\tau)}{d\tau^{2n}}$

Mutual covariation: $B_{\xi^{(k)}\xi^{(l)}}(\tau) = (-1)^k \dfrac{d^{k+l} B_\xi(\tau)}{d\tau^{k+l}}$

Correlation: $K_{\xi^{(n)}}(\tau) = (-1)^n \dfrac{d^{2n} K_\xi(\tau)}{d\tau^{2n}}$

Mutual correlation: $K_{\xi^{(k)}\xi^{(l)}}(\tau) = (-1)^k \dfrac{d^{k+l} B_\xi(\tau)}{d\tau^{k+l}}$

Spectral density: $S_{\xi^{(n)}}(\omega) = \omega^{2n} \cdot S_{\xi^0}(\omega)$

Appendix 21 **Common Simulation Algorithms**

♦ **Simulation Algorithms for Random Variables with Common Distributions**

Hereinafter u is the random variable uniformly distributed at $(0,1)$, $z0G \subset N(0,1)$ is the random variable with standard Gaussian distribution ($m = 0$ and $\sigma = 1$).

1 *Gaussian Distribution*

Density function: $f(z) = \dfrac{1}{\sqrt{2\pi}\,\sigma_z} \cdot \exp\left[-\dfrac{(z - m_z)^2}{2\sigma_z^2} \right]$

Simulation algorithm: $z0G = \sqrt{-2 \cdot \ln(u1)} \cdot \cos(2\pi\, u2)$

or

$$z0G = \sqrt{-2 \cdot \ln(u1)} \cdot \sin(2\pi\, u2)$$

$z_G = m_z + \sigma_z \cdot z0G$

2 *Rayleigh Distribution*

Density function: $f(z) = \dfrac{z}{\sigma_z} \cdot \exp\left(-\dfrac{z^2}{2\sigma_z^2} \right)$ $z \geq 0$

Simulation algorithm: $z_R = \sigma_z \cdot \sqrt{-2 \cdot \ln(u)}$

3 *Ricean Distribution*

Density function:

$$f(z) = \frac{z}{\sigma_z} \cdot \exp\left(-\frac{z^2 + a^2}{2\sigma_z^2} \right) \cdot I_0\left(\frac{az}{\sigma_z^2} \right)   z \geq 0$$

Simulation algorithm:

$$z_{RN} = \sqrt{a^2 - 2\sigma_z^2 \cdot \ln(u1) - 2a\sigma_z \sqrt{-2 \cdot \ln(u1)} \cdot \cos(2\pi\, u2)}$$

4 *Exponential Distribution*

Density function: $f(z) = \lambda \cdot \exp(-\lambda z)$

Simulation algorithm: $z_E = \dfrac{-\ln(u)}{\lambda}$

5 *Log-Normal Distribution*

Density function: $f(z) = \dfrac{1}{z\sqrt{2\pi}\,\sigma_z} \cdot \exp\left[-\dfrac{(\ln(z) - m_z)^2}{2\sigma_z^2} \right]$

Simulation algorithm: $z_L = \exp(z_G)$

6 *Chi-Squared Distribution* (with $2k$ degrees of freedom)

Density function: $f(z) = \dfrac{1}{2^k \cdot \Gamma(k)} \cdot z^{k-1}\, e^{-\frac{z}{2}}$

Simulation algorithm: $z_{Ch} = -2\ln\left(\displaystyle\prod_{i=1}^{k} u_i \right)$

7 *Student Distribution* (with n degrees of freedom)

Density function: $f(z) = \dfrac{\Gamma\left(\dfrac{n+1}{2}\right)}{\sqrt{n\pi}\cdot\Gamma\left(\dfrac{n}{2}\right)\cdot\left(1 + \dfrac{z^2}{n}\right)^{\frac{n+1}{2}}}$

Simulation algorithm: $z_S = \dfrac{z0G}{\sqrt{z_{Ch}}} \cdot \sqrt{n}$ for z_{Ch} with $2k = n$.

8 *Cauchy Distribution*

Density function: $f_1(z) = \dfrac{a}{\pi\left[a^2 + (y - b)^2\right]}$

Simulation algorithm: $z_C = a \cdot \tan\left[\pi\left(u - \frac{1}{2}\right)\right] + b$

9 *Gamma Distribution*

Density function: $f_1(z) = \dfrac{\beta^a}{\Gamma(\alpha)} z^{\alpha-1} \cdot e^{-\beta z}, \quad z > 0$

Simulation algorithm: $z_\gamma = -\ln\left(\displaystyle\prod_{k=1}^{n} \dfrac{u_k}{\beta}\right); \; \alpha = n = \text{integer}.$

10 *Beta Distribution*

Density function: $f_1(z) = \dfrac{\Gamma(p+m)}{\Gamma(p)\Gamma(m)} z^{p-1} \cdot (1-z)^{m-1}, \quad 0 \le z \le 1$

Simulation algorithm: $z_\beta = \dfrac{z_\gamma}{\left(z_\gamma + z0G\right)}$

11 *F-Distribution*

Density function: $f_1(z) = \dfrac{\Gamma(\alpha_1 + \alpha_2) \cdot \beta^{\alpha_1}}{\Gamma(a_1)\Gamma(\alpha_2)} \cdot \dfrac{z^{\alpha_1 - 1}}{(1 + \beta z)^{\alpha_1 + \alpha_2}}$

Simulation algorithm: $z_F = z1_\gamma / z2_\gamma$,

where $z1_\gamma \in \Gamma(\alpha_1, \beta_1) \quad z2_\gamma \in \Gamma(\alpha_2, \beta_2)$

♦ **Simulation Algorithms for the Stationary Gaussian Random Processes with Common Correlation Functions**

1 *The process* $\xi_1(t)$ *with correlation function* $K_1(\tau) = \sigma_\xi^2 \cdot e^{-\alpha|\tau|}$

Simulation algorithm: $z_n = a_1 \cdot z_{n-1} + b_1 \cdot z0G_n$

Parameters: $a_1 = e^{-\alpha \Delta t}; \; b_1 = \sigma_\xi \cdot \sqrt{1 - e^{-2\alpha \Delta t}}$

2 *The process* $\xi_2(t)$ *with correlation function* $K_2(\tau) = \sigma_\xi^2 \cdot e^{-\alpha|\tau|} \cdot \cos \beta \tau$

Simulation algorithm: $z_n = a_1 z_{n-1} + a_2 z_{n-2} + b_1 \cdot z0G_n + b_2 \cdot z0G_{n-1}$

$a_1 = 2 \cdot e^{-\alpha \Delta t} \cdot \cos(\beta \Delta t);\ a_2 = -e^{-2\alpha \Delta t};\ b_1 = \sqrt{-c_0/\vartheta_1};\ b_2 = -\vartheta_1 \cdot b_1$

$\vartheta_1 = -\chi + sign(\chi) \cdot \sqrt{\chi^2 - 1};\ \chi = c_1/(2c_0);\ c_0 = a_{12}R_{12} - R_{11} \cdot a_{22};$

$c_1 = R_{11}(a_{22})^2 - 2R_{12}a_{12}a_{22} + R_{22}(a_{12})^2 + R_{11};\ a_{11} = \rho s_1;\ a_{12} = (\rho/\beta) \cdot \sin \beta_\Delta;$

$a_{21} = -\rho \alpha \overline{\alpha} (1 + \overline{\beta}^2) \sin \beta_\Delta;\ a_{22} = \rho s_2;$

$R_{11} = 1 - \rho^2 + 2\overline{\alpha}^2 \rho^2 \sin^2 \beta_\Delta \left(\sqrt{1 + \overline{\beta}^2} - 1 \right);$

$R_{12} = R_{21} = \alpha \rho^2 \left[\cos \beta_\Delta + \sin \beta_\Delta \left(\sqrt{\alpha^2 + 1} - \overline{\alpha} \right) \right]^2 - \alpha;$

$R_{22} = \alpha^2 \left\{ \rho^2 \left[2s_1^2 \left(\sqrt{1 + \overline{\beta}^2} - 1 \right) - 1 - \overline{\beta}^2 \right] + 1 + \left(1 - \sqrt{1 + \overline{\beta}^2} \right)^2 \right\}$

3 *The process* $\xi_3(t)$ *with correlation function*

$$K_3(\tau) = \sigma_\xi^2 \cdot e^{-\alpha|\tau|} \cdot \left(\cos \beta \tau + \frac{\alpha}{\beta} \sin \beta|\tau| \right)$$

Simulation algorithm: $z_n = a_1 z_{n-1} + a_2 z_{n-2} + b_1 \cdot z0G_n + b_2 \cdot z0G_{n-1}$

$a_1 = 2 \cdot e^{-\alpha \Delta t} \cdot \cos(\beta \Delta t);\ a_2 = -e^{-2\alpha \Delta t};\ b_1 = \sqrt{-c_0/\vartheta_1};\ b_2 = -\vartheta_1 \cdot b_1$

$\vartheta_1 = -\chi + sign(\chi) \cdot \sqrt{\chi^2 - 1};\ \chi = c_1/(2c_0);\ c_0 = a_{12}R_{12} - R_{11} \cdot a_{22};$

$c_1 = R_{11}(a_{22})^2 - 2R_{12}a_{12}a_{22} + R_{22}(a_{12})^2 + R_{11};\ a_{11} = \rho s_1;\ a_{12} = (\rho/\beta) \cdot \sin \beta_\Delta;$

$a_{21} = -\rho \alpha \overline{\alpha} (1 + \overline{\beta}^2) \sin \beta_\Delta;\ a_{22} = \rho s_2;$

$$R_{11} = 1 - \rho^2 \left[s_1^2 + \left(1 + \overline{\alpha}^2\right) \sin^2 \beta_\Delta \right];$$

$$R_{12} = R_{21} = 2\rho^2 \alpha \left(1 + \overline{\alpha}^2\right) \sin^2 \beta_\Delta ;$$

$$R_{22} = \beta^2 \left(1 + \overline{\alpha}^2\right) \left\{1 - \rho^2 \left[s_2^2 + \left(1 + \overline{\alpha}^2\right) \sin^2 \beta_\Delta \right]\right\}$$

4 *The process* $\xi_4(t)$ *with correlation function* $K_4(\tau) = \sigma_\xi^2 \left(1 - |\tau| \alpha/2\right) \cdot e^{-\alpha|\tau|}$

Simulation algorithm: $z_n = a_1 z_{n-1} + a_2 z_{n-2} + b_1 \cdot z0G_n + b_2 \cdot z0G_{n-1}$

$$a_1 = 2 \cdot e^{-\alpha \Delta t} ; \quad a_2 = -e^{-2\alpha \Delta t} ; \quad b_1 = \sqrt{-c_0 / \vartheta_1} ; \quad b_2 = -\vartheta_1 \cdot b_1$$

$$\vartheta_1 = -\chi + sign(\chi) \cdot \sqrt{\chi^2 - 1} ; \quad \chi = c_1 / (2c_0) ; \quad c_0 = a_{12} R_{12} - R_{11} \cdot a_{22} ;$$

$$c_1 = R_{11}(a_{22})^2 - 2R_{12}a_{12}a_{22} + R_{22}(a_{12})^2 + R_{11} ; \quad a_{11} = \rho(1 + \alpha_\Delta) ; \quad a_{12} = \rho \Delta t ;$$

$$a_{21} = -\alpha^2 \rho \Delta t ; \quad a_{22} = \rho(1 - \alpha_\Delta);$$

$$R_{11} = 1 + \rho^2 \left[\alpha_\Delta + \alpha_\Delta^2 \left(\sqrt{3} - 2\right) - 1 \right];$$

$$R_{12} = R_{21} = \alpha \left\{ \rho^2 \left[1.5 + \alpha_\Delta \left(\sqrt{3} - 3\right) + \alpha_\Delta^2 \left(2 - \sqrt{3}\right) \right] - 1.5 \right\};$$

$$R_{22} = \alpha^2 \rho^2 \left[\sqrt{3} - 4 + \alpha_\Delta \left(5 - 2\sqrt{3}\right) + \alpha_\Delta^2 \left(\sqrt{3} - \Delta\right) \right] + \alpha^2 \left(4 - \sqrt{3}\right)$$

The following notations are used: $\alpha_\Delta = \alpha \cdot \Delta t$; $\beta_\Delta = \beta \cdot \Delta t$; $\rho = \exp(-\alpha_\Delta)$; $\overline{\alpha} = \alpha/\beta$; $\overline{\beta} = \beta/\alpha$; $s_1 = \cos \beta_\Delta + \overline{\alpha} \sin \beta_\Delta$; $s_2 = \cos \beta_\Delta - \overline{\alpha} \sin \beta_\Delta$

◆ **Simulation Algorithm for a Three-Dimensional Gaussian Markovian Random Process**

Here $z1_G, z2_G, z3_G \in N(0,1)$ are independent random variables with standard Gaussian distribution ($m = 0$ and $\sigma = 1$).

1 *The first component* $\xi_1(t)$

$$z_1(t_{n+1}) = z_1(t_n) \cdot f(\tau) + \sqrt{D_1 \left[1 - f^2(\tau)\right]} \cdot z1_G(t_{n+1}); \quad z_1(t_0) = \sqrt{D_1} \cdot z1_G(t_0)$$

2 *The second component* $\xi_2(t)$

$$z_2(t_{n+1}) = z_2(t_n) \cdot f(\tau) + \rho_{12}\sqrt{D_2\left[1 - f^2(\tau)\right]} \cdot z1_G(t_{n+1})$$

$$+ \sqrt{D_2\left(1 - \rho_{12}^2\right)\left[1 - f^2(\tau)\right]} \cdot z2_G(t_{n+1})$$

$$z_2(t_0) = \rho_{12}\sqrt{D_2} \cdot z1_G(t_0) + \sqrt{D_2\left(1 - \rho_{12}^2\right)} \cdot z2_G(t_0)$$

3 *The third component* $\xi_3(t)$

$$z_3(t_{n+1}) = z_3(t_n) \cdot f(\tau) + \rho_{13}\sqrt{D_3\left[1 - f^2(\tau)\right]} \cdot z1_G(t_{n+1})$$

$$- \left(\rho_{12}\rho_{13} - \rho_{23}\right)\sqrt{\frac{D_3\left[1 - f^2(\tau)\right]}{1 - \rho_{12}^2}} \cdot z2_G(t_{n+1})$$

$$+ \sqrt{\frac{D_3\left[1 - f^2(\tau)\right]\left[1 + 2\rho_{12}\rho_{13}\rho_{23} - \rho_{12}^2 - \rho_{13}^2 - \rho_{23}^2\right]}{1 - \rho_{12}^2}} \cdot z3_G(t_{n+1})$$

$$z_3(t_0) = \rho_{13}\sqrt{D_3} \cdot z1_G(t_0) - \left(\rho_{12}\rho_{13} - \rho_{23}\right)\sqrt{\frac{D_3}{1 - \rho_{12}^2}} \cdot z2_G(t_0)$$

$$+ \sqrt{\frac{D_3\left(1 + 2\rho_{12}\rho_{13}\rho_{23} - \rho_{12}^2 - \rho_{13}^2 - \rho_{23}^2\right)}{1 - \rho_{12}^2}} z3_G(t_0)$$

Appendix 22 Common Algorithms of Applied Statistics

♦ **Algorithms to Estimate Parameters of a Stationary Ergodic Random Process** $x(n\Delta t)$ **with N Samples** ($n = 0, \dots N{-}1$):

1 *Mean*:

$$\hat{m}_\xi = \frac{1}{N} \sum_{n=0}^{N-1} x(n\Delta t)$$

2 *Variance*:

$$\hat{D}_\xi = \frac{1}{N} \sum_{n=0}^{N-1} [x(n\Delta t)]^2 - \frac{1}{N^2} \left[\sum_{n=0}^{N-1} x(n\Delta t) \right]^2$$

3 *Autocorrelation Function*:

$$\hat{K}_\xi(r \cdot \Delta t) = \frac{1}{N-r} \sum_{n=0}^{N-1-r} x(n\Delta t) \cdot x[(n+r)\Delta t] - \frac{1}{(N-r)^2} \sum_{n=0}^{N-1-r} x(n\Delta t) \cdot \sum_{n=r}^{N-1} x(n\Delta t)$$

4 *Spectral Density*:

$$\hat{S}_\xi[\omega(r\Delta t)] = \frac{\hat{K}_\xi(0) \cdot \Delta t}{2\pi} + \frac{1}{\pi} \sum_{r=1}^{N-2} K_\xi(r\Delta t) \cos[\omega(r\Delta t)] \left(1 - \frac{r}{N-1} \right) \Delta t$$

♦ **Algorithms to Estimate the First Four Moments of a Nonstationary Random Process with** M **samples** ($m = 0, \dots M{-}1$):

1 *Moment* $m_1 = m_\xi$

Algorithm: $\hat{m}_\xi(n) = \dfrac{1}{M} \sum_{m=0}^{M-1} x_m(n)$;   Error: $\sigma_{\hat{m}_\xi} = \sqrt{D_\xi / M}$

2 _Moment_ $M_2 = D_\xi$

Algorithm: $\hat{D}_\xi(n) = \dfrac{1}{M-1} \displaystyle\sum_{m=0}^{M-1} \left[x_m(n) - \hat{m}_\xi(n)\right]^2$

Error:

$$\sigma_{\hat{D}_\xi} = \sqrt{\frac{M}{M-1}} \left[\frac{M_4 - M_2^2}{M} - \frac{2\left(M_4 - 2M_2^2\right)}{M^4} + \frac{M_4 - 3M_2^2}{M^3}\right]^{1/2}$$

$$\approx \sqrt{\frac{M_4 - M_2^2}{M}} + o\left(\frac{1}{M}\right)$$

3 _Moment_ M_3

Algorithm: $\hat{M}_3(n) = \dfrac{M}{(M-1)(M-2)} \displaystyle\sum_{m=0}^{M-1} \left[x_m(n) - \hat{m}_\xi(n)\right]^2$

Error: $\sigma_{\hat{M}_3} = \sqrt{\dfrac{M_6 - 6M_2 M_4 - M_3^2 + 9M_2^3}{M}} + o\left(\dfrac{1}{M}\right)$

4 _Moment_ M_4

Algorithm:

$$\hat{M}_4(n) = \frac{M^2 - 2M + 3}{(M-1)(M-2)(M-3)} \times \sum_{m=0}^{M-1} \left[x_m(n) - m_\xi(n)\right]^4$$

$$- \frac{3(2M-3)}{M(M-1)(M-2)(M-3)} \times \left\{\sum_{m=0}^{M-1} \left[x_m(n) - m_\xi(n)\right]\right\}^2$$

Error: $\sigma_{\hat{M}_4} = \sqrt{\dfrac{M_8 - 8M_3 M_5 - M_4^2 + 16M_3^2}{M}} + o\left(\dfrac{1}{M}\right)$

Appendix 23 Greek Alphabet

Letter (Upper/Lower)			Name	Pronunciation
A	α		alpha	al-fuh
B	β		beta	bay-tuh
Γ	γ		gamma	gam-uh
Δ	δ		delta	del-tuh
E	ε		epsilon	ep-sil-on
Z	ζ		zeta	zay-tuh
H	η		eta	ay-tuh
Θ	θ		theta	thay-tuh
I	ι		iota	eye-oh-tuh
K	κ		kappa	kap-uh
Λ	λ		lambda	lam-duh
M	μ		mu	myoo
N	ν		nu	noo
Ξ	ξ		xi	ks-eye
O	o		omicron	om-i-cron
Π	π		pi	pie
P	ρ		rho	row
Σ	σ	ς	sigma	sig-muh
T	τ		tau	tau
Y	υ		upsilon	oop-si-lon
Φ	φ	ϕ	phi	fee
X	χ		chi	k-eye
Ψ	ψ		psi	sigh
Ω	ω		omega	oh-may-guh

ABOUT THE AUTHORS

Dr. Sergey A. Leonov is known both in Russia and the West as a bilingual expert in the field of applied mathematics and electrical engineering. Dr. Leonov was born in St. Petersburg, Russia, in 1953. He received his B.Sc. degree from the Moscow University of Radioengineering, Electronics and Automation (1976); his M. Sc. degree from Kharkov University (1980); and his Ph.D. and D. Sc. (Eng.) degrees from the Moscow Aerospace Institute. Dr. Leonov holds the "All-Russian Honorable" title in the field of science and engineering. He has authored and coauthored over 70 papers and books, both in Russian and English, including *Air Defense Radars* (Voenizdat, 1988), *Radar Test* (Radio I Svayz, 1990), *Russian-English and English-Russian Dictionary on Radar and Electronics* (Artech House, 1993), *Radar Technology Encyclopedia* (Artech House, 1997), and *Handbook of Computer Simulation in Radio Engineering, Communications, and Radar* (Artech House, 2001).

Dr. Alexander I. Leonov is well known in Russia as a scientist in the field of applied mathematics and electrical engineering. He was born in the Tambov region of Russia in 1927. He received his M.Sc. degree from the Leningrad Military University of Communications (1956), and his Ph.D. and D.Sc. (Eng.) degrees from the Moscow Aerospace Institute. Dr. Leonov holds the "All-Russian Honorable" title in the field of science and engineering. He has authored and coauthored over 100 papers and books, in both Russian and English, including *Radar in Anti-Missile Defense* (Voenizdat, 1967), *Monopulse Radar* (Sovetskoe Radio, 1970, 1984; trans. Artech House, 1986), *Modeling in Radar* (Radio I Svayz, 1979), *Radar Test* (Radio I Svayz, 1990), *Radar Technology Encyclopedia* (Artech House, 1997), and *Handbook of Computer Simulation in Radio Engineering, Communications, and Radar* (Artech House, 2001).

INDEX

For further information on these and other Artech House titles, including previously considered out-of-print books now available through our In-Print-Forever® (IPF®) program, contact:

Artech House
685 Canton Street
Norwood, MA 02062
Phone: 781-769-9750
Fax: 781-769-6334
e-mail: artech@artechhouse.com

Artech House
46 Gillingham Street
London SW1V 1AH UK
Phone: +44 (0)20 7596-8750
Fax: +44 (0)20 7630-0166
e-mail: artech-uk@artechhouse.com

Find us on the World Wide Web at:
www.artechhouse.com